Impacts, Monitoring and Management of Forest Pests and Diseases

Impacts, Monitoring and Management of Forest Pests and Diseases

Special Issue Editors

Young-Seuk Park
Won Il Choi

MDPI • Basel • Beijing • Wuhan • Barcelona • Belgrade

Special Issue Editors
Young-Seuk Park
Kyung Hee University
Korea

Won Il Choi
National Institute of Forest Science
Korea

Editorial Office
MDPI
St. Alban-Anlage 66
4052 Basel, Switzerland

This is a reprint of articles from the Special Issue published online in the open access journal *Forests* (ISSN 1999-4907) in 2019 (available at: https://www.mdpi.com/journal/forests/special_issues/pest_disease).

For citation purposes, cite each article independently as indicated on the article page online and as indicated below:

LastName, A.A.; LastName, B.B.; LastName, C.C. Article Title. *Journal Name* **Year**, *Article Number*, Page Range.

ISBN 978-3-03928-166-4 (Pbk)
ISBN 978-3-03928-167-1 (PDF)

Cover image courtesy of Won Il Choi.

Contents

About the Special Issue Editors . . . . **vii**

Preface to "Impacts, Monitoring and Management of Forest Pests and Diseases" . . . . **ix**

Won Il Choi and Young-Seuk Park
Monitoring, Assessment and Management of Forest Insect Pests and Diseases
Reprinted from: *Forests* **2019**, *10*, 865, doi:10.3390/f10100865 . . . . **1**

Won Il Choi, Youngwoo Nam, Cha Young Lee, Byoung Ki Choi, Yu Jin Shin, Jong-Hwan Lim, Sang-Hyun Koh and Young-Seuk Park
Changes in Major Insect Pests of Pine Forests in Korea Over the Last 50 Years
Reprinted from: *Forests* **2019**, *10*, 692, doi:10.3390/f10080692 . . . . **7**

Zbigniew Sierota, Wojciech Grodzki and Andrzej Szczepkowski
Abiotic and Biotic Disturbances Affecting Forest Health in Poland over the Past 30 Years: Impacts of Climate and Forest Management †
Reprinted from: *Forests* **2019**, *10*, 75, doi:10.3390/f10010075 . . . . **22**

Jinyeong Choi, Deokjea Cha, Dong-Soo Kim and Seunghwan Lee
Review of Japanese Pine Bast Scale, *Matsucoccus matsumurae* (Kuwana) (Coccomorpha: Matsucoccidae), Occurring on Japanese Black Pine (*Pinus thunbergii* Parl.) and Japanese Red Pine (*P. densiflora* Siebold & Zucc.) from Korea
Reprinted from: *Forests* **2019**, *10*, 639, doi:10.3390/f10080639 . . . . **39**

David E. Jennings, Jian J. Duan and Paula M. Shrewsbury
Comparing Methods for Monitoring Establishment of the Emerald Ash Borer (*Agrilus planipennis*, Coleoptera: Buprestidae) Egg Parasitoid *Oobius agrili* (Hymenoptera: Encyrtidae) in Maryland, USA
Reprinted from: *Forests* **2018**, *9*, 659, doi:10.3390/f9100659 . . . . **53**

Fu Liu, Chengxu Wu, Sufang Zhang, Xiangbo Kong, Zhen Zhang and Pingyan Wang
Initial Location Preference Together with Aggregation Pheromones Regulate the Attack Pattern of *Tomicus brevipilosus* (Coleoptera: Curculionidae) on *Pinus kesiya*
Reprinted from: *Forests* **2019**, *10*, 156, doi:10.3390/f10020156 . . . . **62**

Beatriz Mora-Sala, Mónica Berbegal and Paloma Abad-Campos
The Use of qPCR Reveals a High Frequency of *Phytophthora quercina* in Two Spanish Holm Oak Areas
Reprinted from: *Forests* **2018**, *9*, 697, doi:10.3390/f9110697 . . . . **76**

Thaissa P. F. Soares, Edson A. Pozza, Adélia A. A. Pozza, Reginaldo Gonçalves Mafia and Maria A. Ferreira
Calcium and Potassium Imbalance Favours Leaf Blight and Defoliation Caused by *Calonectria pteridis* in Eucalyptus Plants
Reprinted from: *Forests* **2018**, *9*, 782, doi:10.3390/f9120782 . . . . **90**

Sufang Zhang, Sifan Shen, Shiyu Zhang, Hongbin Wang, Xiangbo Kong, Fu Liu and Zhen Zhang
Chemosensory Characteristics of Two *Semanotus bifasciatus* Populations
Reprinted from: *Forests* **2019**, *10*, 655, doi:10.3390/f10080655 . . . . **104**

About the Special Issue Editors

Young-Seuk Park (Professor) is Professor in the Department of Biology, Kyung Hee University, Seoul, Republic of Korea. He teaches courses on ecology, ecological modelling and ecological informatics. He is interested in ecological monitoring and the assessment of sustainable ecosystem management based on computational approaches. His research focuses on the effects of global changes on ecosystems and biodiversity maintenance, including the invasion of alien species. He has served as Editor-in-Chief of Korean Journal of Ecology and Environment and Associate Editor of Ecological Modelling since 2019. He has also been associate editor of the International Journal of Limnology since 2009.

Won Il Choi is a senior researcher at the National Institute of Forest Science, Seoul, Republic of Korea. He has researched the population dynamics of forest insect pests using long-term monitoring data based on computational approaches since 2006. His research focuses on the effects of climate change on the population dynamics of forest insect pests and ecology of invasive forest insect pests. In 2014, he won the 'Hyun, Shin-Kyu' research award for young scientists by the Korean Forest Society.

Preface to "Impacts, Monitoring and Management of Forest Pests and Diseases"

Forest pests have diverse negative impacts on forestry economy, ecosystem services, biodiversity, and sustainable ecosystem management. The first step towards effectively managing forest pests would be to monitor their occurrence and assess their impact on forest ecosystems. The monitoring results can provide basic information for effective management strategies. In particular, the long-term monitoring programs can provide information concerning the occurrence patterns of forest pests. The data from monitoring programs can result in the development of new methods for monitoring, assessing impact, and developing management techniques.

This special issue ("Impacts, Monitoring, and Management of Forest Pests and Diseases") aims to share information to assist in the effective management of forest pests, by understanding the responses of forest pests to natural and anthropogenic changes, and discussing new studies on the monitoring, assessment, and management of forest pests. Thirteen papers are included in this special issue, focused on monitoring, assessing, and managing forest pests.

Two articles reviewed long-term changes in forest pests and forests. Choi et al. reviewed changes in major forest pests, specifically insect pests in pine forests in Korea in the last 50 years, and Sierota et al. reported on changes in the health of forests and their biological diversity in Poland in the last 30 years.

Four papers focused on the monitoring of forest pests. Choi et al. revised the taxonomy of the Japanese pine bast scale, Matsucoccus matsumurae (Kuwana), which had previously been described as Matsucoccus thunbergii. Jennings et al. compared two methods (visual survey and bark shifting) for monitoring the establishment of Oobius agrili (Zhang and Huang), which is a parasitoid on the emerald ash borer (Agrilus planipennis (Fairmaire)), an invasive species. Mora-Sala et al. demonstrated the efficiency of qPCR approaches to determine the presence and distribution of Phytophthora sp. in holm oak areas. Liu et al. reported that the initial attack location and aggregation pheromones were important for mediating the aggressive behavior and top-down attack pattern of Tomicus brevipilosus on Pinus kesiya.

Three papers focused on the assessment of forest pests. Soares et al. reported that an imbalance of Ca and K favors the leaf blight and defoliation caused by the fungus in Eucalyptus plants. Zhang et al. studied the differences in the trapping effects of Semanotus bifasciatus between two populations (Beijing and Shandong), based on next-generation sequencing. Szmidla et al. reported that the population size of Melolonthinae and the location of the larvae were related to weather conditions.

Finally, four papers focused on the management of forest pests. Kamczyc et al. showed that the mite abundance was influenced by tree species, which suggests differences in the litter quality. Lalik et al. reported that pitfall traps with attractants of pine twigs and ethanol are useful for monitoring and controlling H. abietis. Baek et al. evaluated the current and future distribution of Ricania shantungensis (Chou and Lu), which is an invasive planthopper and an important pest in agriculture and forestry in Korea. Finally, Lee et al. predicted the potential distribution of the citrus Flatid planthopper, Metcalfa pruinosa (Say), which is an invasive species in many countries. Their results show that factors relating to human activity strongly influence the occurrence and dispersal of the citrus Flatid planthopper, which has a high potential to disperse over the whole of South Korea. We believe that this special issue provides a better understanding of the structures and processes of forest ecosystems, and fundamental information for the effective management of forest pests.

Young-Seuk Park, Won Il Choi
Special Issue Editors

forests

Editorial

Monitoring, Assessment and Management of Forest Insect Pests and Diseases

Won Il Choi [1] and Young-Seuk Park [2,3,*]

[1] Division of Forest Ecology and Climate Change Division, National Forest Research Institute, Dongdaemun, Seoul 02445, Korea; wchoi71@korea.kr

[2] Department of Biology, Kyung Hee University, Dongdaemun, Seoul 02447, Korea

[3] Department of Life and Nanopharmaceutical Sciences, Kyung Hee University, Dongdaemun, Seoul 02447, Korea

* Correspondence: parkys@khu.ac.kr; Tel.: +82-2-961-0946

Received: 23 September 2019; Accepted: 27 September 2019; Published: 3 October 2019

Abstract: Forest pests are one of the most important factors disturbing forest ecosystems, by impacting forestry economy, ecosystem services, biodiversity, and sustainable ecosystem management. Monitoring the occurrence of forest pests offers clues to understand their impacts on the forest ecosystem and develop a sustainable ecosystem management strategy. This special issue is designed to create a better understanding of the changes and impacts of forest pests according to forest changes, caused by natural or anthropogenic causes. There are 13 papers published in this special issue, covering several issues concerning forest pests. Two of the papers reviewed the changes in forest pests in Korea or Poland. The remaining twelve papers covered issues concerning the monitoring, assessment, and management of forest pests. Through this special issue, we expect to contribute towards the improvement of our knowledge of the structures and processes in forest ecosystems relating to forest pests and fundamental information for the effective management of forest pests.

Keywords: climate change; forest ecosystem; forest pests; invasive species; monitoring

---

1. Introduction

Forest pests, including insects and microorganisms, are considered a regulating factor in the nutrient cycling and energy flow in forest ecosystems [1]. Among them, some species can cause severe damage to forest natural resources, resulting in serious changes in both the structures and functions of forest ecosystems. Forest pests have diverse negative impacts on forestry economy, ecosystem services, biodiversity, and sustainable ecosystem management, which are mostly related with alien species [2]. The first step towards effectively managing forest pests would be to monitor their occurrence and assess their impacts. The monitoring results can provide basic information for effective management strategies. For example, in North America the strategy for the gypsy moth, *Lymantria dispar* L., was constructed on the basis of the long-term monitoring of the range expansion of the gypsy moth [3]. In particular, long-term monitoring programs concerning forest pests offer an insight into their outbreaks, which is fundamental for the effective management of ecosystems [4]. The results of long-term monitoring programs can provide information concerning the occurrence patterns of forest pests. The data from monitoring programs can result in the development of new methods for monitoring, assessing impacts, and developing management techniques.

Forest pest monitoring programs have been intensively conducted in many countries, including the USA and European countries [5]. In the USA, the status of major forest pests, such as the gypsy moth, mountain pine beetle (*Dendroctonus ponderosae* (Hopkins)), and southern pine beetle (*Dendroctonus frontalis* (Zimmermann)), along with risk maps for the major insects and diseases, are offered on a web site (https://www.fs.fed.us). Using long-term monitoring data, analyses of the population dynamics relating to an outbreak of forest insect pests are widely attempted [6]. The stratified dispersal of the gypsy moth was elucidated based on long-term monitoring of the dispersal of the moth [3]. Long-term monitoring data of pine-defoliating lepidopteran pests, including *Dendrolimus pini* L., *Hyloicus pinastri* L., and *Bupalus piniarius* L., in pine forests from 1880 to 1940 were recorded and used for clarifying the periodicity in the occurrence of these pests [4]. The occurrence of the pine processionary moth (*Thaumetopoea pityocampa* (Denis and Schiffermuller)), observed from 1981 on permeant plots in France, displayed periodicity in a range of 7 to 11 years [7].

The monitoring program of forest pests in Korea, which has been conducted for both insect pests and pathogens since 1968 [4,8], consists of monitoring for major and occasional forest pests. The major pests selected for monitoring have been changed according to the damage severity of the pests [4]. For example, the pine caterpillar (*Dendrolimus spectabilis* (Butler)) was one of the main insect pests that were monitored from the 1960s to 1980s. However, it was removed from the monitoring list in 2016 due to the decline in its population. Currently, five major pests, including pine wilt disease (PWD) caused by pine wood nematodes; pine needle gall midge (PNGM) (*Thecodiplosis japonensis* (Uchida and Inouye)); black pine bast scale (*Matsucoccus matsumurae* (Kuwana)); Korean oak wilt disease caused by *Raffaelea quercus-mongolicae* (K.H. Kim, Y.J. Choi, and H.D. Shin); and fall webworm (*Hyphantria cunea* (Drury)), are monitored for their distribution, damage areas, or densities.

The occurrence of forest pests is influenced by various intrinsic and extrinsic factors [4]. The effective evaluation of the impacts of pests on the target ecosystems is important for forest ecosystem management. In particular, it is crucial to identify and assess the impacts of invasive species, to control their dispersal and delimitate the invasive species [9,10]. For example, tools have been developed to identify nematode species, for the rapid and exact diagnosis of pine wood nematodes in pine wood [11]. The classification of dispersal types, according to the invasive history of the pine wilt disease, offered insights for developing management strategies [9]. Meanwhile, diverse hazard rating systems have been developed to evaluate the potential distribution area and economic/ecological impacts of the invasive forest species, such as PNGM, black pine bast scale, and PWD [12–14].

An assessment of the relationship between environmental factors and forest pests provides fundamental information on their occurrence and abundance, which is essential for decision making concerning management strategies. Therefore, many studies have been conducted on the impacts of environmental factors on the occurrence of forest pests. For example, Kurz et al. [15] reported that both an increase in the winter temperature, due to recent climate change, and the uniform age structure of forests with older trees caused an outbreak of the mountain pine beetle (*Dendroctonus ponderosae* (Hopkins)) in British Columbia, Canada. The higher precipitation in August and drought during spring were key factors that reduced the population density of the pine caterpillar and PNGW, respectively [8,16]. Temperature is a limiting factor for determining the geographical distribution of *Monochamus alternatus* and *Monochamus saltuarius*, vectors for PWD in Korea [17].

The development of effective methods for managing forest pests is important for reducing the impacts of forest pests. The new methods can be developed based on theoretical approaches, with modelling and simulations for policy decision making, and can be tested and implemented through practical field applications. Information that is required for decision making is provided through hazard ratings, the examination of the potential economic and ecological impacts, the prediction of dispersal patterns for forest pests, etc. Models can simulate natural conditions to understand the causes of occurrence and decline of forest pests and evaluate the influence of various environmental factors.

This special issue ("Impacts, Monitoring, and Management of Forest Pests and Diseases") aims to share information to assist in the effective management of forest pests, by understanding the responses of forest pests to natural and anthropogenic changes, and discussing new studies on the monitoring, assessment, and management of forest pests.

2. Papers in this Issue

The thirteen papers included in this issue focus on monitoring, assessing, and managing forest pests (Table 1). Choi et al. [4] reviewed changes in major forest pests, specifically, insect pests of pine forests in Korea over the last 50 years, and presented the shift in pests from the pine needle gall midge to the pine wilt disease due to changes in the forest structure. Meanwhile, Sierota et al. [18] reported on changes in the health of forests and their biological diversity in Poland over the last 30 years, by considering the climate, species composition, impacts of forest pests, forest management, etc. They concluded that forests in Poland are highly diversified but vulnerable to outbreaks of forest pests.

In this special issue, four papers focused on the monitoring of forest pests. Choi et al. [19] revised the taxonomy of the Japanese pine bast scale, *Matsucoccus matsumurae* (Kuwana), which had been described as *Matsucoccus thunbergii* by Miller and Park [20], based on morphological and DNA sequence analyses. They also updated the occurrence distribution of this species. Jennings et al. [21] compared two methods (visual survey and bark shifting) for monitoring the establishment of *Oobius agrili* (Zhang and Huang), which is a parasitoid on the emerald ash borer (*Agrilus planipennis* (Fairmaire)), an invasive species in Maryland, USA. They concluded that visual surveying was more efficient than bark shifting. Meanwhile, Mora-Sala et al. [22] demonstrated the efficiency of qPCR approaches to determine the presence and distribution of *Phytophthora* sp. in holm oak areas. In a study on the top-down attack pattern of *Tomicus brevipilosus* on *Pinus kesiya*, Liu et al. [23] reported that the initial attack location and aggregation pheromones were important for mediating the aggressive behavior and top-down attack pattern.

Meanwhile, three papers focused on the assessment of forest pests. Soares et al. [24] studied eucalyptus defoliation caused by the fungus *Calonectria pteridis*. They reported that the defoliation varied according to the doses of calcium (Ca) and potassium (K). An imbalance of Ca and K favors leaf blight and defoliation caused by the fungus in Eucalyptus plants, indicating the need for a balanced supply of Ca and K, to reduce the disease severity and defoliation. Zhang et al. [25] studied the differences in the trapping effects of *Semanotus bifasciatus* between two populations (Beijing and Shandong), based on next-generation sequencing, by analyzing the antennal transcriptome of both sexes of *S. bifasciatus* from the two populations. They found that the expression levels of odorant binding proteins, odorant receptors, and sensory neuron membrane proteins in males were lower in the Beijing population than in the Shandong population. Szmidla et al. [26] studied the threat of root-feeding *Melolonthinae* larvae on scots pine plantations, and reported that the population size of *Melolonthinae* and location of the larvae were related to the weather conditions. Therefore, they recommended an assessment during spring instead of autumn, to properly assess the cockchafer threat.

Finally, four papers focused on the management of forest pests. Kamczyc et al. [27] studied *Mesostigmata* mite communities in the decomposed litter of broadleaved and coniferous trees, to evaluate the effects of the re-establishment of mixed or broadleaved forests with native species. The results revealed that the species richness and diversity of the mite community were not affected, but the mite abundance was influenced by the tree species, which suggests differences in the litter quality. They also proposed that the exposition time is an important driver in shaping the mite community during the early stages of litter decomposition. Lalik et al. [28] compared the efficacy of different attractants in pitfall traps to capture the large pine weevil, *Hylobius abietis* (Linnaeus), which is the main pest of coniferous seedlings in Europe. They reported that the pitfall traps with attractants of pine twigs and ethanol are useful for monitoring and controlling *H. abietis*. Baek et al. [29] evaluated the current and future distribution of *Ricania shantungensis* (Chou and Lu), which is an invasive planthopper and an important pest in agriculture and forestry in Korea [30], using CLIMEX and the Maximum Entropy Model (MaxEnt). In the MaxEnt, the most important variables determining the distribution of *R. shantungensis* were the annual mean temperature, mean temperature of the coldest month, maximum temperature in the warmest month, and precipitation of the driest month. The study concluded that the probability of occurrence of this species is higher in western areas than in eastern areas of Korea, with great potential to spread eastward. Finally, Lee et al. [31] predicted the potential distribution of the citrus *Flatid planthopper*, *Metcalfa pruinosa* (Say), which is an invasive species in many countries, based on the gradient of environmental conditions using a random forest model. Their results show that factors relating to human activities strongly influence the occurrence and dispersal of the citrus *Flatid planthopper*, which has a high potential to disperse over the whole of South Korea.

Table 1. List of papers included in the special issue.

Category	Organism	Title	Authors
Review	Insects and nematodes	Changes in major insect pests of pine forests in Korea over the last 50 years	Choi et al. [4]
	Forests	Abiotic and biotic disturbances affecting forest health in Poland over the past 30 years: Impacts of climate and forest management	Sierota et al. [18]
Monitoring	Insect	Review of Japanese pine bast scale, *Matsucoccus matsumurae* (Kuwana) (Coccomorpha: Matsucoccidae), occurring on Japanese black pine (*Pinus thunbergii* Parl.) and Japanese red pine (*P. densiflora* Siebold & Zucc.) from Korea	Choi et al. [19]
	Insect	Comparing methods for monitoring establishment of the emerald ash borer (*Agrilus planipennis*, Coleoptera: Buprestidae) egg parasitoid *Oobius agrili* (Hymenoptera: Encyrtidae) in Maryland, USA	Jennings et al. [21]
	Insect	Initial location preference together with aggregation pheromones regulate the attack pattern of *Tomicus brevipilosus* (Coleoptera: Curculionidae) on *Pinus kesiya*	Liu et al. [23]
	Pathogen	The use of qPCR reveals a high frequency of *Phytophthora quercina* in two Spanish holm oak areas	Mora-Sala et al. [22]
Assessment	Pathogen	Calcium and potassium imbalance favours leaf blight and defoliation caused by *Calonectria pteridis* in eucalyptus plants	Soares et al. [24]
	Insect	Chemosensory characteristics of two *Semanotus bifasciatus* populations	Zhang et al. [25]
	Insect	The spring assessing method of the threat of *Melolontha* spp. grubs for scots pine plantations	Szmidla et al. [26]
Management	Insect	Mite communities (Acari, Mesostigmata) in the initially decomposed 'litter islands' of 11 tree species in scots pine (*Pinus sylvestris* L.) forest	Kamczyc et al. [27]
	Insect	Simple is best: Pine twigs are better than artificial lures for trapping of pine weevils in pitfall traps	Lalik et al. [28]
	Insect	Current and future distribution of *Ricania shantungensis* (Hemiptera: Ricaniidae) in Korea: Application of spatial analysis to select relevant environmental variables for MaxEnt and CLIMEX modeling	Baek et al. [29]
	Insect	Occurrence prediction of the citrus *Flatid planthopper* (*Metcalfa pruinosa* (Say, 1830)) in South Korea using a random forest model	Lee et al. [31]

3. Conclusions

Global change and increases in human activities induce changes in the habitats of various organisms, including plants and animals, resulting in a reduction in ecosystem stability and increased outbreaks of insect pests and pathogens. In many countries, most forest pests are currently invasive species, causing huge negative impacts on both economic and ecological aspects. To minimize the impacts and effectively control them, the appropriate systems for monitoring, assessing, and managing forest insect pests and diseases are essential. We believe that this special issue provides a better understanding of the structures and processes in forest ecosystems and fundamental information for the effective management of forest pests.

Author Contributions: Conceptualization, W.I.C. and Y.-S.P.; writing—original draft preparation, W.I.C. and Y.-S.P.; writing—review and editing, W.I.C. and Y.-S.P.

Funding: This study was supported by the National Institute of Forest Science and R & D Program for Forest Science Technology (FTIS 2017042A00-1823-CA01) provided by Korea Forest Service (Korea Forestry).

Acknowledgments: We would like to thank all contributors in this special issue and all reviewers who provided very constructive and helpful comments to evaluate and improve the manuscripts.

Conflicts of Interest: The authors declare no conflict of interest. The funders had no role in the design of the study; in the collection, analyses, or interpretation of data; in the writing of the manuscript, or in the decision to publish the results.

References

1. Hobbie, S.E.; Villeger, S. Interactive effects of plants, decomposers, herbivores, and predators on nutrient cycling. In *Trophic Ecology: Bottom-Up and Top-Down Interactions across Aquatic and Terrestrial System*; Hanley, T.C., Pierre, K.J.L., Eds.; Cambridge University Press: Cambridge, UK, 2015; pp. 233–259. [CrossRef]

2. Seidl, R.; Thom, D.; Kautz, M.; Martin-Benito, D.; Peltoniemi, M.; Vacchiano, G.; Wild, J.; Ascoli, D.; Petr, M.; Honkaniemi, J.; et al. Forest disturbances under climate change. *Nat. Clim. Chang.* **2017**, *7*, 395. [CrossRef] [PubMed]
3. Sharov, A.A.; Liebhold, A.M. Model of slowing the spread of gypsy moth (Lepidoptera: *Lymantriidae*) with a barrier zone. *Ecol. Appl.* **1998**, *8*, 1170–1179. [CrossRef]
4. Choi, W.I.; Nam, Y.; Lee, C.Y.; Choi, B.K.; Shin, Y.J.; Lim, J.-H.; Koh, S.-H.; Park, Y.-S. Changes in major insect pests of pine forests in Korea over the last 50 years. *Forests* **2019**, *10*, 692. [CrossRef]
5. Potter, K.M.; Conkling, B.L. *Forest Health Monitoring: National Status, Trends, And Analysis 2017*; Southern Research Station: Asheville, NC, USA, 2018.
6. Turchin, P.; Taylor, A.D. Complex dynamics in ecological time series. *Ecology* **1992**, *73*, 289–305. [CrossRef]
7. Li, S.; Daudin, J.J.; Piou, D.; Robinet, C.; Jactel, H. Periodicity and synchrony of pine processionary moth outbreaks in France. *For. Ecol. Manag.* **2015**, *354*, 309–317. [CrossRef]
8. Choi, W.I.; Park, Y.-S. Dispersal patterns of exotic forest pests in South Korea. *Insect Sci.* **2012**, *19*, 535–548. [CrossRef]
9. Choi, W.I.; Song, H.J.; Kim, D.S.; Lee, D.-S.; Lee, C.-Y.; Nam, Y.; Kim, J.-B.; Park, Y.-S. Dispersal patterns of pine wilt disease in the early stage of its invasion in South Korea. *Forests* **2017**, *8*, 411. [CrossRef]
10. Lee, D.-S.; Nam, Y.; Choi, W.I.; Park, Y.-S. Environmental factors influencing on the occurrence of pine wilt disease in Korea. *Korean J. Ecol. Environ.* **2017**, *50*, 374–380. [CrossRef]
11. Cha, D.J.; Kim, D.S.; Lee, S.K.; Han, H.R. A new on-site detection method for *Bursaphelenchus xylophilus* in infected pine trees. *For. Pathol.* **2019**, *49*, e12503. [CrossRef]
12. Park, Y.-S.; Chung, Y.-J. Hazard rating of pine trees from a forest insect pest using artificial neural networks. *For. Ecol. Manag.* **2006**, *222*, 222–233. [CrossRef]
13. Nam, Y.; Koh, S.-H.; Jeon, S.-J.; Youn, H.-J.; Park, Y.-S.; Choi, W.I. Hazard rating of coastal pine forests for a black pine bast scale using self-organizing map (SOM) and random forest approaches. *Ecol. Inform.* **2015**, *29*, 206–213. [CrossRef]
14. Park, Y.-S.; Chung, Y.-J.; Moon, Y.-S. Hazard ratings of pine forests to a pine wilt disease at two spatial scales (individual trees and stands) using self-organizing map and random forest. *Ecol. Inform.* **2013**, *13*, 40–46. [CrossRef]
15. Kurz, W.A.; Dymond, C.C.; Stinson, G.; Rampley, G.J.; Neilson, E.T.; Carroll, A.L.; Ebata, T.; Safranyik, L. Mountain pine beetle and forest carbon feedback to climate change. *Nature* **2008**, *452*, 987–990. [CrossRef] [PubMed]
16. Hyun, J.S. Studies on the prevision for occurrence of pine moth, *Dendrolimus spectabilis* Butler. *Ent. Res. Bull.* **1968**, *4*, 57–80.
17. Kwon, T.S.; Lim, J.H.; Sim, S.J.; Kwon, Y.D.; Son, S.K.; Lee, K.Y.; Kim, Y.T.; Park, J.W.; Shin, C.H.; Ryu, S.B.; et al. Distribution patterns of *Monochamus alternatus* and *M. saltuarius* (Coleoptera: *Cerambycidae*) in Korea. *J. Korean For. Soc.* **2006**, *95*, 543–550.
18. Sierota, Z.; Grodzki, W.; Szczepkowski, A. Abiotic and biotic disturbances affecting forest health in Poland over the past 30 years: Impacts of climate and forest management. *Forests* **2019**, *10*, 75. [CrossRef]
19. Choi, J.; Cha, D.; Kim, D.-S.; Lee, S. Review of Japanese pine bast scale, *Matsucoccus matsumurae* (Kuwana) (Coccomorpha: *Matsucoccidae*), occurring on Japanese black pine (*Pinus thunbergii* Parl.) and Japanese red pine (*P. densiflora* Siebold & Zucc.) from Korea. *Forests* **2019**, *10*, 639. [CrossRef]
20. Miller, D.R.; Park, S.-C. A new species of *Matsucoccus* (Homoptera: *Coccoidae: Margarodidae*) from Korea. *Korean J. Plant. Prot.* **1987**, *26*, 49–62.
21. Jennings, D.E.; Duan, J.J.; Shrewsbury, P.M. Comparing methods for monitoring establishment of the emerald ash borer (*Agrilus planipennis*, Coleoptera: *Buprestidae*) egg parasitoid *Oobius agrili* (Hymenoptera: *Encyrtidae*) in Maryland, USA. *Forests* **2018**, *9*, 659. [CrossRef]
22. Mora-Sala, B.; Berbegal, M.; Abad-Campos, P. The use of qPCR reveals a high frequency of *Phytophthora quercina* in two Spanish holm oak areas. *Forests* **2018**, *9*, 697. [CrossRef]
23. Liu, F.; Wu, C.; Zhang, S.; Kong, X.; Zhang, Z.; Wang, P. Initial location preference together with aggregation pheromones regulate the attack pattern of *Tomicus brevipilosus* (Coleoptera: *Curculionidae*) on *Pinus kesiya*. *Forests* **2019**, *10*, 156. [CrossRef]

24. Soares, T.P.F.; Pozza, E.A.; Pozza, A.A.A.; Mafia, R.G.; Ferreira, M.A. Calcium and potassium imbalance favours leaf blight and defoliation caused by *Calonectria pteridis* in Eucalyptus plants. *Forests* **2018**, *9*, 782. [CrossRef]
25. Zhang, S.; Shen, S.; Zhang, S.; Wang, H.; Kong, X.; Liu, F.; Zhang, Z. Chemosensory characteristics of two *Semanotus bifasciatus* populations. *Forests* **2019**, *10*, 655. [CrossRef]
26. Szmidla, H.; Małecka, M.; Tkaczyk, M.; Tarwacki, G.; Sierota, Z. The spring assessing method of the threat of *Melolontha* spp. grubs for Scots pine plantations. *Forests* **2019**, *10*, 399. [CrossRef]
27. Kamczyc, J.; Dyderski, M.K.; Horodecki, P.; Jagodziński, A.M. Mite communities (Acari, Mesostigmata) in the initially decomposed 'litter islands' of 11 tree species in Scots pine (*Pinus sylvestris* L.). *Forests* **2019**, *10*, 403. [CrossRef]
28. Lalík, M.; Holuša, J.; Galko, J.; Resnerová, K.; Kunca, A.; Nikolov, C.; Mudrončeková, S.; Surový, P. Simple is best: Pine twigs are better than artificial lures for trapping of pine weevils in pitfall traps. *Forests* **2019**, *10*, 642. [CrossRef]
29. Baek, S.; Kim, M.-J.; Lee, J.-H. Current and future distribution of *Ricania shantungensis* (Hemiptera: *Ricaniidae*) in Korea: Application of spatial analysis to select relevant environmental variables for MaxEnt and CLIMEX modeling. *Forests* **2019**, *10*, 490. [CrossRef]
30. Baek, S.; Koh, S.-H.; Lee, J.-H. Occurrence model of first instars of *Ricania shantungensis* (Hemiptera: *Ricaniidae*). *J. Asia-Pac. Entomol.* **2019**, *22*, 1040–1045. [CrossRef]
31. Lee, D.-S.; Bae, Y.-S.; Byun, B.-K.; Lee, S.; Park, J.K.; Park, Y.-S. Occurrence prediction of the citrus flatid planthopper (*Metcalfa pruinosa* (Say, 1830)) in South Korea using a random forest model. *Forest* **2019**, *10*, 585. [CrossRef]

forests

MDPI

Review

Changes in Major Insect Pests of Pine Forests in Korea Over the Last 50 Years

Won Il Choi [1], Youngwoo Nam [2], Cha Young Lee [2], Byoung Ki Choi [3], Yu Jin Shin [1], Jong-Hwan Lim [1], Sang-Hyun Koh [3] and Young-Seuk Park [4,5,*]

1. Division of Forest Ecology and Climate Change Division, National Institute of Forest Science, Dongdaemun, Seoul 02445, Korea
2. Division of Forest Diseases and Insect Pests, National Institute of Forest Science, Dongdaemun, Seoul 02445, Korea
3. Warm Temperate and Subtropical Forest Research Center, National Institute of Forest Science, Seogwipo, Jeju 63582, Korea
4. Department of Biology, Kyung Hee University, Dongdaemun, Seoul 02447, Korea
5. Department of Life and Nanopharmaceutical Sciences, Kyung Hee University, Dongdaemun, Seoul 02447, Korea

* Correspondence: parkys@khu.ac.kr; Tel.: +82-2961-0946

Received: 29 June 2019; Accepted: 12 August 2019; Published: 15 August 2019

Abstract: Understanding the occurrence patterns of forest pests is fundamental for effective forest management from both economic and ecological perspectives. Here, we review the history of the occurrence patterns and causes of outbreaks and declines of pests in Korean pine forests over the last 50 years. During this period, the major pests of pine forests in Korea have shifted from pine caterpillar (*Dendrolimus spectabilis* Butler) to the pine needle gall midge (PNGM, *Thecodiplosis japonensis* (Uchida and Inouye)) and finally to pine wilt disease (PWD) caused by the pine wood nematode (*Bursaphelenchus xylophilus* (Steiner and Buhrer) Nickle). Outbreaks of pine caterpillar, a native species in Korea, have been recorded as far back as 900 years, and it was the most relevant forest pest in Korea until the 1970s. The decline of its importance has been attributed to reforestation and higher levels of subsequent natural enemy activity. The PNGM is an invasive species, first discovered in Korea in 1929, that became widely distributed by 1992 and the major forest pest in the 1980s and 1990s. A suite of parasitic wasps attacking the PNGM contributed at least partially to the decline of PNGM densities. Following the decline of the PNGM, damage from PWD has increased since 2003. These shifts in major forest pests might be related to changes in forest composition and interactions among forest pests. Therefore, a new management strategy for controlling forest pests is required to mitigate the decline of pine forests in Korea.

Keywords: invasive species; natural enemies; pine caterpillar; pine needle gall midge; pine wilt disease

1. Introduction

Understanding forest pest outbreaks is fundamental for the effective economic and ecological management of forest ecosystems. Outbreaks of forest pests are affected by both intrinsic and extrinsic factors. Intrinsic factors are related to the rate of population increase and cyclical population dynamics [1], while extrinsic factors include biological factors such as the interactions between species and the effects of abiotic factors such as temperature and precipitation [2]. Biological interactions include the bottom-up effects of forest composition and tree age and the top-down regulation exerted by natural enemies [2–4]. In addition, anthropogenic factors directly or indirectly influence forest pest dynamics [5]. Long-term monitoring data for pests and their analyses can allow us to understand historical changes in natural forest ecosystems under the influence of both natural phenomenon and anthropogenic activities.

Long-term monitoring programs on forest pests allow us to describe abundance patterns of pests [6,7]. Schwerdtfeger [8] described population changes of three species of pine-needle feeding Lepidoptera (*Dendrolimus pini* L., *Hyloicus pinastri* L. and *Bupalus piniarius* L.) in a pine plantation in Germany between 1880 and 1940. Using these data, Turchin and Taylor [9] showed the periodical occurrence of these species. Meanwhile, Tenow [10] studied the outbreaks of *Oporina autumnata* and *Operophtera* sp. in the Swedish Scandes Mountains on the basis of historical data from 1862 to 1968, and periodic outbreaks of *Operophtera brumata* L. continued until the 1990s [11].

In Korea, the monitoring of forest pests has been conducted for both major and occasional pests such as *Thecodiplosis japonensis* (Uchida and Inouye) (pine needle gall midge (PNGM)) and *Hyphantria cunea* (Drury) (fall webworm) populations since 1968 [12], including the measurement of annual changes of densities, dispersal, and distributions [13]. Choi and Park [12] reported the occurrence and dispersal patterns of these invasive species with their ecology and management histories. However, few studies have examined the long-term changes in forest pest populations in relation to the conditions of the forest ecosystem.

Pine forests are one of the most important forests in Korea because of their dominance and cultural importance. Forests of *Pinus densiflora* Siebold and Zuccarini account for 26% of all Korean forests [14], and wood from *P. densiflora* has been intensively used in Korea since the Goryeo Dynasty (918–1392) [15]. The major forest pests in Korea have occurred in pine forests [12].

Here, we review the changes in the major pests of pine forests in Korea, in particular of stands of *P. densiflora*, over the last 50 years. We consider occurrence histories and ecologies of these pests as well as the environmental factors influencing the pests' population dynamics. Management strategies based on natural enemies and chemical controls are discussed and new directions for forest pest management are proposed.

2. Forest Change and Monitoring Records in Korea

Wood was the main building material in ancient Korea. Houses from different periods reflect the availability of wood for construction. Based on the materials used, Park and Lee [15] estimated that oak trees were the dominant material (57%) from 100 B.C. to 910 (Three Kingdoms Period of Korea), whereas pine trees were the dominant species from 910 to the 1910s (Goryeo and Joseon Dynasties) with increasing dominancy from 71% in the early period to 88% in the late period [15]. The change in forests from oak to pine likely influenced the distribution and occurrence of animals living in the forests [16].

As the pine trees became dominant in Korean forests, the pine caterpillar (PC) (*Dendrolimus spectabilis* Butler) outbreaks were recorded in the Goryeo and Joseon dynasties. After the first record of a PC outbreak in *The History of Goryeo* (Goryeosa) in 1101 (http://www.history.go.kr), more than 50 cases of PC outbreaks were subsequently recorded in the history books of the Goryeo (918–1392) and Joseon (1392–1897) dynasties, such as *The History of Goryeo* (Goryeosa), *The Veritable Records of the Joseon Dynasty* (Joseon Wangjo Sillok, also called as *The Annals of the Chosun Dynasty*) and *Seungjeongwon Ilgi* (http://www.history.go.kr). These historical records qualitatively describe both the occurrence and control of pests and the prevailing forest conditions, but they do so with little quantitative information.

Modern Korean forests are considered a successful example of reforestation, and the annual growing stock and species composition of forests have changed dramatically as a result [17] (Figure 1). After the Korean War in the early 1950s, deforestation was intensive, with the major causes of deforestation being utilization of wood as a fuel source and slash-and-burn agriculture [17]. In particular, Japanese red pines, *P. densiflora*, were intensively used for fuel from the 1950s to the early 1970s, with 10 million m^3 of woods including pine woods consumed annually [18]. The area of slash-and-burn agriculture likewise increased until the early 1970s [19], when it was prohibited by the government new policy for reforestation, and the main domestic fuel sources were changed from wood to coal. Due to this change in forest management policy, the growing stock in forests increased from 5.7 m^3/ha in 1952 to 146.0 m^3/ha in 2015 [14]. Meanwhile, the dominance of coniferous species decreased from 57.3% in 1970 to 41.8% in 2010 (Figure 1a), although *P. densiflora* remained the most dominant coniferous species, accounting for 60% of conifers [14].

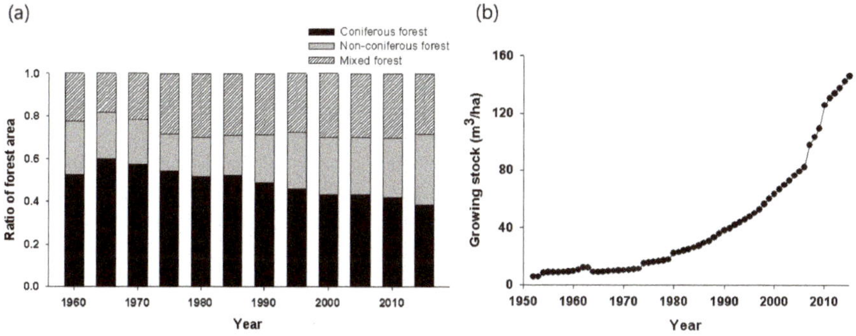

Figure 1. (**a**) Changes of forest composition in Korea from 1960 to 2015 and (**b**) annual changes in forest growing stock (m^3/ha). The graphs were redrawn based on data from Statistical Yearbook of Forestry [14].

3. Monitoring of Korean Forest Pests

The Korea Forest Research Institute (later renamed the National Institute of Forest Science in 2015) initiated the Forest Insect Pests Monitoring Program in 1968. The occurrence of the PC, the major forest pests in 1960s, was checked annually at a series of national long term monitoring sites. The determination of the emergence period and leading edge of dispersal of the PNGM has been done since the mid-1970s. Outbreaks of occasional forest pests and the annual density of fall webworm (*H. cunea*) have also been monitored from the 1970s, and the areas damaged by black bast scale (*Matsucoccus matsumurae* (Kuwana)) and pine wilt disease (PWD) and their dispersal areas have been monitored since the 1980s. Monitoring data have been published in the form of annual reports entitled "Annual Report of Monitoring for Forest Insect Pests and Diseases in Korea" each year since 1968.

Targeted pests of the monitoring program have changed over the years according to the severity of each pest species. Major forest pests such as PWD, the PNGM, black pine bast scale, fall webworm, and Korean oak wilt disease caused by *Raffaelea quercus-mongolicae* (K.H. Kim, Y.J. Choi and H.D. Shin) and its insect vector (*Platypus koryoensis* (Murayama)) have been continuously monitored, although several domestic forest pests such as the PC were excluded in reorganization of monitoring system in 2016. Different factors have been monitored for different pests depending on their characteristics of damage. For PWD, dead pine trees at the forefront of PWD dispersal were sampled to determine if the trees were infected by the pine wood nematode; for the PNGM, the gall density per branch was used as a measure of PNGM density, and its parasitism was assessed by the microscopic observation parasitized larvae of the PNGM. The annual emergence patterns of forest insect pests such as the PNGM, fall webworm and the vector insects such as *Monochamus saltuarius* (Gebler), *Monochamus alternatus* Hope and *P. koryoensis* have been annually surveyed in at least nine sites since 2016. For occasional pests, the occurrence areas and degree of damages have been recorded. Recently, spotted lanternfly (*Lycorma delicatula* (White)), the citrus flatid plant hopper (*Metcalfa pruinose* (Say)) and ricanid plant hopper (*Ricania shantungensis* Chou and Lu) were observed in the field, and their damage areas have gradually increased [13,20].

The total forest area damaged by insect pests in South Korea has dramatically decreased from 421,234 ha in 1968 to 36,217 ha in 2016 (Figure 2). Such a sharp decrease was likely due to increases in both the budget and the manpower deployed to control forest insect pests. The area damaged by the

PC was the highest with 543,244 ha in 1965, from which it gradually decreased to less than 18,500 ha in 1983. Meanwhile, the area damaged by the PNGM was 390,185 ha in 1976 and gradually decreased to 73,206 ha in 2004. After that, the damaged area has increased up to 195,759 ha in 2006 and then decreased to 44,166 ha in 2016, showing annual fluctuations at different regions. These trends display that the PNGM is in the stabilization phase in terms of invasive species dispersal with the population dynamics. The area damaged by black pine bast scale was 5390 ha in 1984, and that area increased to 16,007 ha in 1996. It then gradually decreased to 4906 ha in 2016, indicating the saturation phase in its dispersal as well. Black pine bast scale occurs along the coastal areas where Japanese black pines grow, and the distribution of black pine bast scale has gradually increased [12]. Korean oak wilt disease was first observed in Korea in 2004, and the damage area increased to 4087 ha by 2008. Then the damaged area decreased with fluctuation to 2081 ha in 2016. The area damaged by PWD was 72 ha in 1988 and was less than 100 ha until 1999. However, it increased to 7811 ha in 2005, indicating rapid dispersal since 2000 [14]. The damaged areas increased up to 11,550 ha in 2013 and decreased to 6325 ha in 2016 due to intensive control measurements [14].

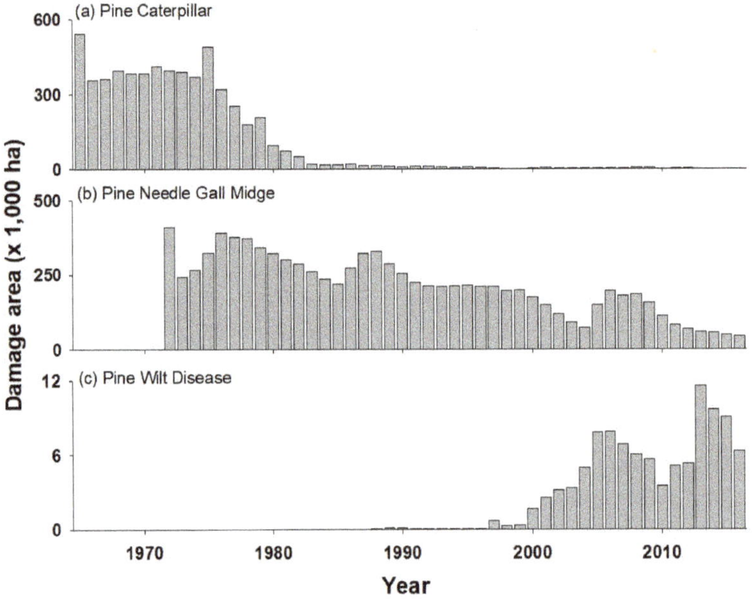

Figure 2. Changes of area damaged by forest pests in Korea. (**a**) Pine caterpillar, (**b**) pine needle gall midge, and (**c**) pine wilt disease. The graphs were redrawn based on data from Statistical Yearbook of Forestry [14].

It is worth noting that these three species have different impacts on pine trees and that the damaged area was not coincided with the number of dead trees. The defoliation of pine trees caused by the PC was rarely resulted in the death of trees, whereas pine trees infested by the PNGM displayed less than 30% of tree mortality in the outbreak area [21]. Meanwhile, pine trees infested by the PWD have mostly died [22]. Therefore, PWD has more serious impacts on forest ecosystems than other two species, although the overall damaged area is relatively smaller than others.

4. Changes in Major Pests in Pine Forests

During the last 50 years, the major forest pests in South Korea have been pine-feeding insects, and shifted from the PC via the PNGM to PWD. To understand these processes, we reviewed their ecologies and occurrence histories and the effects of environmental factors on the population dynamics of these species. In addition, we considered management strategies with natural enemies and chemical controls. Following the review of Choi and Park [12] on the dispersal patterns of invasive species in Korean forests, we revised the occurrence and dispersal patterns of major insect pests in pine forests in South Korea.

4.1. Pine Caterpillar (PC)

4.1.1. Occurrence History and Ecology

The occurrence of the pine caterpillar (PC; *D. spectabilis*) had been documented in ancient literature, as well as in the pest monitoring reports. After its first recorded occurrence in 1101, the PC was documented in the Goryeo Dynasty and Joseon Dynasty (http://www.history.go.kr). PC outbreaks were also reported from the 1900s to 1950s. These records confirm that the PC has been a major forest pest in Korea for a very long time [23]. The PC is native in areas of the Asian Far East, such as Korea, China, Japan and Russia.

The PC was a major forest pest in Korea from 1950s to the early 1970s, after which its occurrence dramatically decreased to the point that it is now considered an occasional pest with very low abundance. Since 2000, the PC has occurred mainly in limited areas on islands or along roads rather than in forested areas [24]. The occurrence of the PC on islands has been reported consistently [13].

Before the 1970s, the PC was regarded as a univoltine species throughout South Korea, with no temporal or spatial variation [24–27]. However, the voltinism of the PC population in the central region of the Korean Peninsula was reported to have changed from univoltinism to bivoltinism in late 1990s and early 2000s [24,28]. Choi et al. [24] reported the host switching of the PC from the Japanese red pine to the pitch pine (*Pinus rigida* (Mill.)) by comparing with the results of Park and Kim [29], who noted a preference of the PC for *P. densiflora* over other pines, including *P. rigida* and *P. thunbergii*, under laboratory conditions in 1973.

The body length of PC adults is about 30 mm for males and 40 mm for females, with a wing span ranging from 50 to 67 mm in males and 64–88 mm in females [27]. Eggs are reddish brown and blue brown ovals, and their diameter is about 2 mm; larvae are dark yellow-grey with an irregular pattern of dark orange and light gray [27]. The life cycle of the PC was reported in detail by Choi et al. [24]. Pine trees heavily defoliated by the PC have typically died [27].

4.1.2. Environmental Factors and Management

The lower developmental temperature thresholds of the PC are 3.2 °C and 0.91 °C for eggs and larvae, respectively. The minimum and the optimal developmental temperatures for ≥4th instar PC larvae are 7.6 °C and 29.8 °C, respectively; for pupae, the minimum and optimal development temperatures are 12.5 °C and 30.4 °C, respectively [24].

Though the PC was a major forest pest in 1950s and 1960s, there were few studies on its population dynamics [30]. Hyun [30] reported that mortality of young larvae ranged from 70.9% to 93.4%, and precipitation in August was the main factor influencing larval mortality. Similarly, Kokubo [31] reported that precipitation was the main mortality factor in Japan for the younger larval stages, with mortality of the first and the second larval stages of 70%–80%.

Outbreaks of *Dendrolimus* spp. including *D. spectabilis* in Shandong province, China, have been found to be affected by drought [32]. Based on occurrence data for the PC from 1992 to 2012, outbreaks of the PC increased under drought conditions, and long-term or prolonged drought had a greater positive impact on PC outbreaks than short-term drought.

The egg mortality of the PC is caused mainly by parasitic wasps. In the late 1960s, the parasitism by *Trichogramma dendrolimi* Matsumura averaged 15.7%, while that by *Anastatus bifasciatus* (Geoffroy) was extremely low [30]. By 1999, the importance of these two egg parasitoids had reversed, with 0.01%–9.5% by *Trichogramma* sp. and *Anastatus* sp., while larval parasitism ranged from 1.6% to 22.8% for *Aleiodes dendrolimi* (Matsumura) and 8.0%–18.5% for an unidentified tachinid fly. Pupal parasitism by the tachinid fly ranged from 13.5% to 41.3%, and PC populations have been found to be density-independently regulated by summer rainfall (especially in August) and parasitic insects [30,33].

Insect pathogens are also an important factor affecting PC abundance, often being the factor that terminates outbreaks, acting as a density-dependent source of mortality [30]. One such important pathogen has been the fungus *Beauveria bassiana* (Bals.) [34].

Suppression by natural enemies appears to have increased during over a period of reforestation. Generally, reforestation increase relative humidity, resulting in a presumably more favorable condition for insect-pathogenic fungi. For example, outbreaks of the gypsy moth in southern New England from 2015 to 2017 were attributed to unusual spring drought because activities of fungal pathogen, *Entomophaga maimaiga* Humber, Shimazu, and Soper were inhibited due to dry condition [35]. Reforestation may also promote increases in the populations of natural enemies such as parasitoids—both the number of species and their abundance—due to an increasing availability of alternative hosts [36,37].

Before the development of chemical pesticides and augmentative biological control methods, the manual collection of larvae by soldiers or citizens was employed, mainly during the Joseon Dynasty (1392–1897) (http://www.history.go.kr). Pesticides have also been widely used for the control of the PC [33], while in contrast, biological control using an egg parasitoid such as *T. dendrolimi* through the rearing and releasing the parasitoid was not attempted until the late 1990s and 2000s [33]. However, parasitoid releases were stopped around 2000 because of the decline of PC abundance.

4.2. *Pine Needle Gall Midge (PNGM)*

4.2.1. Occurrence History and Ecology

The invasive pine needle gall midge (PNGM) (*T. japonensis*) was a major forest insect pest in South Korea in the 1980s and 1990s, when it caused serious damage to *P. densiflora* and *P. thunbergii*. The first occurrence of the PNGM in Korea was reported in Seoul and Mokpo in 1929, possibly by two independent introductions. Meanwhile, the PNGM was first reported in Japan in 1901 without a report of the pest's origin [38].

After the first report of the occurrence in 1929, the PNGM spread to Busan (in the southern part of South Korea) in 1936 and to Danyang (in the middle part of South Korea) in 1964 [39]. By 1985, the PNGM occurred in more than 80% of South Korea and was found throughout the country by 1996 [40]. The dispersal of the PNGM in Korea was classified as Type 2 based on Shigesada et al. [41], moving at about 1.2–2.2 km/year during the early invasion phase [12] and accelerating to 5.2–8.2 km/year by the 1960s. For detailed information on the pest's dispersal in Korea, see Lee et al. [40] and Choi and Park [12].

The larvae of PNGMs overwinter in soil, and adults emerge from pupae in the soil from late May to early July [12]. Female adults lay eggs in the pine needles of pine trees, and larvae form galls at the bases of the pine needles, thus damaging their host pine trees through loss of new pine needles [42]. This results in a reduction of tree growth and an increase in tree mortality.

The PNGM reached outbreak levels five to seven years after invasion [43], and outbreaks have shown a periodicity of 10–12 years [44]. During outbreaks, the mortality of pine trees reached 30% [21] and then gradually decreased, probably due to increase in its parasitoids [27,45]. The area damaged by the PNGM was the highest (351,679 ha) in 1971 and then gradually decreased to 73,206 ha by 2004 [14]. This might have been caused by several factors such as PNGM management and the development of natural enemies [12,46].

4.2.2. Environmental Factors and Management

The minimum and optimal developmental temperatures for the PNGM are 5.9 °C and 27.0 °C, respectively, while the lower and the upper lethal threshold temperatures for post-diapause larvae through adults are 6.1 °C [47] and 30.0 °C [45], respectively. Soil water content in spring influences

PNGM density. The emergence rate increases with increasing soil water content, while it decreases with the decreasing water content [48]. A field life table analysis showed that population size was decreased due to a lower soil water content in spring [49]. This indicated that post-diapause mortality was a key factor to determine population size of the PNGM [12]. Choi et al. [50] determined the most influential environmental factors for population dynamics of the PNGM through spatial synchrony analysis using 20 years of monitoring data at 67 sites. They showed that differences in maximum temperature and precipitation were strongly related with variation in spatial synchrony, indicating that these density-independent factors potentially contributed to a fluctuation of the PNGM population.

Besides these abiotic environmental factors, population dynamics are also influenced by biological interactions such as competition, predator–prey, and host–parasite dynamics [51,52]. Park and Chung [42] developed models for the risk assessment of pine trees from the PNGM using two different artificial neural networks such as a self-organizing map and a multilayer perceptron, and they revealed that the crown density of pine trees highly influences the survival status of pine trees infested by the PNGM. Four species—*Inostemma matsutama* Yoshida and Hirashima, *Inostemma seoulis* Ko, *Inostemma hockpari* Ko and *Platygaster matsutama* Yoshida—parasitize the PNGM in Korea [53]. There are no studies on the origin of these species, although we infer that they co-invaded with their host. Of these, *I. seoulis* was first reported as a parasitoid of the PNGM by Ko [45], while *I. matsutama* and *P. matsutama* were later described by Yoshida and Hirashima [54]. Meanwhile, Ko [55] described *I. hockpari* as a new species in 1980. These species have different distribution patterns with different abundances [56]; *I. seoulis* and *P. matsutama* were widely distributed across Korea by 1985, while *I. matsutama* and *I. hockpari* displayed a relatively local distribution [56].

Parasitoids appear to play little role in the control of the PNGM in the early stages of its invasion of a new area, but the parasitoids become more important with time [52,53]. Two species, *I. seoulis* and *I. matsutama*, are the most common in the early phase of a PNGM invasion, while *P. matsutama* is found in the later, stabilized PNGM population [51]. During the invasive process or in the early stage of invasion, interspecific competitions between *I. matsutama* and *I. seoulis* and between *P. matsutama* and *I. seoulis* are minimal [53,57]. The interactions among three parasitoids are affected by differences in their phenology; *I. matsutama* and *I. seoulis* have different phenologies because of difference in their thermal biology [58], resulting in a weak direct competition. Choi et al. [51] reported that the differences in establishment sequence and competitive ability among the parasitoids in the invasion process of their host determined the parasitoid community's structure and dynamics.

Several methods were developed to manage the PNGM in South Korea, including trunk injection of insecticide, spray of insecticide on soil, silviculture, and the release of natural enemy [12]. The trunk injection of insecticide applied insecticides such as phosphamidon and acetamiprid to trunk of pine trees infested by the PNGM. This was effective but labor intensive and costly. Insecticides such as carbofuran and imidacloprid were sprayed on the soil where the PNGM occurred. Silviculture methods such as thinning were applied to reduce the tree mortality caused by the PNGM by improving the health of pine forests. The release of natural enemies such as parasitoids (*I. matsutama*, *I. seoulis*, *I. hockpari* and *P. matsutama*) was used to reduce the PNGM population level [12,56].

4.3. Pine Wilt Disease (PWD)

4.3.1. Occurrence History and Ecology

Pine wilt disease (PWD) caused by the pine wood nematode (PWN) (*Bursaphelenchus xylophilus* Nickle) is a serious pest for pine trees in many countries, including South Korea, Japan, China, Taiwan, and Portugal [12,46,59–61]. Pine trees in Asia and Europe are susceptible to the PWN, and the tree mortality of *P. densiflora* and *P. thunbergii* can reach 100% when infected by the PWN [12]. The PWN, originally from North America [62], is a tree-parasitic nematode with a body length ranging from 0.6–1.0 mm [63]. It develops to an adult through egg and 1st–4th juveniles. Specific dispersal juveniles dauer larvae enter the body of vectors such as *M. alternatus* and *M. saltuarius* through the tracheal system when the vectors are larvae in the host tree. The female nematode, after mating, lays about 100 eggs.

PWD in South Korea was firstly reported at Mt. Geumjong in Busan in 1988 [12]. Its initial dispersal speed was relatively low at 1.1–1.2 km/year [64,65]. However, dispersal was accelerated to 13.8 km/year within 10 years [12]. These results show that the dispersal patterns of PWD are Type 2 based on Shigesada et al. [41]. Choi et al. [46] estimated the annual dispersal distance in South Korea to be 0.37 km, based on the annual changes of areas damaged by PWD in the early stages of invasion, and the annual dispersal distances were less than 1.0 km in most (88.8%) of damaged areas; they also showed that "jumping" was the dominant dispersal method in the early stages of the invasion, while later dispersal was due to the expansion of existing colonies. Dispersal speed increased with increasing human population density, suggesting human-mediated dispersal [66,67].

PWD damage occurred only on 72 ha in 1988 and less than 100 ha until 1999. Initially, the infested trees were removed to control PWD. A large number of infested pine trees (1598) were removed at Mt. Geumjong in 1989, and 14, 24, and 21 trees were removed in 1990, 1991 and 1992, respectively [13]. Meanwhile, 10, two and eight pine trees were cut to control PWD at outside of Mt. Geumjong in 1990, 1991, and 1992, respectively. These numbers indicate that PWD likely invaded Korea several years before the first report in 1988, and the initial control activities were not effective to minimize the dispersal of PWD in Korea because pine trees killed by PWD were found out of Mt. Geumjong within three years after the first report.

Meanwhile, the damaged area greatly increased to 7811 ha in 2005 [14], indicating that PWD had expanded its range significantly from 2000 to 2005 [12]. Nearly 1.7 million trees were cut to control PWD from May 2014 to April 2015, and the number of infected pine trees in 2017 decreased to 686,000 trees due to this intensive control effort [14].

The host plants of PWN in Korea include the Japanese red pine (*P. densiflora*), the Japanese black pine (*P. thunbergii*) and the Korean white pine (*Pinus koraiensis* Sieb. and Zucc.) [63]. The PWN vectors in Korea are *M. alternatus* and *M. saltuarius*, which both overwinter as larvae or pupae in a pupal chamber near the bark surface. Adults of *M. alternatus* emerge from late May to early August, and the female lays about 100 eggs [27], whereas adults of *M. saltuarius* emerge in mid-April to late May, earlier than *M. alternatus* [27,68]. PWN does not have the ability to disperse from an infected tree to a new tree outside of its vector. PWN individuals are transmitted from beetle vectors to pine trees during the maturation feeding of the adult beetles [12].

Togashi [69] evaluated the dispersal ability of *M. alternatus* by releasing 756 beetles, and 75.5% of them were captured within 100 m of the release point; it was estimated that *M. alternatus* could disperse 7.1–37.8 m per week in pine stands. The dispersal distances of *M. saltuarius* over its entire life span was estimated to be 2.71 and 1.93 km for males and females, respectively, based on a flight mill experiment [70]. The contributions of human-mediated dispersal of PWD were reported in China [71], Japan [72] and Korea [46,59]. Due to human-mediated dispersal, actual PWD dispersal distance was longer than the dispersal capacity of the vector insects.

4.3.2. Environmental Factors and Management

The development and distribution of PWD vectors are dependent on temperature with geographical variation. The lower threshold temperature for *M. alternatus* development in southern Japan and Taiwan is above 12 °C, whereas in northern Japan it is less than 13 °C [73]. The optimal development temperature of *M. saltuarius* is 20 °C, and the lower threshold temperature for development is estimated to be 10.1 °C for both sexes [74]. In Korea, the lower threshold, optimal, and upper threshold temperatures for post-diapause development of *M. saltuarius* are 8.3, 32.2 and 38.7 °C, respectively [75]. Park et al. [76] successfully predicted the spring emergence of *M. alternatus* based on a threshold temperature of 11.9 °C. These differences in lower threshold temperatures may explain the geographical distribution of two vectors in Korea. *M. saltuarius* is the most abundant in the central to northern areas of South Korea, while *M. alternatus* is found in the southern part of the country [68].

The transmission efficiency of PWN was increased when the nematodes and their vectors were exposed to temperatures in the range of 16 °C–25 °C [77], suggesting that optimal temperature positively affects the biological performance of both the nematodes and their vectors as well as their interactions.

The effect of temperature, precipitation and landscape features was evaluated on the level of risk of PWD occurrence [69]. A MB (*Monochamus/Bursaphelenchus*) index was developed [78] by summing the difference between the monthly average temperature and 15 °C when the monthly average temperatures exceeds 15 °C for one year [67]. The index displayed large differences between the PWD occurrence area in the field and the area predicted with the index in Korea, but the predictability of the index was improved by the modification of the threshold temperature [79]. Meanwhile, Park, Chung and Moon [22] developed a hazard rating model for pine trees and pine stands to PWD, showing that large trees have a higher risk rating than smaller ones because their high heights and a large crown volume induce a high probability to be exposure to vector beetles. They also revealed that pine stands at low altitude and south-facing slopes had high risk rating, thus indicating that the occurrence of PWD is highly associated with geographical factors.

Futai [80] demonstrated the importance of asymptomatic carrier trees in the spread of PWD. Meanwhile, Nguyen et al. [65] developed a spatial model to characterize the dispersal patterns of PWD in South Korea; they showed that asymptomatic carriers were important in the PWD dispersal and should be considered in the PDW management.

Information on the natural enemies of PWD vectors is limited even though there was over 30 years after the first occurrence of PWD in South Korea. A parasitic wasp, *Sclerodermus harmandi* (Buysson 1903), caused 54.7% and 98.6% parasitism in laboratory tests on larvae of *M. alternatus* and *M. saltuarius*, respectively [81]. Other reported natural enemies include the predator beetles *Temnochila japonica* Reitter and *Thanassimus lewisi* Jacobson and the parasitoids *Dolichomitus nakamurai* (Uchida 1928) and *Echthrus reluctator* L. [82]. In China, a total of 97 species of natural enemies have been reported, including 47 insects, 17 microorganism pathogens and 33 predatory birds [83]. Among these, four parasitoids [*Scleroderma guani* Xiao et Wu, *Scleroderma sichuanensis* Xiao, *Ontsira palliates* Cameron and *Dastarcus helophoroides* (Fairmaire)] and one predacious insect [*Cryptalaus berus* (Candeze)] have potential as biological control agents for *M. alternatus*. Generalist natural enemies that affect *Monochamus* populations require further study to assess their impact. Pathogenic fungi such as *Beauveria brongniartii* (Saccardo) Petch have also been reported from *Monochamus* species in Japan and may have some potential biological control agents [84].

Similarly, information on the natural enemies of *Monochamus* beetles in Europe is limited. Until 2004, no specific parasitoid of *Monochamus galloprovincialis* Olivier was reported, although generalist braconids such as *Atanycolus genalis* (Thomson) and *Meteorus corax* Marshall are known to parasitize *M. galloprovincialis* [85]. According to Naves et al. [86], no egg parasitoids were found, and larvae parasitism by three Braconidae such as *Cyanopterus flavator* Fabricius, *Iphiaulax impostor* (Scopoli) and *Coeloides sordidator* Ratzeburg was less than 10%.

Burning and chipping pine trees infected by PWD was the main control method employed in South Korea between 1988 and 2001, together with the aerial application of insecticides in 1989. In 2002, a method for felling and fumigation of infected trees with metam-sodium was developed, and this method has been widely applied [87]. Up to 2015, PWD control was carried out mainly with techniques using chemicals such as felling and fumigation, the trunk injection of nematicides, and the aerial spraying of insecticides [63,87]. Since 2015, the use of chemical applications has been reduced due to the concern over environmental contamination, and the felling and crushing of pine trees infected by PWD has become the main control method. Fumigation was conducted in the area where felling and crushing is not applicable because of limited access to the crushing machine. The trunk injection of nematicides is costly and labor intensive, so it is used only for ornamental or nursery trees [12]. For the biological control of the key vectors, the parasitic wasp *S. harmandi* was selected, but there is yet no report of its successful use.

5. Causes of Changes in Dominance of Insect Pests in Korean Forests

The dominant pine forest pests in Korea over the last 50 years has dramatically changed from the PC via the PNGM to PWD. These changes have possibly been caused by bottom-up effects, reflecting changes in forest vegetation including the dominance of pine trees in forest as well as forest environmental conditions. At the same time, top-down effects by natural enemies might also contribute to the stabilization of outbreaks of the PC and the PNGM. The PC and the PNGM directly damage pine trees by defoliating, whereas PWD causes damage by wilting pine trees through the mutualistic interactions between the invasive nematode and native beetle vectors, suggesting that biological traits of the dominant pests as well as the dominant species of forest pests have changed over the last 50 years. Figure 3 presents a schematic diagram for sequential changes in major pine forest pests in Korea relating to invasion history of species and forest management policy.

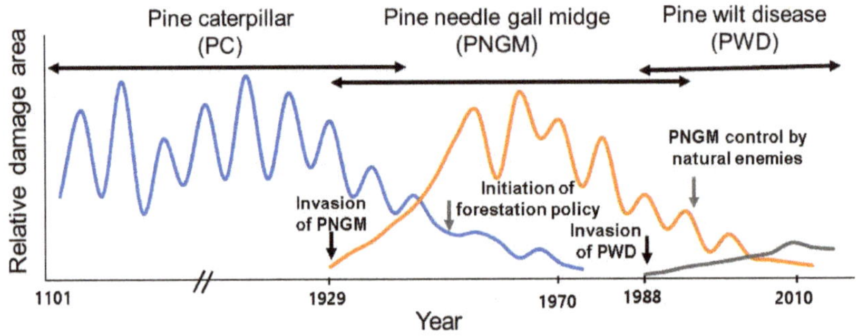

Figure 3. The schematic diagram for sequential changes in major pine forest pests in Korea from pine caterpillar via pine needle gall midge to pine wilt disease. The vertical black arrows indicate the first report of invasive species, and grey arrows indicate the potential causes of decline of pine caterpillar and pine needle gall midge. The first occurrence of pine caterpillar in Korea was reported in 1101 (Goryeo Dynasty) (http://www.history.go.kr).

The causes of PC decline could be explained by (1) the longer periodicity in the outbreak of the PC; (2) an increase in activity of natural enemies, including fungal pathogens probably due to reforestation; or (3) inferior competitiveness with the PNGM. Longer periodicity has possibly contributed to recent decline of the PC in Korea. The second hypothesis is related to bottom-up regulation in a forest ecosystem. From the 1950s to the early 1970s, the dominance of the Japanese red pine was at its peak, and the forests consisted of young trees because Korean forests were severely damaged during the Japanese colonial era (1910–1945) and the Korean War (1950–1953) and then intensively used for fuel. These deforested conditions might have promoted PC outbreaks because the PC prefers younger pine trees, and the activities of fungal pathogen were minimized due to lower humidity in the young forests [88]. The last hypothesis is related to the introduction of the PNGM. The main host plants of both the PC and the PNGM are the Japanese red pine and the Japanese black pine. The period of PC decline coincided with the range expansion of the PNGM. Furthermore, host switching of the PC from the Japanese red pine to the pitch pine was observed in late 1990s and early 2000s [24], suggesting that as the inferior competitor to the PNGM (Kwon T.S, unpublished data), the PC fed on the less preferred pitch pine instead of the Japanese red pine [29].

Meanwhile, the decrease in PNGM occurrence could be explained by (1) an increase in tolerance of pine trees to the PNGM; (2) an increase in activities of natural enemies including exotic parasitoids; or (3) the lower performance of the PNGM under higher temperature, thus showing the influence of climate changes. The outbreak of the PNGM was the highest at seven years after the first local invasions. After the first initial outbreak of the PNGM, the mortality of pine trees decreased and stabilized at a low level. Probably, those pine trees most susceptible to the PNGM died in the first invasion and tolerant individuals survived, or trees that survived the first invasion later acquired resistance to the PNGM. However, there is no evidence to support this hypothesis. The second hypothesis (increase in natural enemies) was tested by several studies [51,53]. Among the four parasitoids of the PNGM, only *I. seoulis* dispersed in similar speed with the PNGM and other parasitoids followed slowly the spread of the PNGM. The interspecific competition between these parasitoids with different phenologies was weak, meaning that more than two parasitoids could act to suppress PNGM populations without strong interspecific competition. Over the long-term, the population of the PNGM decreased with increasing parasitism rates by two or three PNGM parasitoids [51,53]. The last factor is related to the thermal biology of the PNGM, because long-term data analysis suggests that the PNGM is a cool-adapted species [51]. Considering that its adults emerge from late May to early July, an increase in the highest temperature over 30 °C in June due to climate change may have negative impacts on the performance of the PNGM.

6. Perspectives

The occurrence of PWD is still in the expansion phase and the risk of its damage is likely to increase with climate change in South Korea. Meanwhile, there remains very little effect on PWD by natural enemies. Therefore, PWD expansion is likely to increase on the next decades with climate change in South Korea. Considering that outbreaks of the PC and the PNGM decreased due to environmental changes and the activities of natural enemies, however, the outbreak of PWD may decrease due to a reduction in the number of susceptible pine trees or an increase in effective natural enemies of insect vectors. It is likely that mixed species forests can reduce PWD spread. Changes in the major forest pests in pine forests of South Korea over the last 50 years have shown that bottom-up effects, including forest status and biological traits of pest species, likely determined which pest species were dominant, while top-down regulation stabilized outbreaks of the major forest pests. The impacts of the forest pests have increased in terms of tree mortality in the last 50 years. The invasive species have been major threats to forest health, and the influence of climate changes on the performance of forest pests has also been observed. These facts suggest that the decline of pine trees could be accelerated due to forest pests. Therefore, novel management strategies of forest pests in pine trees are needed to sustainably manage pine forests in Korea.

Author Contributions: Conceptualization, W.I.C., J.-H.L., S.-H.K. and Y.-S.P.; data curation, C.Y.L. and Y.J.S.; writing—original draft preparation, W.I.C., Y.N. and B.K.C.; writing—review and editing, W.I.C., Y.N., C.Y.L., B.K.C., Y.J.S., J.-H.L., S.-H.K. and Y.-S.P.

Funding: This study was supported by the National Institute of Forest Science and R&D Program for Forest Science Technology (FTIS 2017042A00-1823-CA01) provided by Korea Forest Service (Korea Forestry).

Acknowledgments: We would like to thank all members who have been involved in the Forest Insect Pests Monitoring Program operated by the National Institute of Forest Science (the former Korea Forest Research Institute) since 1968.

Conflicts of Interest: The authors declare no conflict of interest. The funders had no role in the design of the study; in the collection, analyses, or interpretation of data; in the writing of the manuscript, or in the decision to publish the results.

References

1. Walter, J.A.; Ives, A.R.; Tooker, J.F.; Johnson, D.M. Life history and habitat explain variation among insect pest population subject to global change. *Ecosphere* **2018**, *9*, e02274. [CrossRef]
2. Hentschel, R.; Möller, K.; Wenning, A.; Degenhardt, A.; Schröder, J. Importance of ecological variables in explaining population dynamics of three important pine pest insects. *Front. Plant Sci.* **2018**, *9*, 1667. [CrossRef] [PubMed]
3. Duan, J.J.; Bauer, L.S.; Abell, K.J.; Ulyshen, M.D.; Van Driesche, R.G. Population dynamics of an invasive forest insect and associated natural enemies in the aftermath of invasion: Implications for biological control. *J. Appl. Ecol.* **2015**, *52*, 1246–1254. [CrossRef]

4. Guyot, V.; Castagneyrol, B.; Vialatte, A.; Deconchat, M.; Jactel, H. Tree diversity reduces pest damage in mature forest across Europe. *Biol. Lett.* **2018**, *12*, 20151037. [CrossRef] [PubMed]
5. Speer, J.H.; Swetman, T.W.; Wickman, B.E.; Youngblood, A. Changes in pandora moth outbreak dynamics during the past 622 years. *Ecology* **2001**, *82*, 679–697. [CrossRef]
6. Möller, K.; Hentschel, R.; Wenning, A.; Schröder, J. Improved outbreak prediction for common pine sawfly (*Diprion pini* L.) by analyzing floating 'climatic windows' as keys for changes in voltinism. *Forests* **2017**, *8*, 319. [CrossRef]
7. Li, S.; Daudin, J.J.; Piou, D.; Robinet, C.; Jactel, H. Periodicity and synchrony of pine processionary moth outbreaks in France. *For. Ecol. Manag.* **2015**, *354*, 309–317. [CrossRef]
8. Schwerdtfeger, F. Ueber die Ursachen des Massenwechsels der Insekten. *Ztg. Angeweine Entomol.* **1941**, *28*, 254–303. [CrossRef]
9. Turchin, P.; Taylor, A.D. Complex dynamics in ecological time series. *Ecology* **1992**, *73*, 289–305. [CrossRef]
10. Tenow, O. The ourbreaks of *Oporina autumnata* Bkh. and *Operophtera* spp. (Lep. Geometridae) in the Scandinavian mountain chain and northern Finland 1862–1968. *Zool. Bidr. Upps. Suppl.* **1972**, *2*, 1–107.
11. Tenow, O.; Nilssen, A.C.; Bylund, H.; Hogstad, O. Waves and synchrony in *Epirrita autumnata / Operophtera brumata* outbreaks. I. Lagged synchrony: Regionally, locally and among species. *J. Anim. Ecol.* **2007**, *76*, 258–268. [CrossRef] [PubMed]
12. Choi, W.I.; Park, Y.-S. Dispersal patterns of exotic forest pests in South Korea. *Insect Sci.* **2012**, *19*, 535–548. [CrossRef]
13. Korea Forest Research Institute (KFRI). *Annual Report of Monitoring for Forest Insect Pests and Diseases in Korea*; SeongMunSa: Seoul, Korea, 1968–2018.
14. Korea Forest Service. *Statistical Yearbook of Forestry*; Korea Forest Service: Daejeon, Korea, 2018.
15. Park, W.K.; Lee, K.-H. Changes in the species of woods used for Korean ancient and historic architectures. *J. Archit. Hist.* **2007**, *16*, 9–28.
16. Moon, M.; Kim, S.-S.; Lee, D.-S.; Yang, H.; Park, C.-W.; Kim, H.; Park, Y.-S. Effects of forest management practices on moth communities in a Japanese larch (*Larix kaempferi* (Lamb.) Carrière) plantation. *Forests* **2018**, *9*, 574. [CrossRef]
17. Bae, J.-S.; Joo, R.-W.; Lee, K.-B. *Causes of Forest Degradation and Drivers of Forest Recovery in South Korea*; Upgo MunHwa: Seoul, Korea, 2010.
18. Bae, J.-S.; Lee, K.-B. Impacts of the substitution of firewood for home use on the forest greening after the 1945 liberation of Korea. *J. Korean For. Soc.* **2006**, *95*, 60–72.
19. Lee, K.-B.; Bae, J.-S. Factors of success of the clearance policy for slash-and-burn fields in the 1970s. *J. Korean For. Soc.* **2007**, *96*, 325–337.
20. Lee, D.-S.; Bae, Y.-S.; Byun, B.-K.; Lee, S.; Park, J.K.; Park, Y.-S. Occurrence prediction of the citrus flatid planthopper (*Metcalfa pruinosa* (Say, 1830)) in South Korea using a random forest model. *Forest* **2019**, *10*, 583. [CrossRef]
21. Park, K.N.; Hyun, J.S. Studies on the effects of the pine needle gall midge, Thecodiplosis japonensis Uchida et Inouye, on the growth of the red pine, Pinus densiflora S. et Z. (II)—Growth impact on red pine. *J. Korean For. Soc.* **1983**, *62*, 87–95.
22. Park, Y.-S.; Chung, Y.-J.; Moon, Y.-S. Hazard ratings of pine forests to a pine wilt disease at two spatial scales (individual trees and stands) using self-organizing map and random forest. *Ecol. Inform.* **2013**, *13*, 40–46. [CrossRef]
23. Baek, S.L. The outbreak of pine caterpillar and colonial authorities' response. *Han'guk Munhwa Korean Cult.* **2018**, *81*, 93–123.
24. Choi, W.I.; Park, Y.K.; Park, Y.-S.; Lee, H.P.; Ryoo, M.I. Changes in voltinism in a pine caterpillar *Dendrolimus spectabilis* (Lepidoptera: Lasiocampidae) population: Implications of climatic change. *Appl. Entomol. Zool.* **2011**, *46*, 319–325. [CrossRef]
25. Chung, K.H.; Ryu, J.; Kwon, S.H.; Im, M.S. Field studies on the attractiveness of pine caterpillar moths (*Dendrolimus spectabilis* Butler) to blacklight-traps. *Korean J. Plant Prot.* **1971**, *10*, 43–48.
26. Kim, C.W.; Hyun, J.S. Studies on the control of pine moth, *Dendrolimus spectabilis* (Butler). *Ent. Res. Bull.* **1965**, *1*, 1–109.
27. Shin, S.C.; Choi, K.S.; Choi, W.I.; Chung, Y.J.; Lee, S.G.; Kim, C.S. *An New Illustrated Book of Forest Insect Pests*; Upgo MunHwa: Seoul, Korea, 2008.

28. Kwon, T.-S.; Park, Y.-K.; Oh, K.-S.; Kwon, Y.-D.; Shin, S.-C.; Kim, C.S.; Park, J.D.; Lee, H.-P. Increase in the number of generations in *Dendrolimus spectabilis* (Butler) (Lepidoptera: Lasiocampidae) in Korea. *J. Korean For. Soc.* **2002**, *91*, 149–155.
29. Park, Y.G.; Kim, C.S. Studies on selecting taste of the pine-caterpillar larva: *Dendrolimus spectabilis* Butler on six pine species. *Korean J. Breed.* **1973**, *5*, 27–31.
30. Hyun, J.S. Studies on the prevision for occurrence of pine moth, *Dendrolimus spectabilis* Butler. *Ent. Res. Bull.* **1968**, *4*, 57–80.
31. Kokubo, A. Population fluctuations and natural mortalities of the pine-moth, *Dendrolimus spectalbilis*. *Res. Popul. Ecol.* **1965**, *7*, 23–34. [CrossRef]
32. Bao, Y.; Wang, F.; Tong, S.; Na, L.; Han, A.; Zhang, J.; Bao, Y.; Han, Y.; Zhang, Q. Effect of drought on outbreaks of major forest pests, *Pine Caterpillars* (Dendrolimus spp.), in Shandong Province, China. *Forests* **2019**, *10*, 264. [CrossRef]
33. Park, Y.-K. Changes in Population Biology of Pine Moth, *Dendrolimus spectabilis* (Butler) (Lepidoptera: Lasiocampidae) and Its Biological Control by *Trichogramma dendrolimi* Matsumura (Hymenoptera: Trichogrammatidae). Ph.D. Dissertation, Dongguk University, Seoul, Korea, 2001.
34. Aoki, J. *Beauveria bassiana* (Bals.) Vuill. isolated from some lepidopterous species in Japan. *Jpn. J. Appl. Entomol. Zool.* **1971**, *15*, 222–227. [CrossRef]
35. Pasquarella, V.A.; Elkinton, J.S.; Bradley, B.A. Extensive gypsy moth defoliation in Southern New England characterized using Landsat satellite observations. *Biol. Invasions* **2018**, *20*, 3047–3053. [CrossRef]
36. Moreira, X.; Abdala-Roberts, L.; Rasmann, S.; Castagneyrol, B.; Mooney, K.A. Plant diversity effects on insect herbivores and their natural enemies: Current thinking, recent findings, and future directions. *Curr. Opin. Insect Sci.* **2016**, *14*, 1–7. [CrossRef] [PubMed]
37. Le Borgne, H.; Hebert, C.; Dupuch, A.; Bichet, O.; Pinaud, D.; Fortin, D. Temporal dynamics in animal community assembly during post-logging succession in boreal forest. *PLoS ONE* **2018**, *13*, e0204445. [CrossRef] [PubMed]
38. Soné, K.; Furuno, T. Annual changes in the infestation by the pine needle gall midge, *Thecodiplosis japonensis* Uchida et Inouye (Diptera: Cecidomyiidae), in a pine stand. *J. Jpn. For. Soc.* **1982**, *64*, 301–306.
39. Park, K.-N.; Miura, T.; Hirashima, Y. Outbreaks history and present status of the pine needle gall midge in Korea. *Esakia* **1985**, *23*, 115–118.
40. Lee, B.Y.; Chung, Y.J.; Park, K.N.; Byun, B.H.; Bae, W.I. Distribution of pine needle gall midge, *Thecodiplosis japonensis* Uchida et Inouye (Diptera: Cecidomyiidae), infestations in Korea: A brief history. *FRI J. For. Sci.* **1997**, *56*, 13–20.
41. Shigesada, N.; Kawasaki, K.; Takeda, Y. Modeling stratified diffusion in biological invasions. *Am. Nat.* **1995**, *146*, 229–251. [CrossRef]
42. Park, Y.-S.; Chung, Y.-J. Hazard rating of pine trees from a forest insect pest using artificial neural networks. *For. Ecol. Manag.* **2006**, *222*, 222–233. [CrossRef]
43. Park, K.N.; Hyun, J.S. Studies on the effects of the pine needle gall midge, *Thecodiplosis japonensis* Uchida et Inouye, on the growth of the red pine, *Pinus densiflora* Siebold et Zuccarini (I): Changes of gall formation rate. *J. Korean For. Soc.* **1983**, *61*, 20–26.
44. Chung, Y.-J.; Park, Y.-S.; Lee, B.-Y.; Chon, T.-S. Dynamic patterns of the infestation of pine needle gall midge, *Thecodiplosis japonensis* (Diptera: Cecidomyiidae), in endemic regions of Korea. *FRI J. For. Sci.* **1998**, *59*, 64–69.
45. Ko, J.H. *Studies on the Isostasius seoulis Ko, the Larval Parasite of the Pine Gall-Midge (Thecodiplosis japonensis Uchida et Inouye). III: The Life History*; Research Report; Rural Development: Suwon, Korea, 1966.
46. Choi, W.I.; Song, H.J.; Kim, D.S.; Lee, D.-S.; Lee, C.-Y.; Nam, Y.; Kim, J.-B.; Park, Y.-S. Dispersal patterns of pine wilt disease in the early stage of its invasion in South Korea. *Forests* **2017**, *8*, 411. [CrossRef]
47. Nam, Y.; Choi, W.I. An empirical predictive model for the spring emergence of *Thecodiplosis japonensis* (Diptera: Cecidomyiidae): Model construction and validation on the basis of 25 years field observations data. *J. Econ. Entomol.* **2014**, *107*, 1136–1141. [CrossRef] [PubMed]
48. Chung, Y.J.; Hyun, J.-S. Studies on the major factors affecting the population of the overwintered pine needle gall midge, *Thecodiplosis japonensis* Uchida et Inouye. *Korean J. Plant Prot.* **1986**, *25*, 1–9.
49. Ryoo, M.I.; Chun, Y.S. Population ecology of pine gall midge (*Thecodiplosia japonensis*): Revisited. *Nat. Resour. Res.* **1996**, *4*, 118–129.

50. Choi, W.I.; Ryoo, M.I.; Chung, Y.J.; Park, Y.-S. Geographical variation in the population dynamics of *Thecodiplosis japonensis*: Causes and effects on spatial synchrony. *Popul. Ecol.* **2011**, *53*, 429–439. [CrossRef]
51. Choi, W.I.; Jeon, M.J.; Park, Y.S. Structural dynamics in the host-parasitoid system of the pine needle gall midge (*Thecodiplosis japonensis*) during invasion. *PeerJ* **2017**, *5*, e3610. [CrossRef] [PubMed]
52. Park, Y.-S.; Chung, Y.-J.; Chon, T.-S.; Lee, B.-Y.; Lee, J.-H. Interactions between pine deedle gall midge, *Thecodiplosis japonensis* (Diptera: Cecidomyiidae), and its parasitoids in newly invaded areas. *Korean J. Appl. Entomol.* **2001**, *40*, 301–307.
53. Jeon, M.-J.; Choi, W.I.; Choi, K.-S.; Chung, Y.-J.; Shin, S.-C. Population dynamics of Thecodiplosis jsponenste (Dlptera: Cecidomyiidae) under influence of parasitism by Inostemma matsutama and Inostemma seoulis (Hymenoptera: Platygastridae). *J. Asia-Pac. Entomol.* **2006**, *9*, 269–274. [CrossRef]
54. Yoshida, N.; Hirashima, Y. Systematic studies on proctotrupoid and chalcidoid parasites of gall midges injurious to *Pinus* and *Cryptomeria* in Japan and Korea (Hymenoptera). *Esakia* **1979**, *14*, 113–133.
55. Ko, J.H. A new species of *Inostemma* (Hymenoptera: Platygasteridae), a larval parasite of the pine gall midge, *Thecodiplosis* sp. (Diptera: Cecidomyoodae). *Korean J. Plant Prot.* **1980**, *19*, 35–38.
56. Jeon, M.-J.; Lee, B.-Y.; Ko, J.-H.; Miura, T.; Hirashima, Y. Ecology of *Platygaster matsutama* and *Inostemma seoulis* (Hymenoptera: Platygastridae), egg-larval parasites of the pine needle gall midge, *Thecodiplosis japonensis* (Diptera, Cecidomyiidae). *Esakia* **1985**, *23*, 131–143.
57. Soné, K. Ecology of host-parasitoid community in the pine needle gall midge, *Thecodiplosis japonensis* Uchida et Inouye (Diptera, Cecidomyiidae). *J. Appl. Entomol.* **1986**, *102*, 516–527. [CrossRef]
58. Son, Y.; Chung, Y.-J.; Lee, J.-H. Differential thermal biology may explain the coexistence of *Platygaster matsutama* and *Inostemma seoulis* (Hymenoptera: Platygastridae) attacking *Thecodiplosis japonensis* (Diptera: Cecidomyiidae). *J. Asia-Pac. Entomol.* **2012**, *15*, 465–471. [CrossRef]
59. Lee, D.-S.; Nam, Y.; Choi, W.I.; Park, Y.-S. Environmental factors influencing on the occurrence of pine wilt disease in Korea. *Korean J. Ecol. Environ.* **2017**, *50*, 374–380. [CrossRef]
60. Togashi, K.; Shigesada, N. Spread of the pinewood nematode vectored by the Japanese pine sawyer: Modeling and analytical approaches. *Popul. Ecol.* **2006**, *48*, 271–283. [CrossRef]
61. Mota, M.; Braasch, H.; Bravo, M.A.; Penas, A.C.; Burgermeister, W.; Metge, K.; Sousa, E. First report of *Bursaphelenchus xylophilus* in Portugal and in Europe. *Nematology* **1999**, *1*, 727–734.
62. Kiritani, K.; Morimoto, N. Invasive insect and nematode pests from north America. *Glob. Environ. Res.* **2004**, *8*, 75–88.
63. Shin, S.C. Pine Wilt Disease in Korea. In *Pine Wilt Disease*; Zhao, B.G., Futai, K., Sutherland, J.R., Takeuchi, Y., Eds.; Springer: Tokyo, Japan, 2008.
64. Lee, S.D.; Park, S.; Park, Y.-S.; Chung, Y.-J.; Lee, B.-Y.; Chon, T.-S. Range expansion of forest pest populations by using the lattice model. *Ecol. Model.* **2007**, *203*, 157–166. [CrossRef]
65. Nguyen, T.V.; Park, Y.-S.; Jeoung, C.-S.; Choi, W.-I.; Kim, Y.-K.; Jung, I.-H.; Shigesada, N.; Kawasaki, K.; Takasu, F.; Chon, T.-S. Spatially explicit model applied to pine wilt disease dispersal based on host plant infestation. *Ecol. Model.* **2017**, *353*, 54–62. [CrossRef]
66. Robinet, C.; Roques, A.; Pan, H.; Fang, G.; Ye, J.; Zhang, Y.; Sun, J. Role of human-mediated dispersal in the spread of the pinewood nematode in China. *PLoS ONE* **2009**, *4*, e4646. [CrossRef]
67. Hirata, A.; Nakamura, K.; Nakao, K.; Kominami, Y.; Tanaka, N.; Ohashi, H.; Takano, K.T.; Takeuchi, W.; Matsui, T. Potential distribution of pine wilt disease under future climate change scenarios. *PLoS ONE* **2017**, *12*, e0182837. [CrossRef]
68. Kwon, T.S.; Lim, J.H.; Sim, S.J.; Kwon, Y.D.; Son, S.K.; Lee, K.Y.; Kim, Y.T.; Park, J.W.; Shin, C.H.; Ryu, S.B.; et al. Distribution patterns of *Monochamus alternatus* and *M. saltuarius* (Coleoptera: Cerambycidae) in Korea. *J. Korean For. Soc.* **2006**, *95*, 543–550.
69. Togashi, K. Vector-Nematode relationships and epidemiology in pine wilt disease. In *Pine Wilt Disease*; Zhao, B.G., Futai, K., Sutherland, J.R., Takeuchi, Y., Eds.; Springer: Tokyo, Japan, 2008; pp. 162–183.
70. Kwon, H.J.; Jung, J.-K.; Jung, C.; Han, H.; Koh, S.-H. Dispersal capacity of *Monochamus saltuarius* on flight mills. *Entomol. Exp. Appl.* **2018**, *166*, 420–427. [CrossRef]
71. Hu, S.J.; Ning, T.; Fu, D.Y.; Haack, R.A.; Zhang, Z.; Chen, D.D.; Ma, X.Y.; Ye, H. Dispersal of the japanese pine sawyer, *Monochamus alternatus* (Coleoptera: Cerambycidae), in mainland China as inferred from molecular data and associations to indices of human activity. *PLoS ONE* **2013**, *8*, e57568. [CrossRef] [PubMed]

72. Kawai, M.; Shoda-Kagaya, E.; Maehara, T.; Zhou, Z.; Lian, C.; Iwata, R.; Yamane, A.; Hogetsu, T. Genetic structure of pine sawyer *Monochamus alternatus* (Coleoptera: Cerambycidae) populations in Northeast Asia: Consequences of the spread of pine wilt disease. *Environ. Entomol.* **2006**, *35*, 569–579. [CrossRef]
73. Nakamura-Matori, K. Vector-host tree relationships and the abiotic environment. In *Pine Wilt Disease*; Zhao, B.G., Futai, K., Sutherland, J.R., Takeuchi, Y., Eds.; Springer: Tokyo, Japan, 2008; pp. 144–161.
74. Jikumaru, S.; Togashi, K. Effects of temperature on the post-diapause development of *Monochamus saltuarius* (Gebler) (Coleoptera: Cerambycidae). *Appl. Entomol. Zool.* **1996**, *31*, 145–148. [CrossRef]
75. Jung, C.S.; Koh, S.-H.; Nam, Y.; Ahn, J.J.; Choi, W.I. A forecasting model for predicting the spring emergence of *Monochamus saltuarius* (Coleoptera: Cerambycidae) on Korean white pine, *Pinus koraiensis*. *J. Econ. Entomol.* **2015**, *108*, 1830–1836. [CrossRef] [PubMed]
76. Park, C.G.; Kim, D.S.; Lee, S.M.; Moon, Y.S.; Chung, Y.J.; Kim, D.-S. A forecasting model for the adult emergence of overwintered *Monochamus alternatus* (Coleoptera: Cerambycidae) larvae based on degree-days in Korea. *Appl. Entomol. Zool.* **2014**, *49*, 35–42. [CrossRef]
77. Jikumaru, S.; Togashi, K. Temperature effects on the transmission of *Bursaphelenchus xylophilus* (Nemata: Aphelenchoididae) by *Monochamus alternatus* (Coleoptera: Cerambycidae). *J. Nematol.* **2000**, *32*, 110–116.
78. Taketani, A.; Okuda, M.; Hosoda, R. The meteorological analysis on the epidemic mortality of pine trees, with special reference to the effective accumulated temperature. *J. Jpn. For. Soc.* **1975**, *57*, 169–175.
79. Korea Forest Research Institute (KFRI). *System Design and Structure Analyses for Pine Wilt Disease Control*; KFRI: Seoul, Korea, 2010.
80. Futai, K. Role of asymptomatic carrier trees in epidemic spread of pine wilt disease. *J. For. Res.* **2003**, *8*, 253–260. [CrossRef]
81. Hong, J.-I.; Koh, S.-H.; Chung, Y.J.; Shin, S.S.; Kim, G.-H.; Choi, K.-S. Biological characteristics of *Sclerodermus harmandi* (Hymenoptera: Bethylidae) parasitized on cerambycid. *Korean J. Appl. Entomol.* **2008**, *47*, 133–139. [CrossRef]
82. Kim, J.K.; Won, D.-S.; Park, Y.C.; Koh, S.-H. Natural enemies of wood borers and seasonal occurrence of major natural enemies of *Monochamus slatuarius* on pine trees. *J. Korean For. Soc.* **2010**, *99*, 439–445.
83. Zhang, Y.; Yang, Z. Studies on the natural enemies and biocontrol of *Monochamus alternatus* Hope (Coleoptera: Cerambycidae). *Plant Prot.* **2006**, *32*, 9–14.
84. Shimazu, M. Potential of the Cerambycid-Parasitic Type of *Beauveria brongniartii* (Deuterpmycotina: Hyphomycetes) for microbial control of *Monochamus alternatus* Hope (Coleoptera: Cerambycidae). *Appl. Entomol. Zool.* **1994**, *29*, 127–130. [CrossRef]
85. Petersen-Silva, R.; Pujade-Villar, J.; Naves, P.; Sousa, E.; Belokobylskij, S. Parasitoids on *Monochamus galloprovincialis* (Coleoptera, Cerambycidae), vector of the pine wood nematode, with identification key for Palearctic region. *ZooKeys* **2012**, *251*, 29–48. [CrossRef] [PubMed]
86. Naves, P.; Kenis, M.; Sousa, E. Parasitoids associated with *Monochamus galloprovincialis* (Oliv.) (Coleoptera: Cerambycidae) within the pine wilt nematode-affected zone in Portugal. *J. Pest Sci.* **2005**, *78*, 57–62. [CrossRef]
87. Kwon, T.-S.; Shin, J.H.; Lim, J.-H.; Kim, Y.-K.; Lee, E.J. Management of pine wilt disease in Korea through preventative silvicultural control. *For. Ecol. Manag.* **2011**, *261*, 562–569. [CrossRef]
88. Kamata, N. Outbreaks of forest defoliating insects in Japan, 1950–2000. *Bull. Entomol. Res.* **2002**, *92*, 109–117. [CrossRef] [PubMed]

forests

Review

Abiotic and Biotic Disturbances Affecting Forest Health in Poland over the Past 30 Years: Impacts of Climate and Forest Management †

Zbigniew Sierota [1,*], Wojciech Grodzki [2] and Andrzej Szczepkowski [3]

1 Department of Forestry and Forest Ecology, Warmia and Mazury University in Olsztyn, Pl. Łódzki 2, 10-727 Olsztyn, Poland

2 Department of Mountain Forests, The Forest Research Institute, ul. A. Fredry 39, 30-605 Kraków, Poland; w.grodzki@ibles.waw.pl

3 Department of Forest Protection and Ecology, Division of Mycology and Forest Phytopathology, Warsaw University of Life Sciences—SGGW, ul. Nowoursynowska 159, 02-776 Warsaw, Poland; andrzej_szczepkowski@sggw.pl

* Correspondence: zbigniew.sierota@uwm.edu.pl; Tel.: +48-89-523-49-47

† To mark the centenary of Poland's Independence.

Received: 29 December 2018; Accepted: 18 January 2019; Published: 21 January 2019

Abstract: The current nature of forest management in Poland reflects its history and more than 100 years of economic activity affecting forests since independence in 1918. Before that time, different forest management models were used, related to the nature of the Prussian economy in the north of the country, the Russian economy in the central-eastern part, and the Austrian economy in south-eastern Poland. The consequence of these management models, as well as the differing climate zones in which they were used, resulted in varied forest health. Since the end of World War II, forest coverage within Poland's new borders has increased from 20.8% to currently 29.6%, mainly as a result of afforestation of wastelands and former agricultural lands. This paper describes changes in the health of forests and their biological diversity in Poland in the context of weather extremes, species composition, forest management, the forest industry, and damage from insects and pathogenic fungi over the last 30 years.

Keywords: forestry models; climate change; forest management; abiotic and biotic disturbances; forest health

---

1. Introduction

In order to assess and understand the reasons for the present state of Poland's forests, it is important to be aware of the historical conditions in which forest management in Poland was shaped. Historical changes in the management of forests, their location, and vegetation variability were important predisposing factors leading to declining forest health in Poland and other parts of Central Europe in the 1970s. [1]. Since 1795, and during the nineteenth century the area of the Federal Kingdom of Poland and the Grand Duchy of Lithuania were, at different times, divided between three neighboring countries – Germany in the west (with Prussia in the north-east), Austria-Hungary in the south-east, and Russia in the central and eastern areas [2]. This split also impacted forest policy, forest management, and stand protection up until Polish independence in 1918.

2. Models of Forest Management in the Nineteenth Century

Silviculture and forest planning processes were first developed, and most strongly implemented, in Germany and Prussia, prior to the eighteenth century. Bans on removing manorial forests were introduced, and the concept of forest conservation was promoted in 1713 by the publication of H. von Carlowitz's "Silvicultura oeconomica", which promoted the principles of sustainable forest management [3]. In the Austro-Hungarian Empire, after F. Hohenzollern's Act for Silesian forests (1777), local governments often regulated forest management—for example in the Beskidy mountain region (southern Poland, western Carpathians) at the end of the nineteenth century, there were already forest companies who worked in district forests owned by local communities and the Polish aristocracy [4].

The forest industry was historically a source of employment and consumer goods, but at the same time it harmed both people and the environment. Industrial factors caused reductions in forest area, which resulted in a significant loss of natural fir, oak, beech, and sycamore forests. Forest loss was most severe in the Sudetes and in the western Carpathians (Southern Poland) [5–7]. The depletion of natural forests, both in mountainous regions and in northern and north-eastern areas (Prussia), was a consequence of the rapid development of the industrial economy in the second half of the nineteenth century. Similar forest loss occurred in all regions of Europe, often much earlier than in Poland [3,8]. The theories developed at the beginning of the nineteenth century by German forestry (e.g., W. Pfeil, M. Faustmann), were founded on the principle of obtaining maximum economic income from the forest over its rotation, which is not necessarily the same as harvesting the largest volume of wood [3,9]. As a result, there was considerable tolerance of the market for forest management by land owners, which included clearcutting and the reduction of diverse forest species. In place of the original multi-species, multi-storey forests with a complex physical and ecological structure, forests were managed as "regulated" monocultures. In Prussia and Hessia, from 1819, forests were often harvested on a 100-year cycle, "regulated" using a harvesting cycle where a fixed percentage of the forest was harvested. If the forest was renewed at all, regeneration was mainly with spruce, a fast-growing tree in climatic conditions. Planted trees usually originated from seeds of non-local origins, and were rarely from local populations [10–12]. It was then believed that properly guided harvesting would provide forest owners with a steady maximum income from the forest, and at the same time harmonize the supply of wood to meet industry demand. However, forest practices prior to the twentieth century did not take into account the possible impact of such processes.

As the demand for timber grew rapidly in the nineteenth century, the availability of easily accessible forests for logging dwindled. Planned forest management was increasing, forest offices were being established, and forestry education was developing. This was especially the case in Germany, Austria-Hungary, and Prussia [1]. In place of simply harvesting wood without regard to long-term sustainability of wood supply, silvicultural models of wood production were introduced that drastically changed how forests were harvested, taking into account the age class composition of forests and the need to conserve tree species. It seems that, even today, this modern model for organizing the spatial and temporal structure of managed stands continues to have a positive effect on the development of forest resources and has enabled the management of multiple societal objectives for forests [13]. Only a few writers, such as J.Ch. Hundeshagen and O. Hagen, sensing the looming crisis facing forests, sought to protect them from overharvesting [14]. They had in mind both the impact of forests on moderating climate and reducing soil erosion, as well as the need to preserve forests for future generations, which was expressed in 1804 by L.G. Hartig, one of the co-creators of modern forestry. His vision became the famous slogan of the Second Ministerial Conference on Forest Protection in Europe—Helsinki 1993: "for future generation" [15].

Meanwhile, fire protection practices and the first forest fire insurance were created (1860 in southern Poland, 1862 in the west), which took on special significance after large forest fires occurred in 1863 (3.5 thousand hectares burned in the west) [16]. The planned forest economy began to be implemented, mainly thanks to the work of Schwappach (1908) and Wiedemann (1925, 1948), resulting in table-based stand growth models [17].

A slightly different approach to forest management and forest protection prevailed in areas under Russian rule after 1795. First of all, national and private property were nationalized which resulted in forests no longer being associated with the privileged classes in Poland. In 1892, forest management resulted from the new provisions of the Empire, and the existing provisions of the Kingdom of Poland (beginning in 1898), with new laws for forest protection being introduced along with punishment for their violation [16]. Nevertheless, illegal logging for timber exports and resulting deforestation by landowners still took place, worsening the health of forests [1]. Only government and municipal forests were managed by state forest services, and, thanks to that, they received better management [18].

3. Changes in Forest Cover and Forest Management Strategies in Poland

Wars repeatedly took place on the territory of Poland—from the Napoleonic Wars (1815), to World War I (1914 to 1918), and World War II (1939 to 1945), which dramatically reduced the area and health of forests. These conflicts caused a significant decrease in forest area, which occurred over the 120 years of Polish occupation, until 1918. In areas under Russian control and in mountainous regions (Sudetes, Carpathians) [6,19,20], forest cover decreased from 30.9% (3.8 million ha) in 1815, to 19.2% (2.3 million ha) in 1913 [18]. Over the country as a whole, at the end of the eighteenth century forest cover was about 40%, which decreased to 23% in 1921 and 22.2% in 1937 [21]. The period from 1939, especially as a result of World War II, not only saw massive deforestation and illegal logging in the areas occupied by Nazi-Germany and Soviet Russia, but also local forest losses occurred due to droughts, fires, and insect damage. For example, the extended bark beetle (Scolytinae) outbreak that started just at the end of the war, affected spruce forests in the Sudetes and Western Carpathians [20,22]. Bark beetles, together with outbreaks of Nun moth *Lymantria monacha* (Linnaeus, 1758), affected forests in the region of Mazury in north-eastern Poland [23]. These events caused a decline in the forest cover of Poland within its new borders to 20.8% in 1946 [24].

Forestry changed considerably in the post-war period as it came to be recognized for its contributions to the national economy and was increasingly being seen as an element of the natural environment. The main task of the newly created State Forests (established in 1924), was to rebuild the national forest estate. After World War II and the establishment of the communist government, all forests with an area exceeding 25 hectares were nationalized [25]. Furthermore, afforestation was undertaken on large areas of vacant land, including wasteland, abandoned unproductive agricultural land, dunes, and pastures [26]. Between 1946 and 1970 over 2 million hectares of forest plantations were established, mainly of pine, spruce, and birch, increasing forest cover to 28%.

The above-mentioned afforestation was treated as forecrop, which should re-shape or restore forest soil in a given area. Afforestation on such a large scale was an innovative venture in Europe, with previously unknown scientific premises for tree breeding. However, Poland did not avoid actions that we would today call 'mistakes'—even-aged monocultures covering large areas were created. Many trees were affected by root deformation from poor planting practices, and there was insufficient attention paid to controlling cockchafer (*Melolontha* spp.) larvae [27]. Furthermore, the importance of mycorrhizae for tree health was unknown [28,29]. The result of these practices was that, after 15 to 20 years, many of these plantations were under physiological stress, which contributed to a massive outbreak of root rot (*Heterobasidion* spp.) [26] At the same time, there were increasing outbreaks of defoliating and wood boring insects [30,31], especially in stands growing on rich sites [32]. Altogether, these factors caused significant economic losses and ecosystem changes [5,33].

The years from 1950 to 1970 mark the period of the Poland's socialist economy, in which forest supplied wood was particularly vital to forest-related industry. Forest produced wood enabled the manufacture of stamping mills for mines, railway sleepers, power poles, as well as for the furniture and sawmill industries. Forest stands, however, managed in accordance with the then binding regulations and principles of tree breeding and forest management, were subject to forced summer harvesting, resulting in a thinning of every four to six rows in plantations, and widespread spraying to control insect pests with insecticides that were harmful to the environment [34,35]. Dead wood, although now recognized for its ecological importance, was then often removed as it was considered as wastage and economically negligent.

New concepts, such as 'biodiversity', 'environmental protection', and 'sustainable development' appeared in Poland towards the end of the 1980s, after the change of political system. This marked the transition from forestry that was mainly focused on the production of raw materials to a semi-ecological forestry approach [36–40]. New concepts in Polish forestry, such as "forestry in accordance with nature", "semi-natural forest breeding", and "protection of natural resources in forests" were in line with a growing recognition of such principles elsewhere following the World Commission on Environment and Development (WCED) Report (1987), the Earth Summit (Agenda 21, 1992), and the Helsinki Resolution (1993) [15]. As a result, in many regions of the country, a large program of forest conversion was undertaken to replace conifer monocultures with mixed species stands, using species mixtures amenable to habitat requirements, care for biodiversity, and limitation of chemical methods for biological and ecological activities [41]. As a result, by 2016, forest area had increased to 9.2 million ha (29.5% of the country's area) and, according to the goals of the National Program for Increasing Afforestation, by 2050 forest cover should increase to 33% of the country [42].

4. Abiotic and Biotic Disturbances in Forests

In addition to harvesting by the forest industry, including logging by neighboring countries, Polish forests have been subjected to droughts, hurricanes, fires, fungal diseases, and dieback caused by atmospheric pollution [5,10]. Brazdil [43] notes that from the end of nineteenth century in the north of the present day Czech Republic, it was not only human activity that changed species diversity of natural mountain forests, and introduced more economically preferred species, mainly Norway spruce, but also sulfur dioxide (SO_2) pollution from industrial regions from the south-west. In former East Germany (GDR), for many years, 0.8 million ha of lowland stands were degraded due to industrial emissions, mainly of SO_2 [44]. A similar situation occurred for many years in the adjacent forests of Poland [45,46]. Similarly, the so-called "Black Triangle" (the Izerskie and Ore Mountains in the western Sudetes) was exposed to very high levels of industrial pollution [10,47,48], that indirectly—by the weakening of trees—is one of the most important factors stimulating bark beetle outbreaks [49]. The dieback of forest plantations increased through the 1970s and 1980s in Central and Eastern Europe, and in many papers it was described as 'Waldsterben', or 'forest decline' [49–52].

The present characteristics of forests differ in various parts of Poland that were once under German (north-western), Russian (north-eastern), and Austro-Hungarian (south-eastern) control, in ways that can be traced to each area's former economy. Specific forest health problems include forest insect outbreaks that are related to the dominant tree species in each region.

In the north-western region, Scots pine dominated stands have been, and still are, affected mainly by defoliating insects, such as the Nun moth, *L. monacha*. Outbreaks of Nun moth occurred for several decades. In the period 1946 to 75, insecticide treatments during six *L. monacha* outbreaks were applied on a total area of 7.3 million ha of Polish land [53]. The largest *L. monacha* outbreak took place between 1978 and 1984 [54,55]—in 1982 alone over 2.3 million ha of forest were sprayed with pyrethroids [53]. During the 1992 to 1994 outbreak, insecticide was applied to about 0.7 million ha [35] (Figure 1). After the largest *L. monacha* infestation, cyclical damage to forests also occurred due to minor outbreaks of often co-occurring sawflies (Diprionidae) as well as pine defoliators, such as *Panolis flammea* (Denis and Schiffermüller, 1775), *Bupalus piniaria* (Linnaeus, 1758), *Dendrolimus pini* (Linnaeus, 1758) [35] (Figure 1). However, in recent years the use of insecticides has been strongly reduced due to European legal regulations [35]. Weakened trees were prone to subsequent attacks by bark and wood-boring insects, such as *Tomicus minor* (Hartig, 1834), *Ips acuminatus* (Gyllenhal, 1827), and *Phaenops cyanea* (Fabricius, 1775) [30].

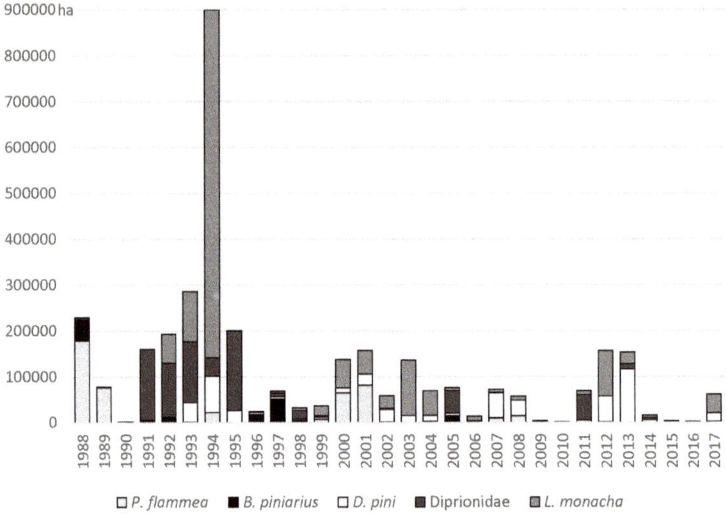

Figure 1. Area of aerial chemical insecticide treatments against pine defoliators in Poland between 1988 and 2017 ([35], modified).

Forests in north-eastern Poland are more diversified, but still contain a high proportion of Scots pine, Norway spruce, and oak. Extended outbreaks of bark beetles, mainly *Ips typographus* (Linnaeus, 1758), occur in spruce-dominated stands [56]. Cyclical *I. typographus* outbreaks occur in the area of the Białowieża Forest, due to the large areas of Norway spruce that occur there [57]. These outbreaks cause deep controversies concerning the active control of insect populations, especially during the most recent outbreak that started in 2012 [58]. Another problem is the occurrence of wood boring insects in oak stands, such as *Agrilus* spp.—especially *A. biguttatus* (Fabricius, 1776) [59], that cause oak decline throughout Europe [60].

The third region, formerly under the control of Austria-Hungary, is located in the south-east, mostly in mountainous terrain, where specific site conditions shape the natural features of forests. As mentioned earlier, due to former (more than 150 years ago) approaches to forest management, stands, especially in the western part of the region, are dominated by Norway spruce that was artificially planted from non-local seed origins. Such species composition has strong effects on forest health problems [8]. The main biotic agents responsible for forest decline in this region are those affecting Norway spruce: *Armillaria ostoyae* (Romagn.) Herink [61] and bark beetles, mainly *I. typographus* [62–64].

In 2003 to 2010 an extended bark beetle outbreak resulted in partial or total deforestation of mountain slopes in the Beskidy Mountains [8,63], where the yearly number of processed logs from trees infested by bark beetles reached more than 1 million m^3 in 2007 to 2008 (Figure 2). In recent years, a new bark beetle outbreak has taken place, started by water stress in Norway spruce after the drought of 2015, affecting stands in the western Carpathian and Sudetes mountains (Figure 2). The ongoing drought conditions, combined with already high bark beetle populations, make the risk of a further outbreak extremely high [63].

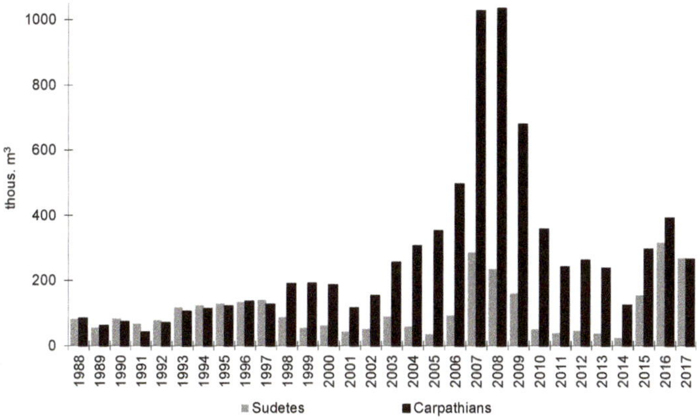

Figure 2. Volume of harvested spruce logs affected by bark beetles from trees in the Carpathian and Sudetes mountains between 1988 and 2017.

Outbreaks of damaging insects often co-occurred with increased incidences of fungal diseases (Figure 3). In the early 1950s, there were high levels of damage in spruce stands caused by *Armillaria* spp. (65 thousand hectares affected). In the following decades, Scots pine, the main forest tree species in Poland, was affected by foliar diseases. For example, in 1996, 950,000 hectares of Scots pine stands were affected by *Lophodermium pinastri* (Schrad.) Chevall., *L. seditiosum* Minter, Staley and Millar, *Gremmeniella abietina* (Lagerb.) M. Morelet (=*Ascocalyx abietina* (Lagerb.) Schläpf.-Bernh.), *Cenangium ferruginosum* Fr., and *Dothichiza pithyophila* (Corda) Petr. [5]. Stands in which these diseases occurred largely occupied former agricultural land, where the root rot *Heterobasidion annosum sensu lato* was endemic. The forest area affected by this root rot grew annually, reaching 5% (200 thousand hectares) of afforested land in 1997 in Poland [26]. In the past 30 years, local forest decline has been repeatedly recorded—silver fir decline in the 1970s (caused mainly by air pollution) [46], oak decline [65], ash dieback caused by *Hymenoscyphus fraxineus* (T. Kowalski) Baral, Queloz and Hosoya (*Chalara fraxinea* T. Kowalski) [66,67], and recently, dieback of Norway spruce stands caused by drought, root pathogens, and bark beetles [63]. Presently, dieback is considered to be caused by a combination of several factors [68,69].

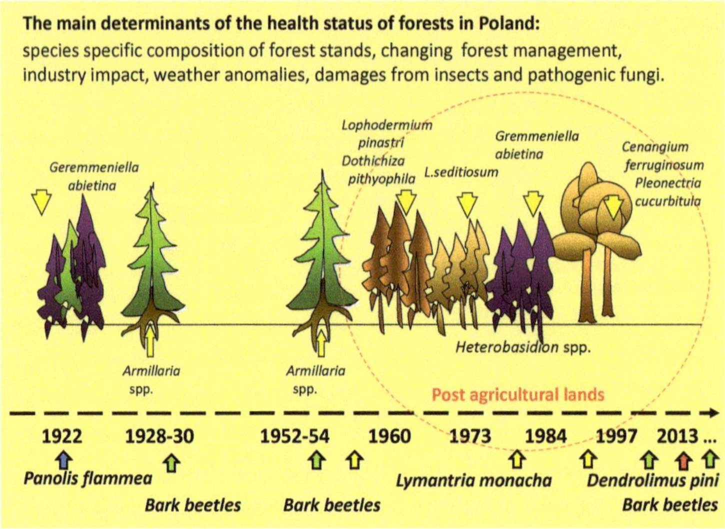

Figure 3. Main fungal diseases and insect pests causing forest mortality since 1922 in Poland [70].

5. Impact of Climatic and Economic Change on Forest Ecosystems in the past and Now

Natural disasters affecting forests cause large quantities of dead organic matter to accumulate, thereby promoting populations of wood boring insects. After World War I (1923 and 1924), outbreaks of defoliating insects (e.g., *Panolis flammea*) occurred on several hundred thousand hectares of State forests, located mainly to the west of the Vistula River [1]. After World War II in forests heavily damaged by war and occupying forces, the range and intensity of insect damage increased significantly, causing severe damage to Scots pine forests in northern and western Poland.

The large-scale regeneration of forests was not only a serious logistical problem, but also a tree seed supply and tree breeding problem, because the appropriate genetic source of specific seeds was needed for the stands that had to be replanted [41,71]. In lowland areas, it was easier to reconstruct natural species composition, whereas in mountainous regions, due to the protective functions of forests, such reconstruction was a complicated process [72]. Forest decline in the Izerskie Mountains in the 1980s resulted in the planting of diverse forest species, which today form young multi-species stands [73] (Figure 4), although specific problems caused by *Ips cembrae* (Heer, 1836) affected young larch trees [74]. An unexpected problem was the decline of older stands in the southern part of the Beskidy Mountains, which were considered more resistant to pollution and biotic damage [12,75,76], although their advanced age was expected to make them more vulnerable to bark beetle attacks [64]. Unfortunately, forest decline has affected even the famous spruces of Istebna, known for their high quality (e.g., "Anderson's spruce" a single tree with a volume of 11 m^3), [12,77], many of which have now died, reducing the amount of spruce in this region by 30%. This is attributed to protecting spruces to an over-mature age, making them more susceptible to root rot diseases, repeated insect outbreaks, as well as to damage by strong winds and breakage by heavy snowfall [64].

Figure 4. Forest stands in the Izerskie Mountains: Decline (1992) and recovery (2005). Photo by W. Grodzki.

Increasingly warmer winters, a lack of winter snowfall, and reduced rainfall during the growing season, have resulted in mortality of many different tree species. Of note, is the dieback of spruce in Poland, observed for about the past 15 years, due to long-term drought and reduced soil moisture levels. Mortality of ash has occurred for many reasons, including the fungus *Hymenoscyphus fraxineus*, responsible for crown dieback, while other deciduous species have been affected by several root pathogens (e.g., *Phytophthora* spp., *Armillaria* spp., *Heterobasidion* spp.) [5].

The weather in different seasons and years was different across the country, due to climatic conditions, as well as to influence of large lakes and forest areas in the north-east, of lowlands with limited forest area in central Poland, and of high mountains in the south. In 1992, extreme drought occurred in central and southern Poland, whereas in the north the growing season was wet, whereas the growing seasons in 2014 and 2015 were very dry throughout the country (Figure 5).

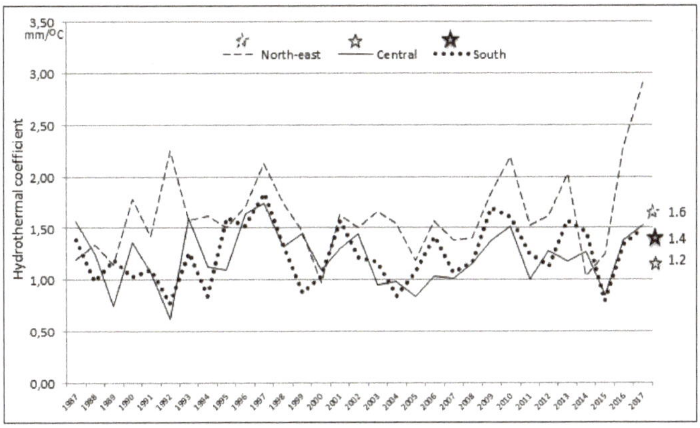

Figure 5. Values of Sielianinov's hydrothermal coefficient, K ($K = P \times 10/\Sigma t$) for three regions of Poland during the last 30 years (lines) and multi-year average (stars), where P is the sum of precipitation and Σt is the sum of average temperatures during the period. Ranges of K values for the period are interpreted as extremely dry (0 to 0.4), very dry (0.41 to 0.70), quite dry (0.71 to 1.0), dry (1.1 to 1.3), optimal (1.31 to 1.6), moist (1.61 to 2.0), wet (2.1 to 2.5), and very wet (>2.5).

A few years were unique in the history of Polish forests—three large-scale fires in August 1992 (burning ~20 thousand hectares in total), widespread flooding of the Nysa and Odra rivers in the west in July 1997 (affecting an area 665.8 thousand ha), high winds in the Pisz Old Forest in the north-east in July 2002 (12 thousand ha), and a hurricane in northern Poland in August 2017 (7.5 million m^3 of blowdown trees). Only the growing seasons of 1997, 2001 to 2002, and 2010 were characterized by favorable weather conditions for forests, without any large scale insect or fungus infestation, although subsequent years were affected by the major disturbances mentioned above [5].

6. What Might the Future Hold for Polish Forests?

Projections for climate change in the twenty-first century, both for Europe as a whole and Poland in particular, predict increases in air temperature, and thus soil temperature, in all seasons [78–80]. In addition, changes in precipitation and increasing concentrations of CO_2 in the atmosphere as a result of greenhouse gas emissions and other pollutants may alter tree species composition. The predicted warming along with nitrogen accumulation from pollution in habitats may mean gradual northward and higher elevational shifting of tree ranges, which will promote deciduous tree species at the expense of the area currently occupied by coniferous species [81–84]. In this part of Europe, the participation and importance of native deciduous trees - beech, birch, linden, and oaks—will probably increase. Non-native species may also appear, including invasive species adapted to warmer temperatures, such as white robinia (*Robinia pseudoacacia* L.), wild black cherry (*Prunus serotina* Ehr.), and red oak (*Quercus rubra* L.) [85]. One should expect a reduction of coniferous species, especially spruce, maybe fir, and, to a lesser extent, pine [46,86].

In the coming years, one of the most important effects of climate change will be its impact on the resilience and health of existing and new forest stands. The increasing frequency and intensity of extreme weather, such as droughts, long periods of high temperature, strong winds, heavy rains, and floods, increase the susceptibility of trees to physical damage, disease, and pests. In addition, climate change can affect the developmental cycles of pathogenic organisms, which in turn may increase threats to trees [70,87,88]. Some species, e.g., endophytes that are of marginal significance to date, such as *Ips acuminatus*, may gain in importance and become serious threats as pathogens/pests [35,89]. It should be remembered that alien and invasive species may appear more frequently [90], as demonstrated by a number of recent examples, such as *Obolodiplosis robiniae* (Haldeman, 1847), *Coleotechnites piceaella* (Kearfott, 1903), *Phylonorycter issikii* (Kumata, 1963) [91].

Changing environmental conditions (e.g., increased soil and air temperature, milder and snowless winters), but also free trade and the long-range shipping of goods favor the emergence of an increasing number of non-native species of fungi in Poland from warmer regions of the world. An example of pathogens of great threat to forest trees in Poland, which until recently were only present in southern Europe, are representatives of the genus *Phytophthora* and *Sphaeropsis sapinea* (Fr.) Dyko and B. Sutton—both may cause dieback of trees [92–94]. One can also expect the spread of fungi affecting trees, responsible for root and stem rot of living trees in Poland, e.g., *Ganoderma resinaceum* Boud. or *Perenniporia fraxinea* (Bull.) Ryvarden [95–98]. An example of the spread (from south to north) of exotic fungal species in Polish forests are saprotrophic ground fungi, members of the stinkhorn (*Phallaceae*) family: Pink stinkhorn *Mutinus ravenelii* (Berk. and M.A. Curtis) E. Fisch and octopus stinkhorn *Clathrus archerii* (Berk.) Dring [99]. To date, there is no evidence of the negative impact of alien stinkhorns on native species of mycobiota or on other elements of forest ecosystems. However, it cannot be ruled out that in the future they may compete with indigenous saprothropic, or even mycorrhizal fungi, exerting some negative (direct or indirect) impact on plants, including trees. Intentional or natural regeneration of non-native woody plants made possible by changing climatic conditions may result in the appearance of non-native species of mycorrhizal fungi, which could affect native mycobiota and displace native species of ectomycorrhizal mushrooms [100].

Under warming conditions with more intense droughts, many insect species can be expected to increase. Thermophilic species with a southerly distribution will increase. Significant damage may be caused by species considered thus far as pests of secondary importance, such as *I. acuminatus* [87,89] and *I. cembrae* [74]. An increase in the frequency of severe drought is expected [101], which, when combined with the emergence of insect species that have been found in Poland (e.g., *Anoplophora glabripennis* (Motschulsky, 1853) [102], *Xylosandrus germanus* (Blandford, 1894) [103], *Gnathotrichus materiarius* (Fitch, 1858) [104], can lead to widespread tree mortality.

Potential increases in deciduous species (e.g., beech, oaks) may fail to materialize if trees are affected by disease, due to the occurrence of phytophthorosis of these species, as well as fungi of the genus *Neonectria*—causing bark damping and beetroot cracks, the activity of *Armillaria* spp., and wood boring beetles of the *Agrilus* genus, as recently observed in oak stands [60,65,105–110]. On moist sites that are exposed to drought, hydric tree species (ash, alder, elm) will be exposed to adverse abiotic and biotic factors that may lead to a drastic decline in their condition or even large-scale dieback, for example due to *Phytophthora* spp. [93], *Ophiostoma novo-ulmi* Brasier, or dying ash, for which *Hymenoscyphus fraxineus* (*Chalara fraxinea*) is responsible [66,67].

Over the next few decades, coniferous species (*Pinus sylvestris, Picea abies*) will be strongly exposed to insect pests and pathogenic factors. In particular, pine monocultures established in the past on former farmland soils, as well as spruce monocultures in lower montane zones, are likely to see increased infection by root pathogens (*Heterobasidion* spp., *Armillaria* spp.), which will require replanting to alter species composition [41,111]. In drought conditions, the negative impact of fungi that produce rhizomorphs (including *Armillaria*) will increase. Since rhizomorphs are capable of growing through the soil for long distances (many meters), they are better able to survive drought conditions by transporting water, carbohydrates, and minerals [105,112–114]. In addition to changing tree species planted in such areas, activities should be undertaken to eliminate food sources of root pathogens (e.g., removing stumps and major roots of dead trees in affected areas). Control of root rot fungi is also recommended using a proven biological method of introducing native strains of *Phlebiopsis gigantea* (Fr.) Jülich, which on contact competes with the spread of hyphae of some root rot fungi [26,41,115–117].

Pine in fertile habitats produces root system tiles, which in the case of water shortage and weakness by other abiotic factors, leads to an increase in susceptibility to attacks by bark and wood boring insects, e.g., *Phaenops cyanea, Ips acuminatus* [118]. Weakened pine trees of practically any age, due to disturbances in water management can be exposed to attacks by pathogens causing dieback (*Gremmeniella abietina, Cenangium ferruginosum, Sphaeropsis sapinea, Pleonectria cucurbitula* (Tode) Hirooka, Rossman and P. Chaverri) [35,70,119].

Spruce is sensitive to high temperatures and drought, both of which have occurred in the last two decades. As a result, there has been a high rate of spruce mortality, especially in the south of Poland. Bark beetles such as *Ips typographus* may be attracted to spruce trees whose roots are infected by *Armillaria ostoyae*, as the fungus can result in the foliage releasing certain chemical compounds [61,63,64,105]. In the coming years, biotic and abiotic factors affecting spruce forest health are likely to persist, and even greater rates of tree decline can be expected.

Climate change may increase the frequency, intensity, and area affected by natural disturbances (fires, wind storms, and snowstorms) in forest ecosystems. The large number of dead trees will be conducive to the development of numerous species of saprotrophic fungi and insects associated with heavily damaged forests.

Already today, forest management faces the need to improve stress prevention and management for improved health in post-conflict forest areas. It is advisable to choose strategies based on scientific principles. An important task in the future will be monitoring and cross-border control of harmful pathogenic insects and fungi. Or indeed: *'Omnia subiecta sunt naturae'* ? (Everything is subject to nature)—Democritus (460 to 375 years b. Ch.).

7. Conclusions

Forests in Poland, mostly growing in a climatic transition zone, are strongly diversified but vulnerable to disease and insect outbreaks. Although specific traits resulting from different management in the past still affect forest characteristics (especially species composition and forest structure) and related threats caused by abiotic and biotic factors, the overall health and vitality of

Polish forests is good. Political and economic transformation have reduced the negative impact of the forest industry on forest health, and, due to new forest policies, the forest area has been gradually increasing. As 80% of the forest area is nationalized and managed by the State Forests Holding, which is primarily focused on the sustainable use of forest resources, the future health of Polish forests looks promising.

Author Contributions: All authors (Z.S., W.G. and A.S.) substantially conceived the idea, contributed in conceptualization, resources, writing original draft, writing review and editing the text; Z.S. supervision.

Funding: This paper was partially supported from the statutory funds of the Forest Research Institute in Sękocin Stary, Poland, the Warmia and Mazury University in Olsztyn, Poland, and of Warsaw University of Life Sciences—SGGW, Poland.

Acknowledgments: The authors would like to thank the anonymous reviewers for their valuable comments and suggestions.

Conflicts of Interest: Authors declare any personal circumstances or interest that may be perceived as inappropriately influencing the representation or interpretation of reported research results.

References

1. Broda, J. *Dzieje Najnowsze Leśnictwa w Polsce 1918–2006. Tom I, II [Recent Forestry History in Poland. Vol. I, II]*; Polskie Towarzystwo Leśne: Warszawa, Poland, 2007.
2. Butterwick, R. Between Anti-Enlightenment and enlightened Catholicism: Provincial preachers in late-eighteenth-century Poland-Lithuania. In *Peripheries of the Enlightenment*; University of Oxford: Oxford, UK, 2008; Volume 1, pp. 201–228.
3. Klocek, A. *Państwowa Administracja Oraz Gospodarka Leśna w Wybranych Krajach [State Administration and Forest Management in Selected Countries]*; Centrum Informacyjne Lasów Państwowych: Warszawa, Poland, 2006.
4. Kawecki, W. *Lasy Żywiecczyzny, Ich Teraźniejszość i Przyszłość (Zarys Monograficzny) [Żywiec Region Forests, Their Present and Future (Monographic Outline)]*; Prace Roln. Leśne 35. Polska Akademia Umiejętności: Kraków, Poland, 1939.
5. Sierota, Z. *Gdy las Choruje [When the Forest Is Ill]*; Centrum Informacyjne Lasów Państwowych: Warszawa, Poland, 2011.
6. Zoll, T. Podstawowe zagadnienia zagospodarowania lasów górskich w Sudetach [Basic issues of mountain forest management in the Sudetes]. *Sylwan* **1958**, *102*, 9–33.
7. Sitková, Z.; Hlásny, T.; Kulla, L.; Grodzki, W.; Šrámek, V. The Beskids—Region of interest. In *Spruce Forests Decline in the Beskids*; Hlásny, T., Sitková, Z., Eds.; National Forest Centre—Forest Research Institute Zvolen & Czech University of Life Sciences Prague & Forestry and Game Management Research Institute Jíloviště-Strnady: Zvolen, Slovakia, 2010; pp. 9–13.
8. Hlásny, T.; Grodzki, W.; Šrámek, V.; Holuša, J.; Kulla, L.; Sitková, Z.; Turčáni, M.; Rączka, G.; Strzeliński, P.; Węgiel, A. Spruce forests decline in the Beskids. In *Spruce Forests Decline in the Beskids*; Hlásny, T., Sitková, Z., Eds.; National Forest Centre—Forest Research Institute Zvolen & Czech University of Life Sciences Prague & Forestry and Game Management Research Institute Jíloviště-Strnady: Zvolen, Slovakia, 2010; pp. 15–31.
9. Kant, S.; Alavalapati, J.R.R. *Handbook of Forest Resource Economics*; Routledge: New York, NY, USA, 2014.
10. Bytnerowicz, A.; Szaro, R.; Karnosky, D.; Manning, W.; McManus, M.; Musselman, R.; Muzika, R.-M. Importance of international research cooperative programs for better understanding of air pollution effects on forest ecosystems in Central Europe. In *Effects of Air Pollution on Forest Health and Biodiversity in Forests of the Carpathian Mountains*; Szaro, R.C., Oszlányi, J., Godzik, B., Bytnerowicz, A., Eds.; NATO Science Series, Series I: Life and Behavioural Sciences; IOS Press: Amsterdam, The Netherlands, 2002; Volume 345, pp. 13–20.
11. Barzdajn, W.; Ceitel, J.; Modrzyński, J. Świerk w lasach polskich—Historia, stan, perspektywy [Spruce in Polish forests—History, condition, perspectives]. In *Drzewostany Świerkowe. Stan, Problemy, Perspektywy Rozwojowe [Spruce Stands. Condition, Problems, Development Perspectives]*; Grzywacz, A., Ed.; Polskie Towarzystwo Leśne: Warszawa, Poland, 2003; pp. 5–22.

12. Faber, A.; Sabor, J.; Grzebelus, D. Zróżnicowanie genetyczne wybranych drzewostanów świerka pospolitego (*Picea abies* L. Karst.) na przykładzie rasy istebniańskiej przy wykorzystaniu markerów RAPD [Genetic diversity of Norway spruce (*Picea abies* L. Karst.) of the Polish provenance from Istebna based on the RAPD analysis]. *Zesz. Probl. Post. Nauk Roln.* **2007**, *517*, 277–283.
13. Sturtevant, B.R.; Fall, A.; Kneeshaw, D.D.; Simon, N.P.P.; Papaik, M.J.; Berninger, K.; Doyon, F.; Morgan, D.G.; Messier, C. A toolkit modeling approach for sustainable forest management planning: Achieving balance between science and local needs. *Ecol. Soc.* **2007**, *12*, 7. [CrossRef]
14. Rykowski, K. *O Leśnictwie Trwałym i Zrównoważonym. W Poszukiwaniu Definicji i Miar [On Permanent and Sustainable Forestry. In Search of Definitions and Measures]*; Centrum Informacyjne Lasów Państwowych: Warszawa, Poland, 2006.
15. Borkowski, J. *Ministerialny Proces Ochrony Lasów w Europie [Ministerial Process of Forest Protection in Europe]*; Biblioteczka Leśniczego 242. Wyd. Świat: Warszawa, Poland, 2011.
16. Smyk, G. Zasady wprowadzania i zakres obowiązywania rosyjskich źródeł prawa w Królestwie Polskim po powstaniu styczniowym [Principles of introduction and scope of operation of Russian sources of law in the Kingdom of Poland after the January Uprising]. *Studia z Dziejów Państwa i Prawa Polskiego* **2011**, *14*, 213–239.
17. Beker, C.; Andrzejewski, T. Model wzrostu niepielęgnowanych drzewostanów sosnowych. I. Lokalny model referencyjny Pinus Zielonka [Growth model of unthinned Scots pine stands I. Local reference model Pinus Zielonka]. *Acta Sci. Pol. Silv. Colendar. Rat. Ind. Lignar.* **2013**, *12*, 5–13.
18. Klemantowicz, D. Lasy powiatu sieradzkiego w połowie XIX w. [Forests in Sieradz district in the mid-19th century]. *Biul. Szadk.* **2008**, *8*, 151–160.
19. Capecki, Z. Charakterystyka zdrowotności i zagrożenia lasów karpackich w Polsce [Characteristics of health and threats in the Carpathian forests in Poland]. *Pr. Inst. Bad. Leśn. Ser. A* **1983**, *617*, 27–54.
20. Capecki, Z. Stan sanitarny lasów górskich a gradacje szkodników wtórnych [The sanitary state of mountain forests and outbreaks of secondary pests]. *Sylwan* **1993**, *137*, 61–68.
21. Poławski, Z.F. Zmiany użytkowania ziemi w Polsce w ostatnich dwóch stuleciach [Land use changes in Poland during last two centuries]. *Teledetekcja Środowiska* **2009**, *42*, 69–82.
22. Capecki, Z. Zagrożenie lasów sudeckich przez szkodniki na tle szkód spowodowanych przez huragany i okiść [Threats to Sudety forests by pests on the background of damage caused by hurricanes and snow tufts]. *Sylwan* **1969**, *113*, 57–64.
23. Michalski, J. Gradacje kornikowe w ostatnim 50-leciu w drzewostanach świerkowych Polski [Bark beetle outbreaks in the last 50 years in spruce stands of Poland]. In *Biologia Świerka Pospolitego [Biology of Norway Spruce]*; Boratyński, A., Bugała, W., Eds.; Bogucki Wydawnictwo Naukowe: Poznań, Poland, 1998; pp. 468–508.
24. Smykała, J. Historia, rozmiar i rozmieszczenie zalesień gruntów porolnych w Polsce w latach 1945–1987 [History, area and distribution of afforestation of post-agricultural lands in Poland in the years 1945–1987]. *Sylwan* **1990**, *134*, 1–7.
25. *Dekret Polskiego Komitetu Wyzwolenia Narodowego z Dnia 6 Września 1944 r. o Przeprowadzeniu Reformy Rolnej [Decree Polish Committee of National Liberation about Implementing the Land Reform]*; PKWN: Lublin, Poland, 1944.
26. Sierota, Z. *Heterobasidion annosum* on Poland's former agricultural lands—Scope of menace and prevention. *Sci. Res. Essays* **2013**, *8*, 2298–2305. [CrossRef]
27. Woreta, D.; Sukovata, L. Survival and growth of the *Melolontha* spp. grubs on the roots of the main forest tree species. *Leś. Pr. Bad./For. Res. Pap.* **2014**, *75*, 375–383. [CrossRef]
28. Sierota, Z. An analysis of the root rot spread in a Scots pine stand growing in post-agricultural land. *Fol. For. Pol. Ser. A For.* **1997**, *39*, 27–37.
29. Małecka, M.; Hilszczańska, D. Influence of resting and pine sawdust application on chemical changes in post-agricultural soil and the ectomycorrhizal community of growing Scots pine saplings. *Leś. Pr. Bad./For. Res. Pap.* **2015**, *76*, 265–272. [CrossRef]
30. Kolk, A. Impact of bark beetles on forest management in Poland in 1981–1990. *J. Appl. Entomol.* **1992**, *114*, 425–430. [CrossRef]
31. Lipa, J.J.; Kolk, A. The recent situation of the gypsy moth (*Lymantria dispar*) and other Lymantriids in Poland. *Bull. OEPP/EPPO Bull.* **1995**, *25*, 623–629. [CrossRef]
32. Capecki, Z. Specyfika zdrowotności lasów wschodniej części Karpat [Specific of forest health in the eastern part of the Carpathians]. *Sylwan* **1999**, *143*, 81–88.

33. Sierota, Z. Choroby infekcyjne—Ocena występowania i wpływ na gospodarkę leśną [Infectious diseases—Occurrence and impact on forest management]. *Sylwan* **1998**, *142*, 21–37.
34. Sierota, Z.; Małecka, M. Ocena zmian w drzewostanie sosnowym na gruncie porolnym po 30 latach od wykonania pierwszych cięć pielęgnacyjnych bez zabiegu ochronnego przeciw hubie korzeni [Assessment of changes in a pine stand established on post-agricultural land after 30 years from first improvement cutting without application of protective measures against root rot]. *Sylwan* **2003**, *147*, 19–26.
35. Skrzecz, I.; Perlińska, A. Current problems and tasks of forest protection in Poland. *Fol. For. Pol. Ser. A For.* **2018**, *60*, 161–172. [CrossRef]
36. Geszprych, M. Specyfika nadzoru i sfera wartości w prawie leśnym [Specific character and the sphere of values in forestry law]. *Studia Lubuskie* **2009**, *5*, 159–181.
37. Häusler, A.; Scherer-Lorenzen, M. *Sustainable Forest Management in Germany: The Ecosystem Approach of the Biodiversity Convention Reconsidered*; Bundesamt für Naturschutz—Skripten 51: Bonn, Germany, 2001.
38. Płotkowski, L. Gospodarka leśna w badaniach ekonomiki leśnictwa [Forest management as a subject of forestry economics research]. *Roczniki Nauk Rolniczych Seria G* **2010**, *97*, 110–120.
39. Brzeziecki, B. Podejście ekosystemowe i półnaturalna hodowla lasu (w kontekście zasady wielofunkcyjności [Ecosystem approach and close-to-nature silviculture (in context of forest multifunctionality principle)]. *Studia i Materiały CEPL* **2008**, *3*, 41–54.
40. Brang, P.; Spathelf, P.; Larsen Bo, J.; Bauhus, J.; Boncčina, A.; Chauvin, C.; Drössler, L.; García-Güemes, C.; Heiri, C.; Kerr, G.; et al. Suitability of close-to-nature silviculture for adapting temperate European forests to climate change. *Forestry* **2014**, *87*, 492–503. [CrossRef]
41. Sierota, Z. (Ed.) *Zmiany w Środowisku Drzewostanów Sosnowych na Gruntach Porolnych w Warunkach Przebudowy Częściowej oraz Obecności Grzyba Phlebiopsis gigantea [Changes in Scots Pine Stands Environment on Former Farmlands in Conditions of a Partial Reconstruction and Presence of Phlebiopsis gigantea]*; IBL: Sękocin Stary, Poland, 2011.
42. *Raport o Stanie Lasów w Polsce [The State Forests in Poland]*; Directorate-General of the State Forests: Warszawa, Poland, 2016.
43. Brázdil, R. Meteorological extremes and their impacts on forests in the Czech Republic. In *The Impacts of Climate Variability on Forests*; Lecture Notes in Earth Sciences 74; Beniston, M., Innes, J.L., Eds.; Springer: Berlin/Heidelberg, Germany, 1998.
44. Heinsdorf, D. The role of nitrogen in declining scots pine forests (*Pinus sylvestris*) in the lowland of East Germany. *Water Air Soil Pollut.* **1993**, *69*, 21–35. [CrossRef]
45. Badea, O.; Tanase, M.; Georgeta, J.; Anisoara, L.; Peiov, A.; Uhlirova, H.; Pajtik, J.; Wawrzoniak, J.; Shparyk, Y. Forest health status in the Carpathian Mountains over the period 1997–2001. *Environ. Pollut.* **2004**, *130*, 93–98. [CrossRef]
46. Barzdajn, W. Adaptacja różnych pochodzeń jodły pospolitej (*Abies alba* Mill.) do warunków Sudetów [Adaptation of different Silver fir (*Abies alba* Mill.) provenances to the conditions of the Sudetes]. *Leśn. Pr. Bad./For. Res. Pap.* **2009**, *70*, 49–58.
47. Percy, K.E. Is air pollution an important factor in forest health. In *Effects of Air Pollution on Forest Health and Biodiversity in Forests of the Carpathian Mountains*; Szaro, R.C., Oszlányi, J., Godzik, B., Bytnerowicz, A., Eds.; NATO Science Series, Series I: Life and Behavioural Sciences; IOS Press: Amsterdam, The Netherlands, 2002; Volume 345, pp. 23–42.
48. Paschalis, P.; Zajączkowski, S. (Eds.) *Protection of Forest Ecosystems. Selected Problems of Forestry in Sudety Mountains*; Grant GEF 05/21685 POL; Forest Biodiversity Protection Project: Warszawa, Poland, 1996.
49. Baltensweiler, W. "Waldsterben": Forest pests and air pollution. *Zeitschrift für Angewandte Entomologie* **1985**, *99*, 77–85. [CrossRef]
50. Schutt, P.; Cowling, E.B. Waldsterben, a general decline of forests in central Europe: Symptoms, development and possible causes. *Plant Dis.* **1985**, *69*, 548–558.
51. Schultze, E.D. Air pollution and forest decline in a spruce (*Picea abies*) forest. *Science* **1989**, *244*, 776–783. [CrossRef] [PubMed]
52. Kandler, O.; Innes, J.L. Air pollution and forest decline in Central Europe. *Environ. Poll.* **1995**, *90*, 171–180. [CrossRef]

53. Głowacka, B. The control of the nun moth (*Lymantria monacha* L.) in Poland: A comparison of two strategies. In *Population Dynamics, Impacts and Integrated Management of Forest Defoliating Insects*; McManus, M.L., Liebhold, A.M., Eds.; USDA Forest Service, General Technical Report NE-247; Northeastern Research Station: Hamden, CT, USA, 1998; pp. 108–115.
54. Schönherr, J. Nun moth outbreak in Poland 1978–1984. *Zeitschrift für Angewandte Entomologie* **1985**, *99*, 73–76. [CrossRef]
55. Śliwa, E. Przebieg masowego pojawu brudnicy mniszki (*Lymantria monacha* L.) i jej zwalczanie w Polsce w latach 1978–1985 oraz regeneracja aparatu asymilacyjnego w uszkodzonych drzewostanach [Course of the mass appearance of the nun moth (*L. monacha* L). and its control in Poland in the years 1978–1985 and the regeneration of the assimilation apparatus in damaged stands]. *Prace Inst. Bad. Lesn.* **1989**, *710*, 1–120.
56. Grodzki, W.; Michalski, J. Historia gradacji kornika drukarza [The history of bark beetle outbreaks]. In *Kornik Drukarz Ips typographus (L.) i Jego Rola w Ekosystemach Leśnych [The Spruce Bark Beetle of Ips typographus (L.) and Its Role in Forest Ecosystems]*; Grodzki, W., Ed.; Centrum Informacyjne Lasów Państwowych: Warszawa, Poland, 2013; pp. 109–125.
57. Michalski, J.; Starzyk, J.R.; Kolk, A.; Grodzki, W. Zagrożenie świerka przez kornika drukarza—*Ips typographus* (L.) w drzewostanach Leśnego Kompleksu Promocyjnego "Puszcza Białowieska" w latach 2000–2002 [Threat of Norway spruce caused by the bark beetle *Ips typographus* (L.) in the stands of the Forest Promotion Complex "Puszcza Białowieska" in 2000–2002]. *Leś. Pr. Bad./For. Res. Pap.* **2004**, *3*, 5–30.
58. Brzeziecki, B.; Hilszczański, J.; Kowalski, T.; Łakomy, P.; Małek, S.; Miścicki, S.; Modrzyński, J.; Sowa, J.; Starzyk, J.R. Problem masowego zamierania drzewostanów świerkowych w Leśnym Kompleksie Promocyjnym "Puszcza Białowieska" [Problem of a massive dying-off of Norway spruce stands in the 'Białowieża Forest' Forest Promotional Complex]. *Sylwan* **2018**, *162*, 373–386.
59. Hilszczański, J.; Sierpiński, A. *Agrilus* spp. the main factor of oak decline in Poland. In Proceedings of the IUFRO Working Party 7.03.10 Workshop, Gmunden, Austria, 11–14 September 2006; pp. 121–125.
60. Moraal, L.G.; Hilszczański, J. The buprestid beetle, *Agrilus biguttatus* (F.) (Col., Buprestidae), a recent factor in oak decline in Europe. *J. Pest Sci.* **2000**, *5*, 134–138.
61. Żółciak, A.; Lech, P.; Małecka, M.; Sierota, Z. Opieńkowa zgnilizna korzeni a stan zdrowotny drzewostanów świerkowych w Beskidach [Armillaria root and the health conditon of Norway spruce stands in the Beskids]. *Polska Akademia Umiejętności. Prace Komisji Nauk Rolniczych Leśnych i Weterynaryjnych* **2009**, *11*, 61–71.
62. Grodzki, W. Zagrożenie górskich drzewostanów świerkowych w zachodniej części Beskidów ze strony szkodników owadzich [Threats to mountain spruce stands of the insect pests in the western part of Beskidy mountains]. *Leś. Pr. Bad./For. Res. Pap.* **2004**, *2*, 35–47.
63. Grodzki, W. The decline of Norway spruce *Picea abies* (L.) Karst. stands in Beskid Śląski and Żywiecki: Theoretical concept and reality. *Beskydy* **2010**, *3*, 19–26.
64. Grodzki, W.; Starzyk, J.R.; Kosibowicz, M. Variability of selected traits of *Ips typographus* (L.) (Col.: Scolytinae) populations in Beskid Żywiecki (Western Carpathians, Poland) region affected by bark beetle outbreak. *Fol. For. Pol. Ser. A For.* **2014**, *56*, 79–92. [CrossRef]
65. Oszako, T. Przyczyny masowego zamierania drzewostanów dębowych [Causes of oak stand decline]. *Sylwan* **2007**, *151*, 62–72.
66. Kowalski, T. *Chalara fraxinea* sp. nov. associated with dieback of ash (*Fraxinus excelsior*) in Poland. *For. Pathol.* **2006**, *36*, 264–270. [CrossRef]
67. Kowalski, T. Rozprzestrzenienie grzyba *Chalara fraxinea* w aspekcie procesu chorobowego jesionu w Polsce [Expanse of Chalara fraxinea fungus in terms of ash dieback in Poland]. *Sylwan* **2009**, *153*, 668–674.
68. Manion, P.D. *Tree Disease Concepts*; Prentice-Hall: Englewood Cliffs, NJ, USA, 1991.
69. Innes, J.L. *Forest Health: Its Assessment and Status*; CAB International: Wallingford, UK, 1993.
70. Sierota, Z. Obserwowane i prawdopodobne zmiany występowania mikroorganizmów chorobotwórczych w związku ze zmianami klimatycznymi oraz ocena ich funkcji ekologicznych w ekosystemach leśnych; potencjalne rozprzestrzenianie się gatunków inwazyjnych [Observed and probable changes in the occurrence of pathogenic microorganisms in relation to climatic changes and assessment of their ecological functions in forest ecosystems; the potential spread of invasive species]. In *KLIMAT. Lasy i Drewno a Zmiany Klimatyczne: Zagrożenia i Szanse. Materiały Pierwszego Panelu Ekspertów w Ramach Prac nad Narodowym Programem Leśnym [CLIMATE. Forests and Wood and Climate Change: Threats and Opportunities. Materials of the First Panel of Experts*

as Part of Work on the National Forest Program]; Rykowski, K., Ed.; Instytut Badawczy Leśnictwa: Sękocin Stary, Poland, 2014; pp. 189–198.

71. Zachara, T. Problem szkód w lasach powodowanych przez śnieg i wiatr oraz sposoby przeciwdziałania im [Damage to forests caused by snow and wind and the ways of counteracting it]. *Sylwan* **2006**, *150*, 56–64.
72. Nikolov, C.; Konôpka, B.; Kajba, M.; Galko, J.; Kunca, A.; Janský, L. Post-disaster Forest Management and Bark Beetle Outbreak in Tatra National Park, Slovakia. *Mount. Res. Develop.* **2014**, *34*, 326–335. [CrossRef]
73. Pietruńko, G. Analiza zmian przestrzennych w nadleśnictwach Sudetów Zachodnich w oparciu o wyniki przeprowadzonych prac urządzeniowych [Analysis of spatial changes in the forest districts of the Western Sudetes based on the results of the forest inventory]. *Postępy Techniki w Leśnictwie* **2004**, *89*, 22–28.
74. Grodzki, W.; Guzik, M. Wiatro- i śniegołomy oraz gradacje kornika drukarza w Tatrzańskim Parku Narodowym na przestrzeni ostatnich 100 lat. Próba charakterystyki przestrzennej [Wind- and snowfall, and the spruce bark beetle outbreaks in the Tatra National Park over the last 100 years. Attempt to spatial characteristics]. In *Długookresowe Zmiany w Przyrodzie i Użytkowaniu TPN [Long-Term Changes in Nature and Use of Tatra NP]*; Guzik, M., Ed.; Wyd. Tatrzański Park Narodowy: Zakopane, Poland, 2009; pp. 33–46.
75. Capecki, Z. Rejony zdrowotności lasów zachodniej części Karpat [Forest health regions in western Carpathians]. *Pr. Inst. Bad. Leśn. Ser. A* **1994**, *781*, 61–125.
76. Małek, S.; Martinson, L.; Sverdrup, H. Modelling future soil chemistry at a highly polluted forest site at Istebna in Southern Poland using the "SAFE" model. *Environ. Poll.* **2005**, *137*, 568–573. [CrossRef]
77. Sabor, J. Możliwości zachowania i metody selekcji drzewostanów świerkowych rasy istebniańskiej [Possibilities of preservation and methods of selection of the Istebna breed spruce stands]. *Sylwan* **1996**, *140*, 61–81.
78. Kundzewicz, Z.W.; Kowalczak, P. *Zmiany Klimatu i Ich Skutki [Climate Changes and Their Impact]*; Wydawnictwo Kurpisz SA: Poznań, Poland, 2008.
79. Lindner, M.; Maroschek, M.; Netherer, S.; Kremer, A.; Barbati, A.; Garcia-Gonzalo, J.; Seidl, R.; Delzon, S.; Corona, P.; Kolström, M.; et al. Climate change impacts, adaptive capacity, and vulnerability of European forest ecosystems. *For. Ecol. Manag.* **2010**, *259*, 698–709. [CrossRef]
80. Liszewska, M. Klimat w Polsce w XXI wieku—Prawdopodobne kierunki zmian; perspektywa klimatów lokalnych [Climate in Poland in the 21st century—Likely directions of change; the prospect of local climates]. In *KLIMAT. Lasy i Drewno a Zmiany Klimatyczne: Zagrożenia i Szanse. Materiały Pierwszego Panelu Ekspertów w Ramach Prac nad Narodowym Programem Leśnym [CLIMATE. Forests and Wood and Climate Change: Threats and Opportunities. Materials of the First Panel of Experts as Part of Work on the National Forest Program]*; Rykowski, K., Ed.; Instytut Badawczy Leśnictwa: Sękocin Stary, Poland, 2014; pp. 35–44.
81. Sadowski, M. Przewidywane zmiany klimatu i ich przyrodnicze, społeczne i polityczne konsekwencje [Predicted changes of the climate and their environmental, social, and political consequences]. *Sylwan* **1996**, *140*, 83–103.
82. Ryszkowski, L.; Kędziora, A.; Bałazy, S. Przewidywane zmiany globalne klimatu a lasy i zadrzewienia krajobrazu rolniczego [The predicted effect of global changes on midfield forests and shelterbelts in agricultural landscape]. *Sylwan* **1995**, *139*, 19–32.
83. Zielony, R. Uwarunkowania siedliskowe gospodarki leśnej u progu XXI wieku—Zarys problem [Site conditions of forest management at the turn of the XX and XXI centuries—Outline of the problem]. In *Zgodność Fitocenozy z Biotopem w Ekosystemach Leśnych [Compatibility between Phytocoenosis and Biotope in Forest Ecosystems]*; Zielony, R., Ed.; Fundacja Rozwój SGGW: Warszawa, Poland, 2001; pp. 6–8.
84. Rabasa, S.G.; Granda, E.; Benavides, R.; Kunstler, G.; Espelta, J.M.; Ogaya, R.; Peñuelas, J.; Scherer-Lorenzen, M.; Gil, W.; Grodzki, W.; et al. Disparity in elevational shifts of European trees in response to recent climate warming. *Glob. Chang. Biol.* **2013**, *19*, 2490–2499. [CrossRef]
85. Tokarska-Guzik, B.; Dajdok, Z.; Zając, M.; Zając, A.; Urbisz, A.; Danielewicz, W.; Hołdyński, C. *Rośliny Obcego Pochodzenia w Polsce ze Szczególnym Uwzględnieniem Gatunków Inwazyjnych [Alien Plants in Poland with Particular Reference to Invasive Species]*; Generalna Dyrekcja Ochrony Środowiska: Warszawa, Poland, 2012.
86. Szwagrzyk, J. Prawdopodobne zmiany zasięgów występowania gatunków drzewiastych—Konsekwencje dla hodowli lasu [Probable changes in the ranges of tree species occurrence—Consequences for forest breeding]. In *KLIMAT. Lasy i Drewno a Zmiany Klimatyczne: Zagrożenia i Szanse. Materiały Pierwszego Panelu Ekspertów w Ramach Prac nad Narodowym Programem Leśnym [CLIMATE. Forests and Wood and Climate Change:*

Threats and Opportunities. Materials of the First Panel of Experts as Part of Work on the National Forest Program]; Rykowski, K., Ed.; Instytut Badawczy Leśnictwa: Sękocin Stary, Poland, 2014; pp. 45–54.
87. Jaworski, T.; Hilszczański, J. The effect of temperature and humidity changes on insects development and their impact on forest ecosystems in the context of expected climate change. *Leś. Pr. Bad./For. Res. Pap.* **2013**, *74*, 345–355. [CrossRef]
88. Hilszczański, J. Dynamika populacji owadów oraz ocena ich funkcji ekologicznych w ekosystemach leśnych w związku ze zmianami klimatycznymi [The dynamics of the insect populations and the assessment of their ecological functions in forest ecosystems in relation to climate change]. In *KLIMAT. Lasy i Drewno a Zmiany Klimatyczne: Zagrożenia i Szanse. Materiały Pierwszego Panelu Ekspertów w Ramach Prac nad Narodowym Programem Leśnym [CLIMATE. Forests and Wood and Climate Change: Threats and Opportunities. Materials of the First Panel of Experts as Part of Work on the National Forest Program]*; Rykowski, K., Ed.; Instytut Badawczy Leśnictwa: Sękocin Stary, Poland, 2014; pp. 174–188.
89. Mokrzycki, T.; Plewa, R. *Kornik Ostrozębny—Ips acuminatus (Gyllenhal, 1827) (Coleoptera, Curculionidae, Scolytinae)—Występowanie, Biologia i Znaczenie Gospodarcze w Lasach Polski [Ips acuminatus (Gyllenhal, 1827) (Coleoptera, Curculionidae, Scolytinae)—Occurrence, Biology and Economic Importance in the Forests of Poland]*; Biblioteczka Leśniczego 382. Wyd. Świat: Warszawa, Poland, 2017.
90. Nowakowska, J.A. Tempo rozprzestrzeniania się nowych inwazyjnych szkodników i patogenów drzew w ekosystemach leśnych w Europie na podstawie modelowania komputerowego w latach 2014–2100 [The rate of spread of new invasive pests and tree pathogens in forest ecosystems in Europe based on computer modeling in 2014–2100]. *Leś. Pr. Bad./For. Res. Pap.* **2014**, *75*, 325–326.
91. Kosibowicz, M. Owady inwazyjne w lasach Polski [Invasive insects in the forests of Poland]. *Wszechświat* **2009**, *110*, 51–54.
92. Kowalski, T. *Zamieranie Pędów Sosny [Pine Shoots Dieback]*; Biblioteczka leśniczego 80; Wydawnictwo Świat: Warszawa, Poland, 1997.
93. Orlikowski, L.; Oszako, T. (Eds.) *Fytoftorozy w Szkółkach i Drzewostanach Leśnych [Phytophthorosis in Nurseries and Forest Stands]*; Centrum Informacyjne Lasów Państwowych: Warszawa, Poland, 2009.
94. Mułenko, W.; Piątek, M.; Wołczańska, A.; Kozłowska, M.; Ruszkiewicz-Michalska, M. Plant parasitic fungi introduced to Poland in modern times. Alien and invasive species. In *Biological Invasions in Poland*; Mirek, Z., Ed.; W. Szafer Institute of Botany, Polish Academy of Sciences: Kraków, Poland, 2010; Volume 1, pp. 49–71.
95. Szczepkowski, A. *Perenniporia fraxinea* (Fungi, Polyporales), a new species for Poland. *Pol. Bot. J.* **2004**, *49*, 73–77.
96. Szczepkowski, A. Macromycetes in the Dendrological Park of the Warsaw Agricultural University. *Acta Mycol.* **2007**, *42*, 179–186. [CrossRef]
97. Kreisel, H. Global warming and mycoflora in the Baltic Region. *Acta Mycol.* **2006**, *41*, 79–94. [CrossRef]
98. Wojewoda, W.; Karasiński, D. Invasive macrofungi (*Ascomycota* and *Basidiomycota*) in Poland. In *Biological Invasions in Poland*; Mirek, Z., Ed.; W. Szafer Institute of Botany, Polish Academy of Sciences: Kraków, Poland, 2010; Volume 1, pp. 7–21.
99. Szczepkowski, A.; Obidziński, A. Obce gatunki sromotnikowatych Phallaceae w lasach Polski [Alien species of stinkhorns Phallaceae in forests of Poland]. *Studia i Materiały CEPL w Rogowie* **2012**, *33*, 279–295.
100. Wrzosek, M.; Motiejūnaitė, J.; Kasparavičius, J.; Wilk, M.; Mukins, E.; Schreiner, J.; Vishnevskiy, M.; Gorczak, M.; Okrasińska, A.; Istel, Ł.; et al. The progressive spread of *Aureoboletus projectellus* (Fungi, Basidiomycota) in Europe. *Fungal Ecol.* **2017**, *27*, 134–136. [CrossRef]
101. Breshears, D.D.; Myers, O.B.; Meyer, C.W.; Barnes, F.J.; Zou, C.B.; Allen, C.D.; McDowell, N.G.; Pockman, W.T. Tree die-off in response to global-change type drought: Mortality insights from a decade of plant water potential measurements. *Front. Ecol. Environ.* **2009**, *7*, 185–189. [CrossRef]
102. Białooki, P. *Anoplophora glabripennis* [Motschulsky] [Coleoptera: Cerambycidae]—Pierwsze stwierdzenie w Polsce [*Anoplophora glabripennis* [Motschulsky] [Coleoptera: Cerambycidae]—First confirmation in Poland]. *Ochrona Roślin* **2003**, *47*, 34–35.
103. Mokrzycki, T.; Grodzki, W. Drzewotocz japoński *Xylosandrus germanus* (Bldf.) (Coleoptera: Curculionidae, Scolytinae) w Polsce [*Xylosandrus germanus* (Bldf.) (Coleoptera: Curculionidae, Scolytinae) in Poland]. *Sylwan* **2014**, *158*, 590–594.

104. Mazur, A.; Witkowski, R.; Góral, J.; Rogowski, G. Occurrence of *Gnathotrichus materiarius* (Fitch, 1858) (Coleoptera, Curculionidae, Scolytinae) in South-Western Poland. *Fol. For. Pol. Ser. A For.* **2018**, *60*, 154–160. [CrossRef]
105. Kubiak, K.; Żółciak, A.; Damszel, M.; Lech, P.; Sierota, Z. *Armillaria* Pathogenesis under Climate Changes. *Forests* **2017**, *8*, 100. [CrossRef]
106. Szczepkowski, A. Objawy zamierania buków oraz związek między stopniem uszkodzenia drzew a wybranymi cechami taksacyjnymi drzewostanów [Symptoms of beech decline and the relationship between tree damage level and selected inventory traits of tree stand]. *Sylwan* **2001**, *145*, 85–99.
107. Szczepkowski, A.; Szyndel, M.S. Attempts to detect and to determine the significance of plant viruses in beech tree decline in Poland. *Phytopathol. Pol.* **2001**, *21*, 45–53.
108. Szczepkowski, A.; Tarasiuk, S. Stan zdrowotny zagrożonych zamieraniem drzewostanów bukowych w Polsce [Health status of threatened European beech (*Fagus sylvatica* L.) stands in Poland]. *Acta Sci. Pol. Silv. Colendar. Rat. Ind. Lignar.* **2005**, *4*, 71–85.
109. Tarasiuk, S.; Szczepkowski, A. The health status of endangered oak stands in Poland. *Acta Sci. Pol. Silv. Colendar. Rat. Ind. Lignar.* **2006**, *5*, 91–106.
110. Jung, T. Beech decline in Central Europe driven by the interaction between Phytophthora infections and climatic extremes. *For. Path.* **2008**, *39*, 77–94.
111. Grodzki, W.; Oszako, T. (Eds.) *Current Problems of Forest Protection in Spruce Stands under Conversion*; Forest Research Institute: Warsaw, Poland, 2008.
112. Agerer, R. Exploration types of ectomycorrhizae. *Mycorrhiza* **2001**, *11*, 107–114. [CrossRef]
113. Agerer, R. Fungal relationships and structural identity of their ectomycorrhizae. *Mycol. Prog.* **2006**, *5*, 67–107. [CrossRef]
114. Yafetto, L.; Davis, D.J.; Money, N.P. Biomechanics of invasive growth by *Armillaria* rhizomorphs. *Fungal Genet. Biol.* **2009**, *46*, 688–694. [CrossRef] [PubMed]
115. Pratt, J.E.; Niemi, M.; Sierota, Z.H. Comparison of three products based on *Phlebiopsis gigantea* for the control of *Heterobasidion annosum* in Europe. *Biocontrol Sci. Technol.* **2000**, *10*, 467–477. [CrossRef]
116. Żółciak, A. *Opieńki [Honey Fungi]*; Centrum Informacyjne Lasów Państwowych: Warszawa, Poland, 2005.
117. Sierota, Z.; Nowakowska, J.; Sikora, K.; Wrzosek, M.; Żółciak, A.; Małecka, M. What is important in selecting *Phlebiopsis gigantea* strain for commercial use? *J. Agric. Sci. Technol. B* **2015**, *5*, 55–64. [CrossRef]
118. Tyburski, Ł.; Zaniewski, P.T.; Bolibok, L.; Piątkowski, M.; Szczepkowski, A. Scots pine *Pinus sylvestris* mortality after surface fire in oligotrophic pine forest *Peucedano-Pinetum* in Kampinos National Park. *Fol. For. Pol. Ser. A For.* **2019**, in press.
119. Gierczyk, B.; Szczepkowski, A.; Kujawa, A.; Ślusarczyk, T. Contribution to the knowledge of mycobiota of the Kampinos National Park (Poland). Part 2. *Acta Mycol.* **2019**, in press.

forests

MDPI

Article

Review of Japanese Pine Bast Scale, *Matsucoccus matsumurae* (Kuwana) (Coccomorpha: Matsucoccidae), Occurring on Japanese Black Pine (*Pinus thunbergii* Parl.) and Japanese Red Pine (*P. densiflora* Siebold & Zucc.) from Korea

Jinyeong Choi [1,2], Deokjea Cha [3], Dong-Soo Kim [3] and Seunghwan Lee [1,2,*]

1 Insect Biosystematics Laboratory, Department of Agricultural Biotechnology, Seoul National University, Seoul 151-921, Korea
2 Research Institute of Agriculture and Life Sciences, Seoul National University, Seoul 151-921, Korea
3 Forest Biomaterials Research Center, National Institute of Forest Science, Jinju 52817, Korea
* Correspondence: seung@snu.ac.kr; Tel.: +82-2880-4703; Fax: +82-2873-2319

Received: 11 June 2019; Accepted: 16 July 2019; Published: 29 July 2019

Abstract: *Matsucoccus matsumurae* (Kuwana, 1905), commonly known as Japanese pine bast scale, is a destructive pest on pine trees in North America, East Asia, and Northern Europe. The spread of damage to black pine trees, *Pinus thunbergii* Parl., due to *M. matsumurae* has been reported throughout Southern and some Eastern and Western coastal regions in Korea, under the name *M. thunbergianae*, which was described by Miller and Park (1987). Recently, *M. thunbergianae* was synonymized with *M. matsumurae* by Booth and Gullan (2006), based on molecular sequences and morphological data. However, *M. thunbergianae* is still considered a valid species in Korea. Since supporting data for the synonyms are unavailable in any DNA database (e.g., GenBank and BOLD), we performed morphological and molecular comparisons to review the results of Booth and Gullan (2006) using samples of *M. matsumurae* collected from Japan and topotype materials of *M. thunbergianae* from Korea. Our study supports the opinion of Booth and Gullan (2006), as the morphological features of the adult female and male of *M. thunbergianae* are identical to those of *M. matsumurae*, and DNA sequences (18S and 28S) of *M. thunbergianae* show identical or very low genetic distances with those of *M. matsumurae*. Additionally, regional sampling of Korea produced the first documented occurrence of *M. matsumurae* in Jeju.

Keywords: *Matsucoccus thunbergianae*; black pine bast scale; taxonomy; synonym

1. Introduction

The genus *Matsucoccus* Cockerell [1], belonging to the archaeococcoid family Matsucoccidae, comprises about 38 species worldwide [2]. Except for six fossil species described from Baltic amber, all 32 extant species exclusively occur on *Pinus* spp. in the Holarctic and Neotropical regions [2]. Among them, some species are the most destructive pests on pine trees; for example, *Matsucoccus acalyptus* Herbert on pinyon pine (*Pinus edulis* Engelm.) and single-leaf pinyon (*P. monophylla* Torr. and Frém.); *M. bisetosus* Morrison and *M. vexillorum* Morrison on ponderosa pine (*P. ponderosa* P. and C. Lawson); and *M. matsumurae* (Kuwana) on Chinese pines (*P. tabuliformis* Carrière and *Pinus massoniana* D. Don.), Japanese black pine (*P. thunbergii* Parl.), and red pine (*P. resinosa* Sol. ex Aiton) [3–7].

Females and males of *Matsucoccus* species have sexually dimorphic life cycles after the second-instar nymphs, which are known as “cysts” in the feeding and overwintering stages (Figure 1). The females are neotenic, including three stages (occasionally four in *M. vexillorum* Morrison), whereas the males have five stages, including a prepupa and pupa, and grow functioning wings as adults (for the life cycle details of *Matsucoccus*, see Foldi [5]).

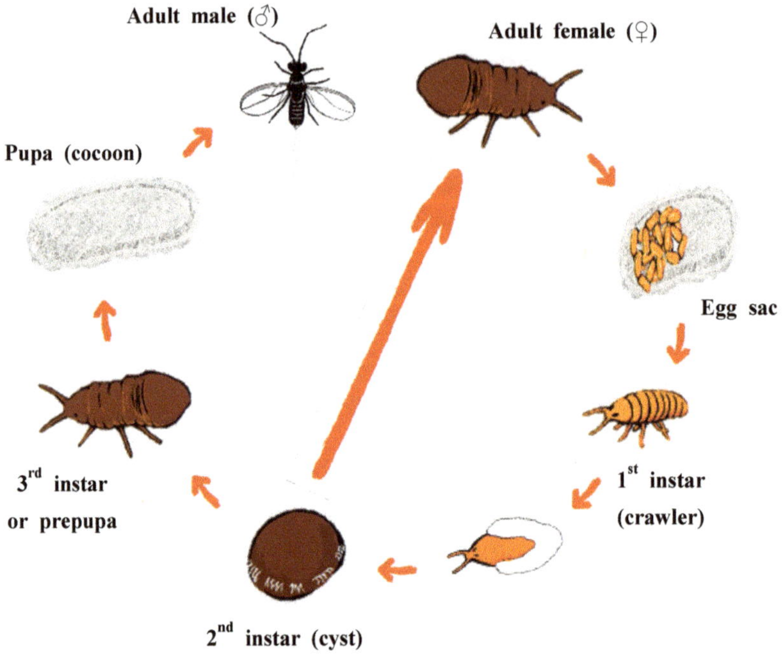

Figure 1. General life cycle of *Matsucoccus* species.

The Japanese pine bast scale, *Matsucoccus matsumurae*, was described by Kuwana [8] as a new species on Japanese black pine (*Pinus thunbergii* Parl.) in Tokyo, Japan, and was mainly found in China, Japan, and Korea. In China, more than 70,000 km^2 of pine forest damage by *M. matsumurae* had been reported each year between the 1970s–1980s despite attempts at chemical control [7,9].

In Korea, there had been no reports of significant damage by *Matsucoccus matsumurae* after the species was first found by Kanda [10], until a heavy infestation was detected on about 12,000 ha of Japanese black pine in the Southwestern coastal area of Korea [11]. This damage was found to be caused by a new species, *M. thunbergianae* Miller & Park [11], which was distinguished based on differences in morphological characteristics and in the number of generations from the congeners, *M. matsumurae* and *M. resinosae* Bean & Godwin. However, several studies implied that *M. matsumurae*, *M. resinosae*, and *M. thunbergianae* could be the same species. For example, the three species showed strong cross-attractiveness to sex pheromones [12,13], the main component of which was identical for the three species [14,15]. Finally, *M. thunbergianae* and *M. resinosae* were synonymized with *M. matsumurae* by Booth and Gullan [16], mainly based on their similarities in morphology and molecular sequences of 18S and 28S rDNA. However, *M. thunbergianae* is still considered a valid species in Korea [6,17–19] despite suggestions that the two species are synonymous, and an evaluation between *M. thunbergianae* collected from Korea and *M. matsumurae* was not possible due to the absence of DNA data from any available database.

In this study, we collected true *Matsucoccus matsumurae* from Fukuoka, Japan, and topotype materials of *M. thunbergianae* from Goheung, Korea, on the Japanese black pine to compare their morphologies and molecular sequences. To examine the spread of *M. matsumurae*, regional sampling was performed in Korea. Based on those data, we provide morphological and molecular characteristics of *M. matsumurae* and its current distribution in Korea.

2. Materials and Methods

2.1. Sample Collection

Eight populations of *Matsucoccus* sp. from Japan and Korea were sampled for morphological and molecular analyses (Figure 2; Table 1). To sample *M. matsumurae*, a population of *Matsucoccus* sp. (assumed to be *M. matsumurae*) on *Pinus thunbergii* was collected from Fukuoka in Japan. To sample *M. thunbergianae*, a population of *Matsucoccus* sp. (assumed to be topotype materials of *M. thunbergianae*) on *P. thunbergii* was collected from Goheung in Korea. For regional sampling in Korea, six populations of *Matsucoccus* sp. on *P. densiflora* or *P. thunbergii* were collected from Buan, Jeju, Pohang, Sacheon, Seoul, and Taean in Korea. Each population was preserved in 95% ethanol and stored at −20 °C for molecular analysis and morphological identification.

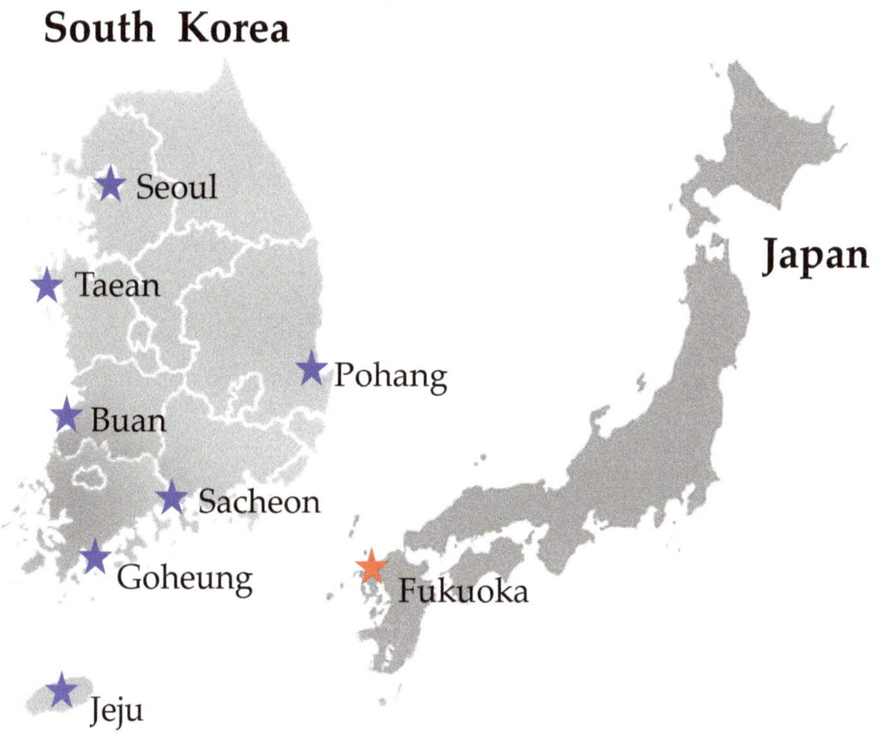

Figure 2. Sampling localities of *Matsucoccus* sp. in Korea and Japan.

Table 1. Collection data of samples used in this study (GPS coordinates; WGS84 coordinate system).

Species	Sex	Host	Locality	GPS		Date	Collector	GenBank Accession No.	
				North	East			18S	28S
Matsucoccus sp.	Female	*Pinus thunbergii*	Goheung, South Korea	34.631599	127.380838	Jan., 2017	D. Cha	-	-
Matsucoccus sp.	Male	*Pinus thunbergii*	Goheung, South Korea	34.631599	127.380838	Jan., 2017	D. Cha	MH574839	MH574783
Matsucoccus sp.	Male	*Pinus thunbergii*	Buan, South Korea	35.596690	126.486645	Jan., 2017	D. Cha	MH574841	MH574785
Matsucoccus sp.	Male	*Pinus thunbergii*	Jeju, South Korea	33.530302	126.718108	Jan., 2017	D. Cha	MH574845	MH574789
Matsucoccus sp.	Male	*Pinus thunbergii*	Pohang, South Korea	36.055884	129.576762	Jan., 2017	D. Cha	MH574843	MH574787
Matsucoccus sp.	Male	*Pinus thunbergii*	Sacheon, South Korea	34.948513	128.050597	Jan., 2017	D. Cha	MH574840	MH574784
Matsucoccus sp.	Male	*Pinus densiflora*	Seoul, South Korea	37.598240	127.024651	Jan., 2017	D. Cha	MH574844	MH574788
Matsucoccus sp.	Male	*Pinus thunbergii*	Taean, South Korea	36.781692	126.132603	Jan., 2017	D. Cha	MH574842	MH574786
Matsucoccus sp.	Female	*Pinus thunbergii*	Fukuoka, Japan	33.580591	130.278336	Jan., 2017	D. Cha	-	-
Matsucoccus sp.	Male	*Pinus thunbergii*	Fukuoka, Japan	33.580591	130.278336	Jan., 2017	D. Cha	MH574846	MH574790

2.2. Morphological Identification

For morphological comparison, the adult females and males of *Matsucoccus* sp. from Fukuoka and Goheung were mounted on glass microscope slides using the methods of Danzig and Gavrilov-Zimin [20]. Morphological descriptions of *Matsucoccus* spp. in Foldi [5] and Morrison [21] were used to identify the adult females and males of slide-mounted specimens. Photomicrographs of the specimens were produced with a digital camera (Infinity3, Lumenera Corporation, Ottawa, Canada) mounted on a compound light microscope (DM 4000B, Leica Microsystems, Wetzlar, Germany). The slide specimens were deposited in either (i) the Insect Biosystematics Laboratory, Seoul National University, Korea (SNU), or (ii) Southern Forest Resources Research Center, National Institute of Forest Science (NIFS).

2.3. Molecular Analyses

Genomic DNA isolation was performed with the DNeasy Blood and Tissue kit (Qiagen, Inc., Dusseldorf, Germany) following the manufacturer's protocols. For molecular comparison, we selected two nuclear ribosomal RNA genes (partial 18S and D2–D3 region of 28S). These two loci were amplified from the total DNA of the adult males of *Matsucoccus* sp. from Fukuoka and Goheung as well as other regions (Buan, Jeju, Pohang, Sacheon, Seoul and Taean). We designed primers based upon Margarodidae spp. 18S and 28S sequences from GenBank. Primer sequences used for the polymerase chain reaction (PCR) are given in Table 2. PCR was conducted with AccuPower PCR PreMix (Bioneer, Daejeon, Korea) in 20 ml including 0.4 μM of each primer, 20 μM dNTPs, 20 μM MgCl2, and 0.05 μg DNA template. The PCRs for 18S and 28S were performed under the following conditions: An initial denaturation step at 95 °C for 3 min, followed by 35 cycles at 95 °C for 30 s, 60 °C for 30 s, and 72 °C for 1 min and a final extraction step at 72 °C for 1 min. PCR products were visualized in 1.5% agarose gel electrophoresis. All amplified samples were purified using the QIAquick PCR purification kit (Qiagen, Inc., Dusseldorf, Germany) and sequenced using an automated sequencer (ABI Prism 3730 XL DNA Analyzer) at Macrogen Inc. (Seoul, Korea). Both strands of each sample were assembled and edited with SeqMan Pro ver. 7.1.0 (DNASTAR, Inc., Madison, WI, USA). The alignment was carried out using MEGAX [22], including sequences of *Matsucoccus macrocicatrices* Richards (used as reference, KF053072 for 18S; KF040573 for 28S) and *Icerya purchasi* Maskell (used as outgroup, AY426078 for 18S; KT199077 for 28S) downloaded from GenBank. The uncertain regions of sequences were removed, and 361 base pairs from *18S* and 462 base pairs from *28S* were used for the analyses. All sequences used in this study were deposited into GenBank under the accession number (from MH574783 to MH574790 for 28S; from MH574839 to MH574846 for 18S; Table 1). Genetic distances were measured in MEGAX using a neighbor-joining tree with the Kimura two-parameter model [23].

Table 2. Primers used in this study.

Gene Regions	Direction	Primer Name	Sequences (5′–3′)	Annealing Temperature
18S	Forward	Matsu_18S_F	CATGTCTAAGTGCAAGCCGG	60 °C
	Reverse	Matsu_18S_R	CCTCATAAGAGTCCCGTATCG	60 °C
28S	Forward	Matsu_28S_F	AAACCACAGCCAAGGGAACG	60 °C
	Reverse	Matsu_28S_R	TTTTCTGACACCTCTCGCTG	60 °C

3. Results

3.1. Morphological Comparison

The morphological characteristics of adult females and males of *Matsucoccus thunbergianae* from Korea are identical to those of *M. matsumurae* from Japan.

3.1.1. Adult Females

Based on the morphological information of *Matsucoccus* spp. in Foldi [5] (Figure 3), adult females of *Matsucoccus* sp. from Fukuoka, Japan, were identified as *M. matsumurae*. In addition, adult females of *Matsucoccus* sp. from Goheung, Korea, are morphologically similar to those of Fukuoka in eight morphological characteristics (Figure 4): (i) Nine-segmented antenna, with bases placed close together, and the scape and pedicel distinctly longer and wider than the associated flagellar segment (Figures 3A and 4B); (ii) two pairs of thoracic spiracles, each with numerous tracheae (Figures 3I and 4E,G); (iii) seven pairs of abdominal spiracles with a structure similar to that of thoracic spiracles (Figures 3G and 4E,F); (iv) femur, tibia, and tarsus, each with a reticulated surface (Figure 3H); (v) a pair of ventral setae present on each abdominal segment III–VII (Figures 3F2 and 4I); (vi) multilocular disc-pores each with 9–13 loculi and about 40–85 pores around the vulva (Figures 3E and 4D); (vii) cicatrices present in transverse rows across abdominal segments III–VII, numbering 180–280 (Figures 3D and 4H); (viii) bilocular tubular ducts in transverse rows on the head, thorax, and abdomen (Figures 3C and 4C).

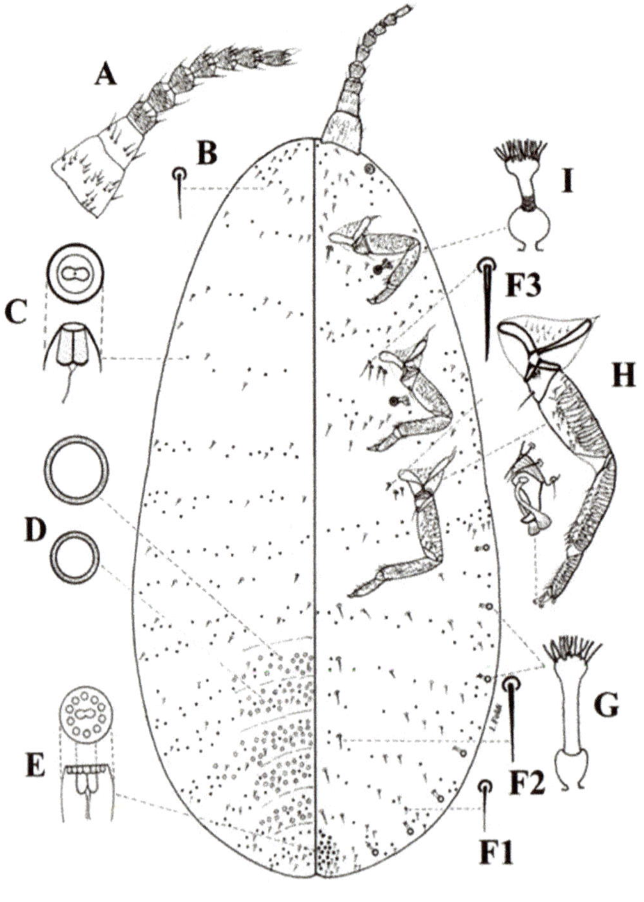

Figure 3. *Matsucoccus matsumurae* (Kuwana, 1905), adult female, from Foldi [5]. **A**, antenna; **B**, dorsal seta; **C**, bilocular tubular duct; **D**, cicatrices; **E**, multilocular disc-pore; **F1–3**, ventral setae; **G**, abdominal spiracle; **H**, leg; **I**, thoracic spiracle. This figure was reproduced with permission of I. Foldi.

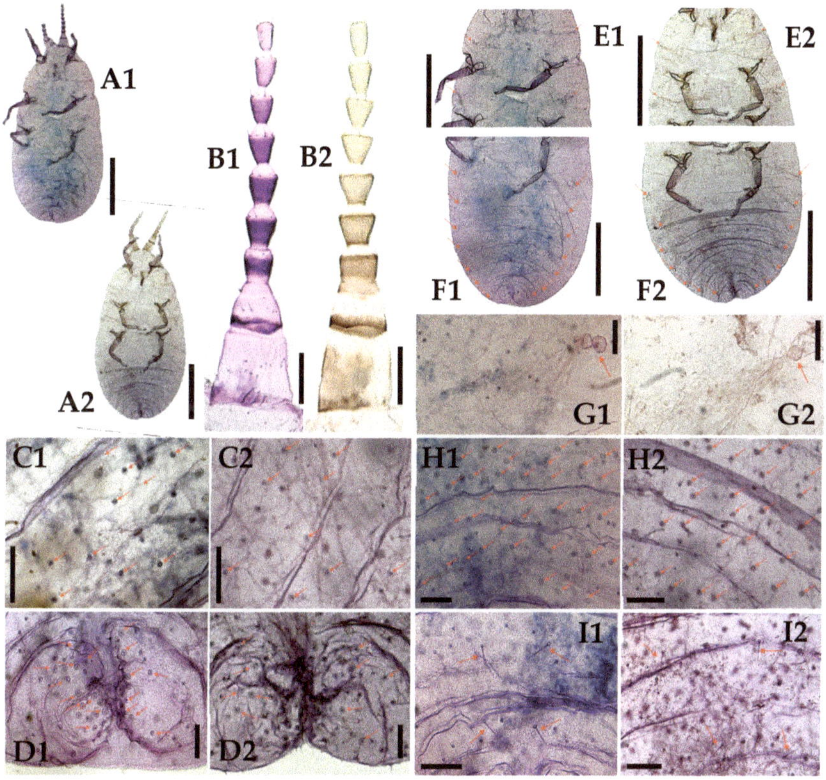

Figure 4. Morphological comparison of the adult females from Goheung (first) and Fukuoka (second). **A**, bodies; **B**, antennae; **C**, bilocular tubular ducts; **D**, multilocular disc-pores; **E**, thoracic spiracles; **F**, abdominal spiracles; **G**, spiracles with numerous tracheae; **H**, cicatrices; **I**, ventral setae on abdomen. Scale lines for A, E, F = 1 mm; B = 100 µm; C, D, G, H, I = 50 µm.

3.1.2. Adult Males

Based on the morphological information of *Matsucoccus* spp. in Morrison [21] (Figure 5), adult males of *Matsucoccus* sp. from Fukuoka were identified as *M. matsumurae.* In addition, adult males of *Matsucoccus* sp. from Goheung were morphologically similar to those of Fukuoka, based on seven morphological characteristics (Figure 6): (i) Ten-segmented antenna, each with short and stout scape and pedicel, but slender and cylindrical in each flagellar segment (Figures 5F and 6F); (ii) a head wider than it is long (transversely elongated), with antennal bases placed close together and large compound eyes (Figures 5C and 6B); (iii) slender legs with reticulated tarsus and stout claws without denticles (Figures 5D and 6G); (iv) wings with a reticulated costal complex that continues to the apex and two main veins extending from the wing base to its apex in the medial area with its under margin directed sharply downward on the basal area (Figure 5I–J and Figure 6C–D); (v) halteres are broadest at the apex, each with about six long and slender setae (Figures 5G and 6H); (vi) abdomen with elongated multilocular tubular pores, clustered in a transversely ovoid area near the apex of the abdomen (Figures 5A and 6I); (vii) the penis sheath is broad at the base and tapers to a rounded tip, while the aedeagus is slender, strongly curved, and protrudes beyond the apex of the penis sheath (Figures 5A and 6E).

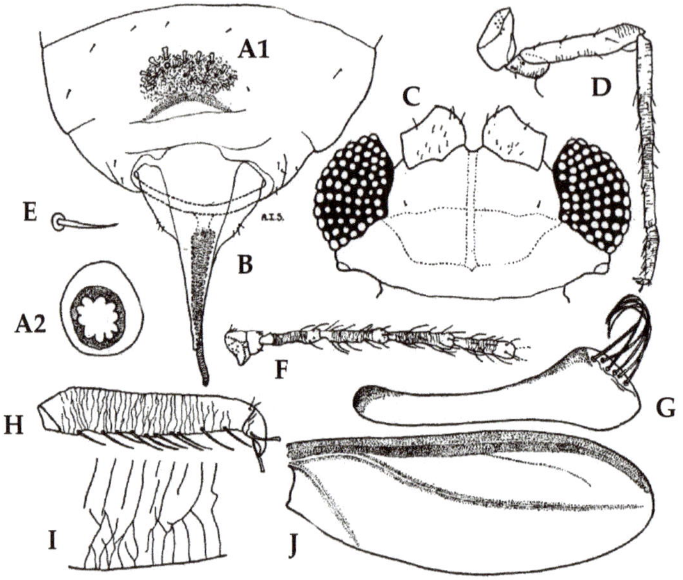

Figure 5. *Matsucoccus matsumurae* (Kuwana, 1905), adult male, from Morison [21]. **A1–2**, multilocular tubular pores near apex of abdomen; **B**, genitalia (aedeagus and penis sheath); **C**, head; **D**, leg; **E**, ventral abdominal seta; **F**, antenna; **G**, halter; **H**, tarsus and claw; **I**, wing venation; **J**, wing.

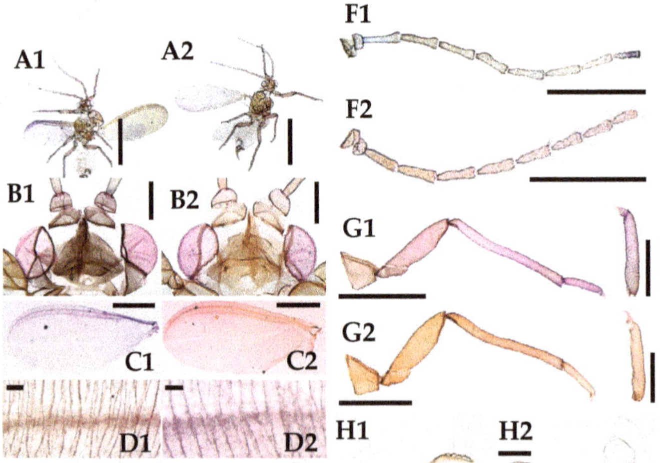

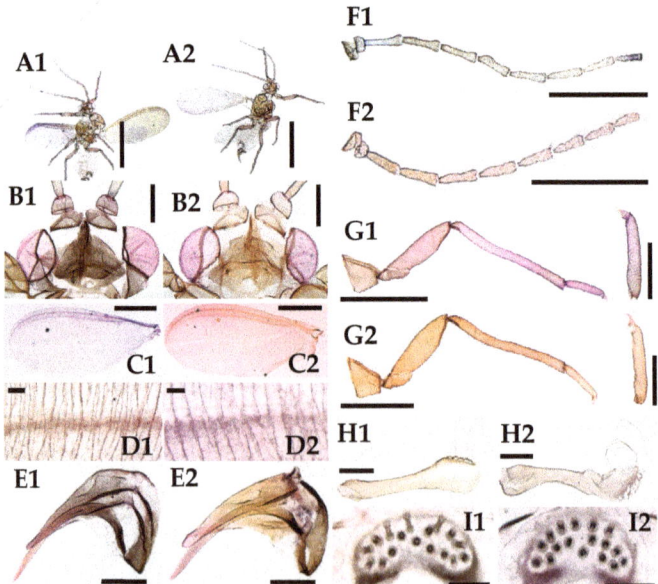

Figure 6. Morphological comparison of the adult males from Goheung (first) and Fukuoka (second). **A**, bodies; **B**, heads; **C**, wings; **D**, reticulated costal complexes; **E**, genitalia (aedeagus and penis sheath); **F**, antennae; **G**, legs, and tarsi and claws; **H**, halteres; **I**, multilocular tubular pores. Scale lines for A = 1 mm; B, E, G (bar beside tarsus and claw) = 100 μm; C = 0.5 mm; D = 10 μm; F = 500 μm; G (bar under entire leg) = 300 μm.

3.2. Molecular Comparison

The DNA sequences (18S rDNA and 28S rDNA) of *Matsucoccus* sp. from Korea showed identical or very small genetic distances compared to those of *M. matsumurae* from Japan. Each neighbor-joining tree and genetic distances of 18S and 28S sequences among the samples are presented in Figure 7 and Tables 3 and 4, respectively.

The 18S sequences of *Matsucoccus* sp. from seven populations in Korea (Buan, Goheung, Jeju, Pohang, Sacheon, Seoul, and Taean) were almost identical to those of *M. matsumurae* from a population in Japan (Fukuoka). The genetic distances ranged from 0% to 0.6% among these samples (Figure 7A; Table 3).

The 28S sequences of *Matsucoccus* sp. from seven populations in Korea (Buan, Goheung, Jeju, Pohang, Sacheon, Seoul, and Taean) were identical to those of *M. matsumurae* from a population in Japan (Fukuoka). No genetic distances were observed among these samples (Figure 7B; Table 4).

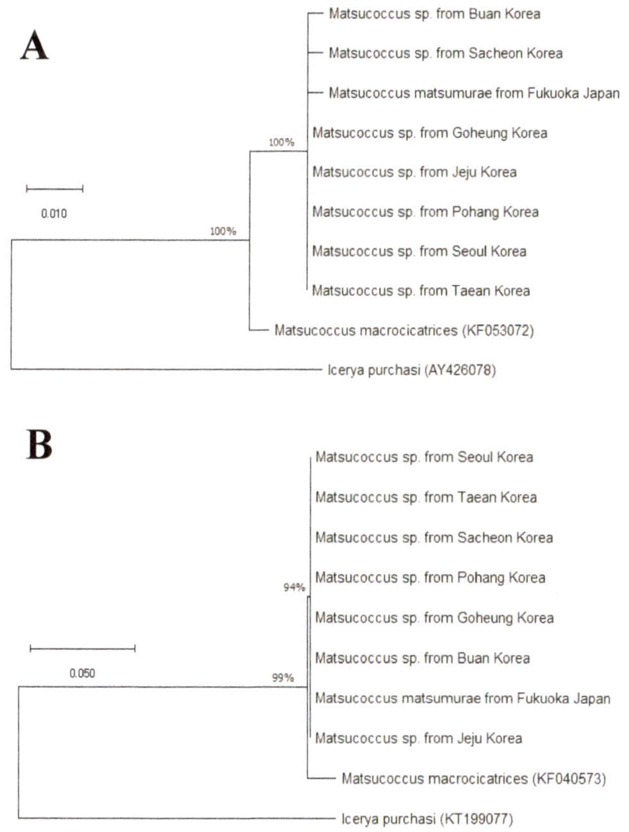

Figure 7. Neighbor-joining (NJ) tree analyses (using Kimura 2-parameter model) of seven regional populations of *Matsucoccus* sp. from Korea and *M. matsumurae* from Japan. **A**. 18S; **B**. 28S.

Table 3. Genetic distances (%) of 18S sequences estimated by Neighbor-joining with Kimura-2 parameter model.

No.		1	2	3	4	5	6	7	8	9	10
1	*Matsucoccus matsumurae* from Fukuoka, Japan										
2	*Matsucoccus* sp. from Buan, Korea	0.6									
3	*Matsucoccus* sp. from Goheung, Korea	0.3	0.3								
4	*Matsucoccus* sp. from Jeju, Korea	0.3	0.3	0.0							
5	*Matsucoccus* sp. from Pohang, Korea	0.3	0.3	0.0	0.0						
6	*Matsucoccus* sp. from Sacheon, Korea	0.6	0.6	0.3	0.3	0.3					
7	*Matsucoccus* sp. from Seoul, Korea	0.3	0.3	0.0	0.0	0.0	0.3				
8	*Matsucoccus* sp. from Taean, Korea	0.3	0.3	0.0	0.0	0.0	0.3	0.0			
9	*Matsucoccus macrocicatrices* (KF053072)	1.7	1.7	1.4	1.4	1.4	1.7	1.4	1.4		
10	*Icerya purchasi* (AY426078)	11.2	11.2	10.8	10.8	10.8	11.2	10.8	10.8	10.2	

Table 4. Genetic distances (%) of 28S sequences estimated by Neighbor-Joining with Kimura-2 parameter model.

No.		1	2	3	4	5	6	7	8	9	10
1	*Matsucoccus matsumurae* from Fukuoka, Japan										
2	*Matsucoccus* sp. from Buan, Korea	0.0									
3	*Matsucoccus* sp. from Goheung, Korea	0.0	0.0								
4	*Matsucoccus* sp. from Jeju, Korea	0.0	0.0	0.0							
5	*Matsucoccus* sp. from Pohang, Korea	0.0	0.0	0.0	0.0						
6	*Matsucoccus* sp. from Sacheon, Korea	0.0	0.0	0.0	0.0	0.0					
7	*Matsucoccus* sp. from Seoul, Korea	0.0	0.0	0.0	0.0	0.0	0.0				
8	*Matsucoccus* sp. from Taean, Korea	0.0	0.0	0.0	0.0	0.0	0.0	0.0			
9	*Matsucoccus macrocicatrices* (KF040573)	1.5	1.5	1.5	1.6	1.5	1.5	1.5	1.5		
10	*Icerya purchasi* (KT199077)	28.7	28.7	28.7	30.4	28.7	28.7	28.7	28.7	29.3	

4. Discussion

The original description of *Matsucoccus thunbergianae* has weak evidence to support it as a new species [11]. This new species was proposed based on the quantitative morphological features (especially in the adult males) and the number of generations per year, which might be considered as autapomorphic features that differentiated it from its congeners, such as *M. matsumurae* and *M. resinosae*. However, the characteristics of *M. thunbergianae* vary considerably according to environmental conditions (e.g., altitude, host plant, locality, and seasons), and all of the quantitative morphological traits have significant overlap among the three species. These points agree well with the opinions of Foldi [5] and Booth and Gullan [16], and are supported by the empirical tests of Boratynsky [24], Rieux [25], Ben-Dov [26], McClure et al. [27], and Miller and Park [11]. In this context, *M. thunbergianae* had been assigned an uncertain taxonomic status before it was synonymized with *M. matsumurae* [16]. As demonstrated by Booth & Gullan [16], the molecular and morphological features of the *Matsucoccus* species occurring on *Pinus thunbergii* and *P. densiflora* in Korea, which have been considered to be *M. thunbergianae*, are identical to those of *M. matsumurae* in this study.

The taxonomic validities of some species in the genus *Matsucoccus* have been controversial. Other congeners of *M. matsumurae*, such as *M. boratynskii* Bodenheimer & Neumark, *M. dahuriensis* Hu & Hu, *M. gallicolus* Morrison, *M. liaoningensis* Tang, *M. pini* (Green), and *M. yunnanensis* Ferris, have very similar morphology and differ in only a few characteristics [5,16]. Although we tentatively identified *Matsucoccus* species from Korea as *M. matsumurae*, further molecular and morphological studies with type specimens of those problematic species are needed, as the case research of Booth & Gullan [16] and this study both suggest that *M. matsumurae*, *M. thunbergianae*, and *M. resinosae* should be considered the same species.

The accurate identification of pest species is essential to establish an effective strategy of pest management. In this study supporting the results of Booth & Gullan [16], the *Matsucoccus* species occurring on *Pinus thunbergii* and *P. densiflora* in Korea should be recognized as *M. matsumurae* instead of *M. thunbergianae*. Until now, the control measures for species recognized as *M. thunbergianae* have been restrictively suggested in Korea; for example, chemical insecticides, such as fenitrothion 50EC and buprofezin 40SC [6], pheromone sticky traps [18], and yellow sticky traps [19]. On the other hand, biological control measures for *M. matsumurae*, such as entomopathogenic fungi, *Lecanicillium fungicola* strain HEB02, *L. lecanii* strain V3.4504 and V3.4505, *Fusarium incarnatum-equiseti* strain HEB01 [7,28], and natural enemies belonging to about 32 species [7], were proposed from China. In addition to this, studies were published on morphological changes in the antenna [29], and the wax glands and wax secretion [30] that might apply to the pest management of this species. Although Korean populations of *M. matsumurae* should be compared with the lineages from China using molecular tools, the results of this research will be useful to control the pests occurring on *P. thunbergii* in Korea.

In this study, *Matsucoccus matsumurae* occurring on *Pinus thunbergii* is reported for the first time from Jeju, Korea. We also observed the recent occurrence of *M. matsumurae on P. densiflora* from Seoul, although this area was recorded as one of the distributions of the species in Korea [31]. According to the results of Lim et al. [17], the occurrence of species under the name of *M. thunbergianae* was confirmed in all Southern coastal regions and some parts of Eastern and Western coastal regions, but it was not discovered in Chungcheongbuk-do, Daejeon, Jeju, or Seoul. In China, *M. matsumurae* also mainly damaged *P. densiflora*, which is widely distributed in Korea, along with *P. thunbergii*. Therefore, monitoring of this pest should be performed to investigate its exact occurrence and damage to *Pinus* species in the extensive regions that have previously been overlooked, especially in the Central and Northern inlands of Korea as well as Jeju.

The origin of Korean populations of *Matsucoccus matsumurae* is unclear, namely whether it is an indigenous species or an invasive species. Although there is no detailed information about the origin of host plants, *Pinus densiflora* and *P. thunbergii*, in the Korean Peninsula, both species are recognized as native *Pinus* species in Korea [32–34]. *P. thunbergii*, the main host plant of *M. matsumurae*, occurs mainly along Southern coastal areas of Korea [34]. According to Kim & Zsuffa [35], a number of

P. thunbergii (ca. 308,624 trees) had been planted for reforestation (ca. 105,863 ha) of South Korea in the period between 1953 and 1990. Based on this evidence, the current outbreak of *M. matsumurae* could not be explained only with "introduction and sudden spread of invasive populations." Moreover, the reproductive adult females of most scale insects, including *M. matsumurae*, have a sedentary lifestyle and very low dispersal ability (they are wingless) except for long-distance migration on the air currents as well as human-mediated transport [35,36]. To understand the possible circumstances concerning the outbreak of *M. matsumurae* in Korea, population genetic analyses using microsatellite markers or double digest restriction-site associated DNA sequencing (ddRAD-seq) are needed for various regional samples from China, Japan, and the USA, as well as Korea.

5. Conclusions

Based on the morphological and molecular evidence, our research corroborates the results of Booth & Gullan (2006) who synonymized *Matsucoccus thunbergianae* with *M. matsumurae*. These results imply the potential use of entomopathogenic fungi and natural enemies that were proposed from China to establish effective pest management in Korea. In addition, the occurrence of *M. matsumurae* is newly reported from Jeju, Korea in this study.

Author Contributions: J.C. and S.L. identified the pest based on morphological and molecular analyses and wrote the first draft of this paper; D.C. and D.-S.K. collected the materials used in this study and carried out the molecular experiments.

Funding: This research was supported by Korea Environment Industry & Technology Institute (KEITI) through Exotic Invasive Species Management Program, funded by Korea Ministry of Environment (MOE) (2018002270005).

Acknowledgments: We would like to acknowledge Imre Foldi (Département Systématique et Evolution, Muséum national d'Histoire naturelle, Paris, France) for giving us permission for reproducing his figure of *Matsucoccus matsumurae*. We also thank Penny J. Gullan (Research School of Biology, The Australian National University, Canberra, Australia) for asking Imre Foldi to get the permission.

Conflicts of Interest: The authors declare no conflict of interest.

References

1. Cockerell, T.D.A. The Japanese Coccidae. *Can. Entomol.* **1909**, *41*, 55–56. [CrossRef]
2. García Morales, M.; Denno, B.D.; Miller, D.R.; Miller, G.L.; Ben-Dov, Y.; Hardy, N.B. ScaleNet: A literature-Based Model of Scale Insect Biology and Systematics. Database 2016. Available online: http://scalenet.info (accessed on 1 October 2018).
3. McKenzie, H.L.; Gill, L.S.; Ellis, D.E. The prescott scale (*Matsucoccus vexillorum*) and associated organisms that cause flagging injury to ponderosa pine in the southwest. *J. Agric. Res.* **1948**, *76*, 33–51.
4. Furniss, R.L.; Carolin, V.M. *Western Forest Insects*; US Department of Agriculture, Forest Service: Washington, DC, USA, 1977; Volume 1339.
5. Foldi, I. The Matsucoccidae in the Mediterranean basin with a world list of species (Hemiptera: Sternorrhyncha: Coccoidea). *Annales de la Société Entomologique de France* **2004**, *40*, 145–168. [CrossRef]
6. Lim, E.; Kim, D.S.; Lee, S.M.; Choi, K.S.; Lee, D.W.; Chung, Y.J.; Park, C.G. Effect of fenitrothion on different life stages of black pine bast scale, *Matsucoccus thunbergianae*. *J. Asia Pac. Entomol.* **2013**, *16*, 55–59. [CrossRef]
7. Liu, W.; Xie, Y.; Yang, Q.; Xue, J.; Tian, F. New Research on *Matsucoccus matsumurae* (Kuwana) (Hemiptera: Matsucoccidae) in China. *Acta Zool. Bulg.* **2014**, *6*, 95–102.
8. Kuwana, S.I. A new *Xylococcus* in Japan. *Insect World* **1905**, *9*, 91–95.
9. Liu, W.; Xie, Y.; Xue, J.; Zhang, Y.; Tian, F.; Yang, Q.; Wu, J.; Tang, X.; Geng, Y.; Zhang, Y.; et al. Morphology, behavior and natural enemies of *Matsucoccus matsumurae* (Homoptera: Matsucoccidae) during development. *Sci. Silvae Sin.* **2015**, *51*, 69–83.
10. Kanda, S. Studies on Coccidae from Corea. *Insect World* **1941**, *45*, 296–303.
11. Miller, D.R.; Park, S.C. A new species of *Matsucoccus* (Homoptera: Coccoidea: Margarodidae) from Korea. *Korean J. Plant Prot.* **1987**, *26*, 49–62.

12. Young, B.L.; Miller, D.R.; McClure, M.S. Attractivity of the female sex pheromone of Chinese *Matsucoccus matsumurae* (Kuwana) to males of *M. matsumurae* in Japan and to males of *M. resinosae* Bean and Godwin in the United States (Margarodidae, Coccoidea, Homoptera). *Contrib. Shanghai Inst. Entomol.* **1984**, *1984*, 1–20.
13. Park, S.C.; West, J.R.; Abrahamson, L.P.; Lanier, G.N.; Silverstein, R.M. Cross-attraction between two species of *Matsucoccus*. *J. Chem. Ecol.* **1986**, *12*, 609–617. [CrossRef] [PubMed]
14. Lanier, G.N.; Qi, Y.T.; West, J.R.; Park, S.C.; Webster, F.X.; Silverstein, R.M. Identification of the sex pheromone of three *Matsucoccus* pine bast scales. *J. Chem. Ecol.* **1989**, *15*, 1645–1659. [CrossRef] [PubMed]
15. Hibbard, B.E.; Lanier, G.N.; Parks, S.C.; Qi, Y.T.; Webster, F.X.; Silverstein, R.M. Laboratory and field tests with the synthetic sex pheromone of three *Matsucoccus* pine bast scales. *J. Chem. Ecol.* **1991**, *17*, 89–102. [CrossRef] [PubMed]
16. Booth, J.M.; Gullan, P.J. Synonymy of three pestiferous *Matsucoccus* scale insects (Hemiptera: Coccoidea: Matsucoccidae) based on morphological and molecular evidence. *Proc. Entomol. Soc. Wash.* **2006**, *108*, 749–760.
17. Lim, E.G.; Lee, S.M.; Kim, D.S.; Kim, J.B.; Lee, S.H.; Choi, K.S.; Park, C.G.; Lee, D.W. The spread of black pine bast scale, *Matsucoccus thunbergianae* (Hemiptera: Margarodidae) in Korea. *Korean J. Appl. Entomol.* **2012**, *51*, 59–65. [CrossRef]
18. Kim, J.; Kim, D.S.; Matsuyama, S.; Lee, S.M.; Lee, S.C.; Park, I.K. Development of a pheromone trap for monitor black pine bast scale, *Matsucoccus thunbergianae* (Hemiptera: Margarodidae). *J. Asia Pac. Entomol.* **2016**, *19*, 899–902. [CrossRef]
19. Lee, C.J.; Kim, D.S.; Chung, Y.H.; Lee, S.M.; Lee, S.J.; Lee, D.W. Monitoring of black pine bast scale, *Matsucoccus thunbergianae* (Homoptera: Margarodidae) using yellow sticky trap. *Korean J. Appl. Entomol.* **2018**, *57*, 143–149.
20. Danzig, E.M.; Gavrilov-Zimin, I.A. *Palaearctic Mealybugs (Homoptera: Coccinea: Pseudococcidae), Part 1: Subfamily Phenacoccinae*; Russian Academy of Sciences, Zoological Institute: St. Petersburg, Russia, 2014.
21. Morrison, H. *A Classification of the Higher Groups and Genera of the Coccid Family Margarodidae (No. 52)*; US Department of Agriculture Technical Bulletin: Washington, DC, USA, 1928.
22. Kumar, S.; Stecher, G.; Li, M.; Knyaz, C.; Tamura, K. MEGA X: Molecular evolutionary genetics analysis across computing platforms. *Mol. Biol. Evol.* **2018**, *35*, 1547–1549. [CrossRef]
23. Kimura, M. A simple method for estimating evolutionary rates of base substitutions through comparative studies of nucleotide sequences. *J. Mol. Evol.* **1980**, *16*, 111–120. [CrossRef]
24. Boratynski, K.L. *Matsucoccus pini* (Green, 1925) (Homoptera, Coccoidea: Margarodidae): Bionomics and external anatomy with reference to the variability of some taxonomic characters. *Trans. R. Entomol. Soc. Lond.* **1952**, *103*, 285–326. [CrossRef]
25. Rieux, R. *Matsucoccus pini* Green (1925) (Homoptera, Margarodidae) dans le Sud-Est de la France. Variations intraspécifiques. Comparaison avec des espèces les plus proches. *Annales de Zoologie Ecologie Animale* **1976**, *8*, 231–263.
26. Ben-Dov, Y. Redescription of *Matsucoccus josephi* Bodenheimer and Harpaz (Homoptera: Coccoidea: Margarodidae). *Isr. J. Entomol.* **1981**, *15*, 35–51.
27. McClure, M.S. Temperature and host availability affect the distribution of *Matsucoccus matsumurae* (Kuwana) (Homoptera: Margarodidae) in Asia and North America. *Ann. Entomol. Soc. Am.* **1983**, *76*, 761–765. [CrossRef]
28. Wang, X.; Xie, Y.; Zhang, Y.; Liu, W.; Wu, J. The structure and morphogenic changes of antennae of *Matsucoccus matsumurae* (Hemiptera: Coccoidea: Matsucoccidae) in different instars. *Arthropod Struct. Dev.* **2016**, *45*, 281–293. [CrossRef] [PubMed]
29. Xie, Y.; Tian, F.; Liu, W.; Zhang, Y.; Xue, J.; Zhao, Y.; Wu, J. The wax glands and wax secretion of *Matsucoccus matsumurae* at different development stages. *Arthropod Struct. Dev.* **2014**, *43*, 193–204. [CrossRef] [PubMed]
30. Paik, W.H. *Illustrated Flora and Fauna of Korea. Insecta (VII)*; No. 22; Min. Education: Seoul, Korea, 1978.
31. Kong, W.S. Species composition and distribution of native Korean conifers. *J. Korean Geogr. Soc.* **2004**, *39*, 528–543.
32. Kong, W.S. Biogeography of native Korean Pinaceae. *J. Korean Geogr. Soc.* **2006**, *41*, 73–93.
33. Kong, W.S.; Lee, S.G.; Park, H.N.; Lee, Y.M.; Oh, S.H. Time-spatial distribution of *Pinus* in the Korean Peninsula. *Quat. Int.* **2014**, *344*, 43–52. [CrossRef]

34. Kim, K.H.; Zsuffa, L. Reforestation of South Korea: The history and analysis of a unique case in forest tree improvement and forestry. *For. Chron.* **1994**, *70*, 58–64. [CrossRef]
35. Gullan, P.J.; Kosztarab, M.P. Adaptations in scale insects. *Annu. Rev. Entomol.* **1997**, *42*, 23–50. [CrossRef]
36. Kondo, T.; Gullan, P.J.; Williams, D.J. Coccidology. The study of scale insects (Hemiptera: Sternorrhyncha: Coccoidea). *Ciencia y Tecnología Agropecuaria* **2008**, *9*, 55–61. [CrossRef]

forests

MDPI

Article

Comparing Methods for Monitoring Establishment of the Emerald Ash Borer (*Agrilus planipennis*, Coleoptera: Buprestidae) Egg Parasitoid *Oobius agrili* (Hymenoptera: Encyrtidae) in Maryland, USA

David E. Jennings [1,*,†], Jian J. Duan [2] and Paula M. Shrewsbury [1]

1 Department of Entomology, University of Maryland, 4112 Plant Sciences Building, College Park, MD 20742, USA; pshrewsb@umd.edu

2 Beneficial Insects Introduction Research Unit, United States Department of Agriculture—Agricultural Research Service, 501 South Chapel Street, Newark, DE 19711, USA; Jian.Duan@ars.usda.gov

* Correspondence: david.e.jennings@gmail.com

† Current address: Vermont Law School, 164 Chelsea Street, South Royalton, VT 05068, USA.

Received: 18 September 2018; Accepted: 19 October 2018; Published: 22 October 2018

Abstract: The emerald ash borer, *Agrilus planipennis* Fairmaire (EAB), is an invasive beetle that has caused widespread mortality of ash trees in North America. To date, four parasitoids have been introduced in North America for EAB biological control, including the egg parasitoid *Oobius agrili* Zhang & Huang (Hymenoptera: Encyrtidae). Monitoring EAB egg parasitism is challenging because female beetles oviposit in bark crevices and EAB eggs and *O. agrili* are small (<1 mm in diameter). Consequently, multiple methods have been developed to recover this parasitoid. Here we compared two methods, visual surveys and bark sifting, used to monitor establishment of *O. agrili* in Maryland, USA. From 2009 to 2015, a total of 56,176 *O. agrili* were released at 32 sites across the state. In 2016, we surveyed nine of the study sites for *O. agrili* establishment using both methods. We compared the amount of time spent searching for eggs separately in each method, and also analyzed the effects of years-post release, total number of parasitoids released, and median month of release, on percent parasitism of EAB eggs, and the percentage of trees per site with parasitized EAB eggs. We found that visually surveying ash trees for EAB eggs was more efficient than bark sifting; the percent parasitism observed using the two methods was similar, but visually surveying trees was more time-efficient. Both methods indicate that *O. agrili* can successfully establish populations in Maryland, and June may be the best month to release *O. agrili* in the state. Future research should investigate EAB phenology in the state to help optimize parasitoid release strategies.

Keywords: ash trees; biological control; Buprestidae; Encyrtidae; invasive species

1. Introduction

Invasive arthropods represent a serious threat to forest ecosystems worldwide [1–3]. In the USA, invasive woodboring insects in particular are increasing in frequency [4]. These insects can cause extensive economic and environmental damage [2,5], yet their management is challenging because they often have cryptic life stages which are difficult to observe and target.

Of the invasive woodboring insects in the USA, emerald ash borer (EAB), *Agrilus planipennis* Fairmaire (Coleoptera: Buprestidae) is especially damaging [6–10]. EAB is native to northeastern Asia and is thought to have been accidentally introduced to North America sometime in the 1990s [11,12]. Management strategies for EAB include trunk injections of pesticides, removal of infested trees, and biological control [11,13,14]. To date, biological control of EAB in North America has involved the release of three parasitoids of EAB larvae (*Tetrastichus planipennisi* Yang (Hymenoptera:

Eulophidae), *Spathius agrili* Yang (Hymenoptera: Braconidae), and *S. galinae* Belokobylskij & Strazanac (Hymenoptera: Braconidae)), and one parasitoid of EAB eggs (*Oobius agrili* Zhang & Huang (Hymenoptera: Encyrtidae)) [13,15].

The establishment and effectiveness of EAB larval parasitoids has been the subject of several studies [16–20], but comparatively little is known regarding the establishment of the egg parasitoid *O. agrili*. Previous research on *O. agrili* has taken place in Michigan [21,22], New York [23], and Kentucky [18], where generally it appears as though populations of this parasitoid are successfully establishing. Nonetheless, monitoring the recovery of *O. agrili* remains especially challenging because of the size of both *O. agrili* and EAB eggs (<1 mm in diameter), as well as the location of the eggs in bark crevices. Consequently, a range of methods has been developed and tested for assessing EAB egg mortality [21,23–28]. These methods include EAB egg sentinel logs [22] and envelopes [27], yellow pan traps [23], visual surveys, and bark sifting [21].

Improving the efficiency of *O. agrili* monitoring is paramount to the EAB biological control program. Although some studies have simultaneously compared multiple methods for monitoring *O. agrili* [21,23], the extent to which results depend on environmental factors such as climate, habitat, and *O. agrili* release protocols remains unclear. Consequently, more data are needed to further refine the *O. agrili* monitoring process. For instance, Abell et al. [21] compared visual surveys and bark sifting in Michigan, finding that the bark sifting method revealed considerably higher *O. agrili* parasitism. Additionally, Parisio et al. [23] compared various methods in New York, including egg sentinel logs and yellow pan traps, and found that yellow pan traps recovered more *O. agrili* than egg sentinel logs. These methods all have different benefits associated with them, including variation in financial costs and labor. Comparing methods within the same study should help to guide practitioners in the field, especially those operating with limited resources.

In the present study, our objective was to monitor *O. agrili* establishment and EAB egg parasitism rates across Maryland and to compare two different methods: visual surveys and bark sifting. These two methods were selected because they are among the most cost-effective to implement, and because of their previous use by Abell et al. [21]. Conducting the research in Maryland enabled us to compare our results to those of Abell et al. [21] in Michigan and determine how *O. agrili* populations respond to different climates and environments in the USA. The results from the present study should help to improve the efficiency of *O. agrili* monitoring.

2. Materials and Methods

2.1. Parasitoid Releases

Oobius agrili released in the present study were obtained from the population maintained at the USDA APHIS EAB Biocontrol Facility in Brighton, MI, USA [29]. Releases of *O. agrili* in Maryland began in the summer of 2009, and by 2015 this parasitoid had been released at 32 sites (Figure 1a). Release methods for *O. agrili* included use of logs and inverted cups attached to trees allowing parasitoids to emerge naturally from EAB eggs, as well as direct releases of adult parasitoids [30]. By 2015, 56,176 *O. agrili* had been released in Maryland overall (Table 1; Supplementary Table S1).

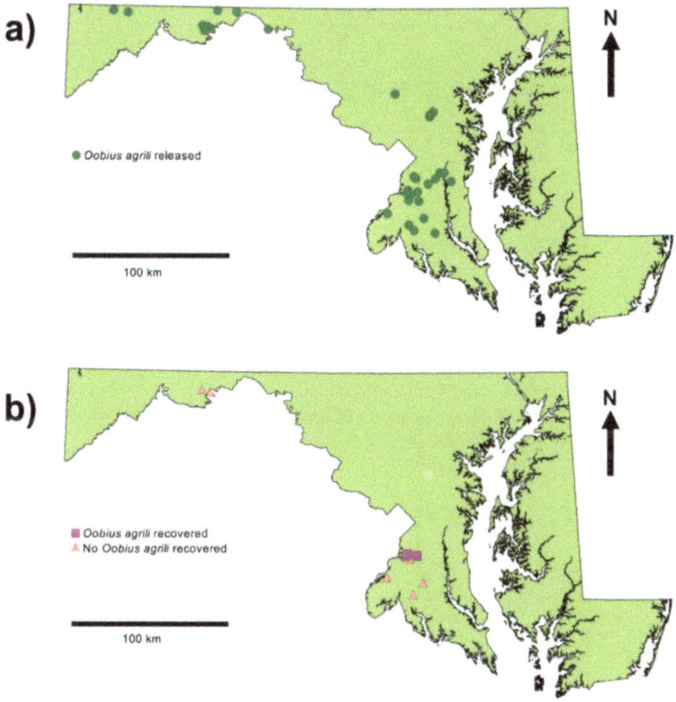

Figure 1. Map of *Oobius agrili* (**a**) release and (**b**) recovery sites in Maryland, USA. Nine of the 32 release sites were sampled for *O. agrili* recovery in the present study. Parasitized eggs were found at three of the sites sampled, and *O. agrili* parasitism was confirmed (through an emerging adult) from one site.

Table 1. Summary of *Oobius agrili* releases in Maryland by year from 2009 to 2015. Shown are number of releases per year, mean ± SE *O. agrili* per release, total number released, and the earliest and latest date of release.

Year	No. of Releases	Mean (±SE)	Total	Earliest	Latest
2009	2	182.0 ± 39.0	364	3 June	1 July
2010	5	145.2 ± 28.5	726	26 May	22 June
2011	16	167.0 ± 6.8	2672	25 May	30 June
2012	10	87.9 ± 36.9	2286	8 June	14 September
2013	10	181.1 ± 75.1	6518	30 May	3 July
2014	24	675.4 ± 96.2	16,210	20 June	11 September
2015	39	702.6 ± 62.0	27,400	18 June	10 September

2.2. Sampling Methods

We surveyed *O. agrili* recovery in late March and early April 2016 at nine of the 32 release sites throughout Maryland (Figure 1b). Recovery sites were selected based on logistics and geography, including urban as well as rural areas. Two different sampling methods were used to monitor *O. agrili* recovery in the present study: visual surveys for EAB eggs on ash trees in the field, and sifting ash bark in the laboratory. Both methods were utilized at all sites.

For the visual survey method, we selected green ash trees (*Fraxinus pennsylvanica* Marshall) that had apparent external signs of EAB infestation (i.e., EAB exit holes, woodpecker damage, epicormic growth, and reduced crown condition). Visual surveys were conducted for 30 min per tree, and on 10 trees per site, for a total of 90 green ash trees (mean diameter at breast height ± standard error

= 16.28 ± 1.05 cm). Using a utility knife, we picked away at the surface layer of bark between approximately 0.5 and 1.5 m high on the trees to expose cracks and crevices where EAB females typically oviposit. We recorded the number of EAB eggs observed and whether or not the eggs were parasitized (indicated by egg discoloration, with eggs turning a dark brown/black, as opposed to their normal light brown color). All EAB eggs were collected and taken to the laboratory where they were stored in environmentally controlled incubation chambers (25 °C, 65% relative humidity, 16:8 light:dark photoperiod). Any parasitoids that emerged from eggs were then identified.

For the bark sifting method, we scraped off a 10 × 20 cm section of bark on each green ash tree using a drawknife. Bark sifting was conducted immediately after visual surveys on a random subsample of five of the visually surveyed trees per site, for a total of 45 trees (mean diameter at breast height ± standard error = 17.88 ± 1.28 cm). We scraped off the same area of bark on each tree to attempt to standardize the amount collected. A small plastic sheet was placed on the ground beneath the tree to catch falling bark, which was then collected in plastic bags and returned to the laboratory for exhaustive inspection using a microscope. The time spent sifting through bark was recorded to enable us to compare the two sampling methods in terms of efficiency. Bark samples were also kept in environmentally controlled incubation chambers to collect and identify any parasitoids that emerged.

2.3. Data Analyses

First, we examined if the recovery sites were independent from each other by using Mantel tests. We then used generalized linear models with binomial error distributions to test the effects of years-post release, total number of parasitoids released, and median month of release, on percent parasitism of EAB eggs, and the percentage of trees per site with parasitized EAB eggs. For this analysis, parasitism from both sampling methods was pooled together. These tests were followed by Tukey HSD tests when there were significant main effects. Lastly, we used a generalized linear model with Gaussian error distribution to compare the amount of time spent searching for eggs separately in each method. For this analysis, we only included trees where both sampling methods had been used. To ensure that all models fit the data ($p > 0.05$), model fits were assessed using Pearson tests. All analyses were conducted using R3.3.2 [31].

3. Results

3.1. Summary

All of the adult parasitoids that emerged in the present study were identified as *O. agrili*. Mantel tests indicated that there were no significant relationships between the distance between sites and egg parasitism (Mantel $r = -0.118$, $p = 0.867$) or trees with egg parasitism (Mantel $r = -0.197$, $p = 0.848$). Therefore, we considered the recovery sites to be independent for subsequent analyses.

We found parasitized EAB eggs at three of the nine sites. At those three sites, mean percent parasitism per tree was 13.03%; mean percent parasitism per tree across all sites was 5.16%. There was a significant effect of the number of years post-release on percent parasitism (LR = 28.48, df = 1, $p < 0.001$; Figure 2a) and the percentage of trees with parasitized eggs (LR = 4.81, df = 1, $p = 0.028$; Figure 2b), with both increasing over time. Mean percent parasitism per tree reached 29.11%, and the percentage of trees with parasitized eggs reached 40%, at the site where *O. agrili* releases had been conducted seven years prior. However, neither the total number of parasitoids released (LR = 0.01, df = 1, $p = 0.916$; Figure 2c), nor median month of release (LR = 0.65, df = 1, $p = 0.421$; Figure 2e), significantly affected percent parasitism. Similarly, there was no significant effect of the total number of parasitoids released (LR = 0.03, df = 1, $p = 0.857$; Figure 2d) or median month of release (LR = 0.89, df = 1, $p = 0.345$; Figure 2f) on the percentage of trees with parasitized eggs.

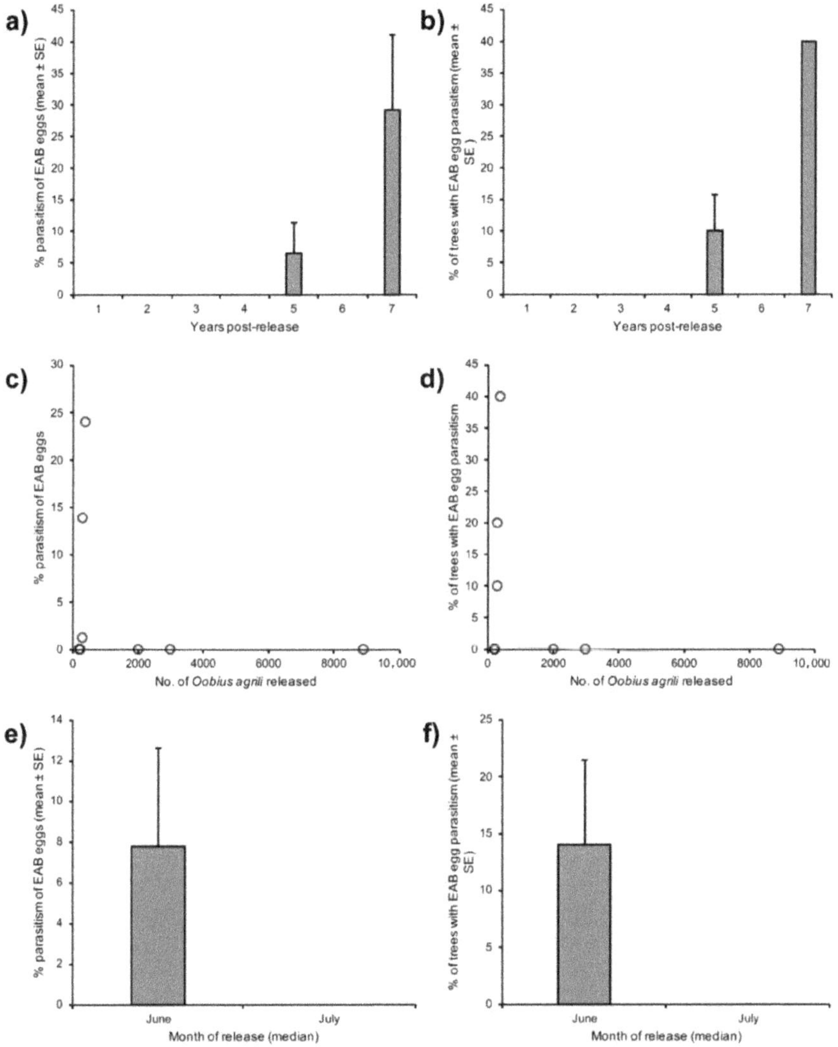

Figure 2. Mean percent parasitism per tree, and mean percentage of trees with parasitized emerald ash borer eggs, by years-post release (**a**,**b**), total number of parasitoids released (**c**,**d**), and median month of release (**e**,**f**), across the nine recovery sites. Data pooled from both survey methods. Black lines represent standard error.

3.2. Comparison of Sampling Method Efficiency

The visual survey method detected 235 EAB eggs, of which 13 were parasitized (5.53%). EAB eggs were found on 62 of the 90 trees (68.89%), and five of those trees had parasitized EAB eggs. The bark sifting method detected 125 EAB eggs, of which five were parasitized (4%). EAB eggs were found on 33 of the 45 trees (73.33%), and three of those trees had parasitized EAB eggs. The mean weight of bark sampled was 4.12 ± 0.34 g.

There were no significant differences in percent parasitism (LR = 0.42, df = 1, $p = 0.518$; Figure 3a) or the percentage of trees with egg parasitism (LR = 0.07, df = 1, $p = 0.798$; Figure 3b) between the two methods. However, there was a difference in time per sample (LR = 19.65, df = 1, $p < 0.001$) when comparing the methods, with almost double the amount of time spent processing (in the laboratory) each bark sample (59.64 ± 6.69 min) compared with the visual survey in the field (30 ± 0 min).

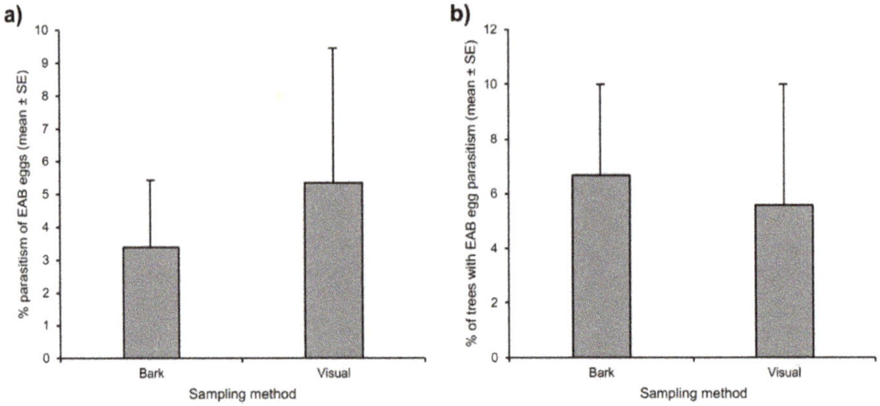

Figure 3. Comparison of results using visual survey and bark sifting methods. Shown are percent parasitism (**a**), and percentage of trees with parasitized emerald ash borer eggs (**b**), across the nine recovery sites. Black lines represent standard error.

4. Discussion

Oobius agrili adults were recovered with both visual survey and bark sifting methods from one site, and parasitized EAB eggs were found at a further two sites. The highest EAB egg parasitism was found at the site where the longest time had passed since the initial release. Generally, rates of *O. agrili* recovery, and parasitism of EAB eggs, were slightly low in comparison with other studies [21,22]. However, percent parasitism at the site sampled seven years post-release was comparable to parasitism found in previous studies elsewhere [21,22]. Thus, these results indicate that it is at least possible for *O. agrili* to successfully establish populations in Maryland.

Parasitism was only detected at sites within Prince George's County, MD. Interestingly, the only other *O. agrili* recovered in Maryland in a separate study was also in Prince George's County [9]. All of the adult parasitoids that emerged in the present study were identified as *O. agrili*. Although adult parasitoids were not collected from all of the parasitized eggs, given the lack of native parasitoids thus far observed attacking EAB eggs [9,21] it is likely that this parasitism can still be attributed to *O. agrili*.

Many factors need to be considered when designing a sampling protocol, such as personnel availability, field conditions, and financial resources. However, if choosing between the two methods used in the present study, visually surveying trees for EAB eggs appears to be a more efficient method than collecting bark samples from the field and processing them in the laboratory. This finding contrasts with the results of Abell et al. [21], who found that the bark sifting method was more effective at detecting EAB egg parasitism. The findings of Abell et al. [21] could be explained by their bark sifting method collecting a larger area of bark than in the present study (10 × 100 cm and 10 × 50 cm compared with 10 × 20 cm used here).

Somewhat surprisingly, the number of parasitoids released, and the median month of release, did not significantly affect *O. agrili* recovery. Indeed, the lack of effect from median month of release was surprising because *O. agrili* was only recovered from sites where June was the median month of release. Erring on the cautious side, we would still suggest that *O. agrili* releases in Maryland and nearby areas take place in June to increase the likelihood of establishment. The total number of parasitoids released may be less important than timing, because *O. agrili* need to be released when EAB oviposition is at its peak, which is June–July in the Midwest [11,32] and likely similar in Maryland.

Another possible explanation for the trends observed in parasitoid recovery could relate to *O. agrili* diapause patterns. If *O. agrili* are released later in the summer there is less time for them to produce multiple generations before entering diapause to overwinter [33,34]. Additionally, *O. agrili* are also sensitive to variation in climate [35], which means that seasonal changes in weather patterns could strongly affect the establishment of this parasitoid.

We appreciate that the present study was relatively limited in the number of sites sampled and methods compared. However, our results demonstrate that *O. agrili* can establish populations in the Mid-Atlantic region of the USA, and highlight the importance of timing for parasitoid releases. Further, the present study shows that although different *O. agrili* monitoring methods may yield similar results, there are clear contrasts in the efficiencies of each method.

5. Conclusions

Egg parasitoids have been implicated in, or deployed for, biological control of other agricultural and forest pests such as brown marmorated stink bug, *Halyomorpha halys* Stål (Hemiptera: Pentatomidae) [36], and eucalyptus longhorned borer, *Phoracantha semipunctata* Fab. (Coleoptera: Cerambycidae) [37]. Biological control of EAB appears to be able to slow ash mortality [38,39]. Even with the comparative lack of research on *O. agrili*, targeting EAB eggs for biological control could provide an additive effect to parasitism by larval parasitoids, and lower the number of EAB larvae boring into trees. Future research should further investigate the phenology of EAB in Maryland [32], with a view to further optimizing biological control release strategies.

Supplementary Materials: The following are available online at http://www.mdpi.com/1999-4907/9/10/659/s1, Table S1: Summary of *Oobius agrili* release data in Maryland by site from 2009–2015. Shown are year of release, number of releases, mean ± SE *O. agrili* per release, total number released, and the earliest and latest date of release. Bold type indicates sites surveyed for recovery, and asterisks indicate sites where *O. agrili* were recovered.

Author Contributions: Conceptualization, D.E.J., J.J.D., and P.M.S.; Data curation, D.E.J.; Formal analysis, D.E.J.; Funding acquisition, J.J.D. and P.M.S.; Investigation, D.E.J., J.J.D., and P.M.S.; Methodology, D.E.J., J.J.D., and P.M.S.; Project administration, J.J.D. and P.M.S.; Supervision, J.J.D. and P.M.S.; Writing—original draft, D.E.J.; Writing—review & editing, D.E.J., J.J.D., and P.M.S.

Funding: This study was supported by the USDA National Institute of Food and Agriculture, McIntire-Stennis Project 1003486, and USDA-ARS Specific Cooperative Agreement (58-1926-167).

Acknowledgments: We thank Dick Bean, Kim Rice, Steve Bell, Aaron Shurtleff, Charles Pickett, Rose Buckner, Sam Stokes, and the late Martin Proctor (all Maryland Department of Agriculture), Mark Beals and Jesse Morgan (both Maryland Department of Natural Resources), and Kristi Larson, Jonathan Schmude, and Phil Taylor (all USDA-ARS), for logistical assistance with this research. We are also grateful to four anonymous reviewers whose comments greatly improved the manuscript.

Conflicts of Interest: The authors declare no conflict of interest.

References

1. Langor, D.W.; DeHass, L.J.; Foottit, R.G. Diversity of non-native terrestrial arthropods on woody plants in Canada. *Biol. Invasions* **2009**, *11*, 5–19. [CrossRef]
2. Lovett, G.M.; Weiss, M.; Liebhold, A.M.; Holmes, T.P.; Leung, B.; Lambert, K.F.; Orwig, D.A.; Campbell, F.T.; Rosenthal, J.; McCullough, D.G.; et al. Nonnative forest insects and pathogens in the United States: Impacts and policy options. *Ecol. Appl.* **2016**, *26*, 1437–1455. [CrossRef] [PubMed]
3. Wan, F.H.; Yang, N.W. Invasion and management of agricultural alien insects in China. *Annu. Rev. Entomol.* **2016**, *61*, 77–98. [CrossRef] [PubMed]
4. Aukema, J.E.; McCullough, D.G.; Von Holle, B.; Liebhold, A.M.; Britton, K.; Frankel, S.J. Historical accumulation of nonindigenous forest pests in the continental United States. *BioScience* **2010**, *60*, 886–897. [CrossRef]
5. Aukema, J.E.; Leung, B.; Kovacs, K.; Chivers, C.; Britton, K.O.; Englin, J.; Frankel, S.J.; Haight, R.G.; Holmes, T.P.; Liebhold, A.M.; et al. Economic impacts of non-native forest insects in the continental United States. *PLoS ONE* **2011**, *6*, e24587. [CrossRef] [PubMed]

6. Gandhi, K.J.K.; Herms, D.A. North American arthropods at risk due to widespread Fraxinus mortality caused by the alien emerald ash borer. *Biol. Invasions* **2010**, *12*, 1839–1846. [CrossRef]
7. Kovacs, K.F.; Haight, R.G.; McCullough, D.G.; Mercader, R.J.; Siegert, N.W.; Liebhold, A.M. Cost of potential emerald ash borer damage in U.S. communities, 2009–2019. *Ecol. Econ.* **2010**, *69*, 569–578. [CrossRef]
8. Kovacs, K.F.; Mercader, R.J.; Haight, R.G.; Siegert, N.W.; McCullough, D.G.; Liebhold, A.M. The influence of satellite populations of emerald ash borer on projected economic costs in U.S. communities, 2010–2020. *J. Environ. Manag.* **2011**, *92*, 2170–2181. [CrossRef] [PubMed]
9. Jennings, D.E.; Duan, J.J.; Bean, D.; Rice, K.A.; Williams, G.L.; Bell, S.K.; Shurtleff, A.S.; Shrewsbury, P.M. Effects of the emerald ash borer invasion on the community composition of arthropods associated with ash tree boles in Maryland, USA. *Agric. For. Entomol.* **2017**, *19*, 122–129. [CrossRef]
10. Klooster, W.S.; Gandhi, K.J.K.; Long, L.C.; Perry, K.I.; Rice, K.B.; Herms, D.A. Ecological impacts of emerald ash borer in forests at the epicenter of the invasion in North America. *Forests* **2018**, *9*, 250. [CrossRef]
11. Herms, D.A.; McCullough, D.G. Emerald ash borer invasion of North America: History, biology, ecology, impacts, and management. *Annu. Rev. Entomol.* **2014**, *59*, 13–30. [CrossRef] [PubMed]
12. Siegert, N.W.; McCullough, D.G.; Liebhold, A.M.; Telewski, F.W. Dendrochronological reconstruction of the epicentre and early spread of emerald ash borer in North America. *Divers. Distrib.* **2014**, *20*, 847–858. [CrossRef]
13. Bauer, L.S.; Duan, J.J.; Gould, J.R.; Van Driesche, R. Progress in the classical biocontrol of *Agrilus planipennis* Fairmaire (Coleoptera: Buprestidae) in North America. *Can. Entomol.* **2015**, *147*, 300–317. [CrossRef]
14. McCullough, D.G.; Mercader, R.J.; Siegert, N.W. Developing and integrating tactics to slow ash (Oleaceae) mortality caused by emerald ash borer (Coleoptera: Buprestidae). *Can. Entomol.* **2015**, *147*, 349–358. [CrossRef]
15. Duan, J.J.; Bauer, L.S.; Van Driesche, R.G.; Gould, J.R. Progress and challenges of protecting North American ash trees from the emerald ash borer using biological control. *Forests* **2018**, *9*, 142. [CrossRef]
16. Duan, J.J.; Bauer, L.S.; Abell, K.J.; Lelito, J.P.; Van Driesche, R. Establishment and abundance of *Tetrastichus planipennisi* (Hymenoptera: Eulophidae) in Michigan: Potential for success in classical biocontrol of the invasive emerald ash borer (Coleoptera: Buprestidae). *J. Econ. Entomol.* **2013**, *106*, 1145–1154. [CrossRef] [PubMed]
17. Hooie, N.A.; Wiggins, G.J.; Lambdin, P.L.; Grant, J.F.; Powell, S.D.; Lelito, J.P. Native parasitoids and recovery of Spathius agrili from areas of release against emerald ash borer in eastern Tennessee, USA. *Biocontrol Sci. Technol.* **2015**, *25*, 345–351. [CrossRef]
18. Davidson, W.; Rieske, L.K. Establishment of classical biological control targeting emerald ash borer is facilitated by use of insecticides, with little effect on native arthropod communities. *Biol. Control* **2016**, *101*, 78–86. [CrossRef]
19. Jennings, D.E.; Duan, J.J.; Bean, D.; Gould, J.R.; Rice, K.A.; Shrewsbury, P.M. Monitoring the establishment and abundance of introduced parasitoids of emerald ash borer larvae in Maryland, USA. *Biol. Control* **2016**, *101*, 138–144. [CrossRef]
20. Johnson, T.D.; Lelito, J.P.; Pfammatter, J.A.; Raffa, K.F. Evaluation of tree mortality and parasitoid recoveries on the contiguous western invasion front of emerald ash borer. *Agric. For. Entomol.* **2016**, *18*, 327–339. [CrossRef]
21. Abell, K.J.; Bauer, L.S.; Duan, J.J.; Van Driesche, R. Long-term monitoring of the introduced emerald ash borer (Coleoptera: Buprestidae) egg parasitoid, *Oobius agrili* (Hymenoptera: Encyrtidae), in Michigan, USA and evaluation of a newly developed monitoring technique. *Biol. Control* **2014**, *79*, 36–42. [CrossRef]
22. Abell, K.J.; Bauer, L.S.; Miller, D.L.; Duan, J.J.; Van Driesche, R.G. Monitoring the establishment and flight phenology of parasitoids of emerald ash borer (Coleoptera: Buprestidae) in Michigan by using sentinel eggs and larvae. *Fla. Entomol.* **2016**, *99*, 667–672. [CrossRef]
23. Parisio, M.S.; Gould, J.R.; Vandenberg, J.D.; Bauer, L.S.; Fierke, M.K. Evaluation of recovery and monitoring methods for parasitoids released against emerald ash borer. *Biol. Control* **2017**, *106*, 45–53. [CrossRef]
24. Bauer, L.S.; Gould, J.R.; Duan, J.J.; Hansen, J.A.; Cossé, A.; Miller, D.; Abell, K.J.; Van Driesche, R.; Lelito, J.P.; Poland, T. Sampling methods for recovery of exotic emerald ash borer parasitoids after environmental release. In Proceedings of the 22nd USDA Interagency Research Forum on Invasive Species, Annapolis, MD, USA, 11–14 January 2011; McManus, K., Gottschalk, K.W., Eds.; United States Department of Agriculture, Forest Service: Morgantown, WV, USA, 2012; pp. 2–4.

25. Duan, J.J.; Bauer, L.S.; Ulyshen, M.D.; Gould, J.R.; Van Driesche, R. Development of methods for the field evaluation of *Oobius agrili* (Hymenoptera: Encyrtidae) in North America, a newly introduced egg parasitoid of the emerald ash borer (Coleoptera: Buprestidae). *Biol. Control* **2011**, *56*, 170–174. [CrossRef]
26. Duan, J.J.; Bauer, L.S.; Hansen, J.A.; Abell, K.J.; Van Driesche, R. An improved method for monitoring parasitism and establishment of *Oobius agrili* (Hymenoptera: Encyrtidae), an egg parasitoid introduced for biological control of the emerald ash borer (Coleoptera: Buprestidae) in North America. *Biol. Control* **2012**, *60*, 255–261. [CrossRef]
27. Jennings, D.E.; Duan, J.J.; Larson, K.M.; Lelito, J.P.; Shrewsbury, P.M. Evaluating a new method for monitoring the field establishment and parasitism of *Oobius agrili* (Hymenoptera: Encyrtidae), an egg parasitoid of emerald ash borer (Coleoptera: Buprestidae). *Fla. Entomol.* **2014**, *97*, 1263–1265. [CrossRef]
28. Jennings, D.E.; Duan, J.J.; Abell, K.J.; Bauer, L.S.; Gould, J.R.; Shrewsbury, P.M.; Van Driesche, R.G. Life table evaluation of change in emerald ash borer populations due to biological control. In *Biology and Control of Emerald Ash Borer*; Van Driesche, R.G., Reardon, R.C., Eds.; United States Department of Agriculture, Forest Service: Morgantown, WV, USA, 2015; pp. 139–151.
29. Bauer, L.S.; Liu, H. *Oobius agrili* (Hymenoptera: Encyrtidae), a solitary egg parasitoid of emerald ash borer from China. In Proceedings of the Emerald Ash Borer and Asian Longhorned Beetle Research and Technology Development Meeting, Cincinnati, OH, USA, 29 October–2 November 2006; Mastro, V., Lance, D., Reardon, R., Parra, G., Eds.; United States Department of Agriculture, Forest Service: Morgantown, WV, USA, 2007; pp. 63–64.
30. Abell, K.J.; Poland, T.; Cossé, A.; Bauer, L. Trapping techniques for emerald ash borer and its introduced parasitoids. In *Biology and Control of Emerald Ash Borer*; Van Driesche, R.G., Reardon, R.C., Eds.; United States Department of Agriculture, Forest Service: Morgantown, WV, USA, 2015; pp. 113–127.
31. R Core Team. *R: A Language and Environment for Statistical Computing*; R Foundation for Statistical Computing: Vienna, Austria, 2016. Available online: http://www.r-project.org (accessed on 15 November 2016).
32. Abell, K.J.; Duan, J.J.; Shrewsbury, P.M. Determining optimal parasitoid release timing for the biological control of emerald ash borer (Coleoptera: Buprestidae). *Fla. Entomol.* **2018**, under review.
33. Hoban, J.; Duan, J.J.; Hough-Goldstein, J. Effects of temperature and photoperiod on the reproductive biology and diapause of *Oobius agrili* (Hymenoptera: Encyrtidae), an egg parasitoid of emerald ash borer (Coleoptera: Buprestidae). *Environ. Entomol.* **2016**, *45*, 726–731. [CrossRef] [PubMed]
34. Larson, K.M.; Duan, J.J. Differences in the reproductive biology and diapause of two congeneric species of egg parasitoids (Hymenoptera: Encyrtidae) from northeast Asia: Implications for biological control of the invasive emerald ash borer (Coleoptera: Buprestidae). *Biol. Control* **2016**, *103*, 39–45. [CrossRef]
35. Wetherington, M.T.; Jennings, D.E.; Shrewsbury, P.M.; Duan, J.J. Climate variation alters the synchrony of host-parasitoid interactions. *Ecol. Evol.* **2017**, *7*, 8578–8587. [CrossRef] [PubMed]
36. Jones, A.L.; Jennings, D.E.; Hooks, C.R.R.; Shrewsbury, P.M. Field surveys of egg mortality and indigenous egg parasitoids of the brown marmorated stink bug, *Halyomorpha halys*, in ornamental nurseries in the mid-Atlantic region of the USA. *J. Pest Sci.* **2017**, *90*, 1159–1168. [CrossRef] [PubMed]
37. Hanks, L.M.; Gould, J.R.; Paine, T.D.; Millar, J.G.; Wang, Q. Biology and host relations of *Avetianella longoi* (Hymenoptera: Encyrtidae), an egg parasitoid of the eucalyptus longhorned borer (Coleoptera: Cerambycidae). *Ann. Entomol. Soc. Am.* **1995**, *88*, 666–671. [CrossRef]
38. Duan, J.J.; Bauer, L.S.; Van Driesche, R.G. Emerald ash borer biocontrol in ash saplings: The potential for early stage recovery of North American ash trees. *For. Ecol. Manag.* **2017**, *394*, 64–72. [CrossRef]
39. Margulies, E.; Bauer, L.; Ibáñez, I. Buying time: Preliminary assessment of biocontrol in the recovery of native forest vegetation in the aftermath of the invasive emerald ash borer. *Forests* **2017**, *8*, 369. [CrossRef]

forests

MDPI

Article

Initial Location Preference Together with Aggregation Pheromones Regulate the Attack Pattern of *Tomicus brevipilosus* (Coleoptera: Curculionidae) on *Pinus kesiya*

Fu Liu, Chengxu Wu, Sufang Zhang, Xiangbo Kong, Zhen Zhang * and Pingyan Wang

Research Institute of Forest Ecology, Environment and Protection, Chinese Academy of Forestry, Beijing 100091, China; liufu2006@163.com (F.L.); muzixuan58@126.com (C.W.); zhangsf@caf.ac.cn (S.Z.); xbkong@sina.com (X.K.); aqwangpy@163.com (P.W.)

* Correspondence: zhangzhen@caf.ac.cn; Tel.: +86-136-7102-2209

Received: 22 January 2019; Accepted: 11 February 2019; Published: 12 February 2019

Abstract: Research Highlights: We found that the initial attack location together with the aggregation pheromones played an important role in mediating the aggressive behavior of *T. brevipilosus* on *P. kesiya*. Background and Objectives: *T. brevipilosus* was identified as an aggressive species, which possesses the ability to kill live, healthy *P. kesiya*. In this scenario, we study the top-down attack pattern of *T. brevipilosus* on *P. kesiya* during the entirety of the reproductive period. Materials and Methods: We investigated the phenology of trunk attack on *P. kesiya* over a period of three years in Pu'er City, China. The hindguts extracts of the females and males *T. brevipilosus* were analyzed by coupled gas chromatography-mass spectrometry (GC-MS). The candidate aggregation pheromone compounds of *T. brevipilosus* were determined through electrophysiology experiments (electroantennographic detection, EAD and electroantennography, EAG), laboratory olfactometer bioassays, and field trapping. Results: we found that the pioneer beetles preferentially infested the crown of *P. kesiya* at the early stage of attack following spring flight with the later arriving beetles selectively attacking the lower area of the trunk to avoid intraspecific competition and better utilize limited resources, which exhibits a top-down attack pattern. During gallery initiation, the beetles release aggregation pheromones to attract conspecifics to conduct a mass attack. The chemical analyses indicated that the hindgut extracts of gallery-initiating beetles contained a larger amount of myrtenol, *cis*-verbenol, *trans*-verbenol, and verbenone. Myrtenol and *trans*-verbenol were identified as candidate aggregation pheromone compounds. In addition, a blend of these two components with *S*-(−)-α-pinene and *S*-(−)-β-pinene attracted more *T. brevipilosus* individuals in a field bioassay. Conclusions: We concluded that the preference for the initial attack location together with the aggregation pheromones played an important role in mediating the top-down attack pattern of *T. brevipilosus* on *P. kesiya*.

Keywords: *Tomicus brevipilosus*; location preference; aggregation pheromones; attack pattern; aggressiveness

1. Introduction

Species of *Tomicus* are noteworthy for their damage to pine forests in Europe, North America, Asia, and North Africa [1–4]. In Southwestern China, these bark beetles are also the main cause of extensive tree damage [5–9]. They attack the fresh shoots of living trees for maturation feeding, which can weaken the host and curtail growth [10]. During the reproductive phase, they infest the trunk of the host, which directly causes death [11–14].

Tomicus brevipilosus (Eggers) is native to Asia and has been reported from China, India, Japan, Korea, and the Philippines [15,16]. This bark beetle is known to infest several *Pinus* species, including *Pinus yunnanensis* Franch. [17], *Pinus koraiensis* Siebold & Zucc., and *Pinus kesiya* Royle ex Gordon [15]. *T. brevipilosus*, together with *other* two *Tomics* species (*Tomicus yunnanensis* Kirkendall & Faccoli, and *Tomicus minor* (Hartig, 1834)) have caused extensive mortality of *P. yunnanensis* in Southwestern China [5]. The life cycle of *T. brevipilosus* contains two phases including a feeding maturation and a reproduction stage. Extensive shoot feeding by adults can cause growth loss and lower a tree's natural resistance, which facilitates trunk attacks [6,13]. Gallery construction and larval feeding are regarded as the direct causes of *P. yunnanensis* tree mortality, particularly when a mass attack occurs on individual trees. The colonization pattern of *T. brevipilosus* varied dramatically, depending on the other two *Tomicus* species that were already present on the tree trunks or not. On trees that were already infested by the other two *Tomicus* species, *T. brevipilosus* colonized areas of the trunk that were not already occupied. When there were no newly infested by the other two *Tomicus* species, *T. brevipilosus* attacked *P. yunnanensis* by itself, infesting the lower parts of the trunk first, and then infesting progressively upward along the trunk into the crown, which exhibits a down-up attack pattern. The ability of *T. brevipilosus* to adjust its infestation pattern in response to other *Tomicus* species likely decreases interspecific competition as well as better utilizes the limited resource. Thus, *T. brevipilosus* appears to be an aggressive bark beetle since it can successfully attack and colonize vigorous *P. yunnanensis* on its own [17].

T. brevipilosus is an oligophagous pest that could cause significant damage to *P. kesiya* pine forests in Pu'er City of China as well [18,19]. The maturation feeding of adults on shoots normally occurs from March through October, which severely injures *P. kesiya* hosts. After completing their sexual development, the adults attack the tree trunk to initiate their reproductive phase. Their longitudinal galleries disrupt the nutrient flow within the phloem tissue, which gradually leads to the death of the *P. kesiya* trees.

Bark beetle aggregation and aggressive behavior are generally mediated by pheromones released by the beetles themselves or by volatiles derived from the host trees [20–22]. *T. piniperda* used host volatiles to locate trees suitable for brood production, the host monoterpenes were effectively lured for this bark beetle [23]. However, Poland et al. concluded that *trans*-verbenol is an aggregation pheromone component for immigrant North American populations of *T. piniperda* [22,24]. *T. destruens* that was known to be attracted to *P. pinea* shoots and the host extracts [25]. *trans*-verbenol was identified from hindgut extracts of both sexes of *T. minor*, which suggested that this beetle might generate aggregation pheromones [26]. Volatiles present in the phloem of infested *P. yunnanensis* could promote aggregation of *T. yunnanensis* [27,28]. In later work, Wang et al. found that both sexes of *T. brevipilosus* showed stronger attraction to damaged shoots than undamaged ones, but the specific components have so far had little reporting [9].

Strategies to reduce tree mortality caused by bark beetle attacks are becoming urgent. An understanding of ecological and behavioral aspects of pheromone production is critical for the development of semiochemical-based strategies to control bark beetles [29]. Application strategies using aggregation pheromones have been successful in monitoring and controlling pest population in Europe and North America, where commercial products exist for major species, *Ips* and *Dendroctonus* [30]. Push-pull tactic have also been successful against *Dendroctonus ponderosae* (Hopkins, 1902) and *Ips paraconfusus* Lanier, 1970 [31,32].

Nowadays, *T. brevipilosus* remains one of the greatest threats to *P. kesiya*, and there are few effective management tactics to control it. The present studies are designed to reveal (1) the regulatory mechanisms of the top-down attack pattern of *T. brevipilosus* on *P. kesiya* during the reproductive period and (2) the semiochemical components of *T. brevipilosus*. One further aim of this study is to utilize effective chemical lures as a pest management strategy to ensure the health of forests.

2. Materials and Methods

2.1. Study Area

The field study reported in this case was carried out in Pu'er City (N 22°54′09.20″, E 101°15′27.87″) at an elevation of 1400 m, and the samples of *T. brevipilosus* were collected on *P. kesiya*, and there is no coexistence of other *Tomicus* species. The studied host trees were originally planted by aerial seeding in the mid-1980s, which ranged from 10 m to 15 m in height and had breast height diameters from approximately 35 cm to 48 cm.

2.2. Insect Collection

Three to five attacked trees were selected at random and cut down every month during the trunk attack period November through January over three years. The pine trunks colonized by *T. brevipilosus* were divided into 1 m-long sections from the base to the top, and the bark was carefully peeled to reveal the beetle numbers and gallery lengths. According to our statistics, the length of *T. brevipilosus* can reach 90 mm during its reproductive phase. The colonization stages were then sampled by the length of the galleries: Stage I (<10 mm): mainly contained female beetles within tunnels where boring had just started, Stage II (10–20 mm): paired males and females that had started mating, Stage III (20–30 mm): mostly paired males and females that had completed their mating behavior and oviposition had occurred, Stage IV (30–50 mm): mostly females in a spawning state (males had gradually left the mating chamber), and Stage V (>50 mm): only females and eggs that were inside a completed gallery system (males had left entirely).

Based on the gallery length classifications, the beetles were individually placed into 5 mL plastic centrifuge tubes with several holes, returned to the laboratory, and then stored at 4 °C for 24 h before testing. Then, the beetles were separated by species and sex using a stereoscope because of the presence of erect hairs, granules, and punctures on the elytral declivity [16].

2.3. Chemicals

The chemicals tested in the study included *S*-(−)-α-pinene (98%), *S*-(−)-β-pinene (98.5%), *cis*-verbenol (95%, enantionmeric excesses ≥50%), *trans*-verbenol (95%, enantionmeric excesses ≥98%), verbenone (≥93%), myrtenal (≥97%), myrtenol (≥95%), (−)-camphene (80%), (+)-camphene (80%), styrene (99%), linalool (99%), terpinen-4-ol (97%), and heptyl acetate (98%), all obtained from Sigma-Aldrich Co. (St Louis, MO, USA). And *n*-hexanol (99.9%, Fisher Scientific Co., Waltham, MA, USA), *R*-(+)-α-pinene (98%), β-phellandrene (85%), (−)-limonene (97%), and (+)-limonene (97%, J&K Scientific Ltd., Beijing, China).

2.4. Hindgut Extracts

The hindguts from the females and males of *T. brevipilosus* were dissected under a microscope (Olympus, SZ61) and placed into 2 mL glass vials containing 200 μL *n*-hexane at 4 °C for 12 h. Then, these extracts were evaporated to 50 μL with nitrogen, transferred into small glass ampoules (100 μL), and stored at −20 °C before testing. Every extract contained five hindguts.

2.5. Coupled Gas Chromatography-Mass Spectrometry (GC-MS)

The component analysis was carried out by using an Agilent gas chromatograph coupled with a mass spectrometry system (TRACE GC 2000). The GC was equipped with a polar HP-5MS column (30 m × 0.25 mm × 0.25 μm) and included an injector temperature set to 230 °C. The oven temperature for the HP-5MS GC column was initially programmed at 45 °C for one minute and then subsequently increased to 105 °C at 2 °C per minute. Lastly, the temperature was increased to 220 °C at 10 °C per minute with helium used as the carrier gas. Heptyl acetate (2 ng/μL *n*-hexane) was then added to the extract as an internal standard. All the samples were analyzed with the same apparatus under the same conditions. Compound identification was based on a comparison of the retention times to those of the synthetic standards, and the reference mass spectra were taken from the NIST11 library (Scientific Instrument Services, Inc., Ringoes, NJ, USA).

2.6. Gas Chromatography–Electroantennographic Detection (GC-EAD)

The GC-EAD system consisted of an Agilent 7890A gas equipped with a HP-5MS column (30 m × 0.25 mm × 0.25 μm, Agilent Technologies, Wilmington, DE, USA) and an electro-antennogram detector (EAD), a 1:1 effluent splitter that allowed simultaneous flame ionization (FID) and electro-antennographic (EAD) detection of the separated volatile compounds. Hydrogen was used as the carrier gas. The injector and detector temperatures were 220 °C and 230 °C. The oven temperature was initially programmed at 60 °C for one minute and then subsequently increased to 100 °C at 5 °C per minute. Lastly, the temperature was increased to 250 °C at 10 °C per minute. Each antenna was prepared by cutting the basic segment carefully and clearing surface debris slightly, which was then inserted into a glass capillary. The glass capillaries housed 0.39 mm silver wires (Sigmund Cohn Corp, Mt. Vernon, NY, USA) and were filled with 0.9% NaCl saline solution. Electrodes were connected to a combi Probe (PRG-3, Syntech, Buchenbach, Germany). A reference electrode was inserted into the base of the excised beetle. The tip of the recording electrode was removed so its opening matched the diameter of the antennal club, and then one side of the antennal club was laid flat against the opening so its entire surface was in contact with the saline [33]. The purified and humidified supplemental airflow (400 mL/min) was supplied continuously. The FID and EAD signals were processed with Syntech software (GcEad version 4.6). In addition, 20 individuals of *T. brevipilosus* housed for 24 h in 3 cm centrifuge tube. Then all hindguts were dissected under a microscope (Olympus, SZ61) and placed into 2 mL glass vials containing 200 μL *n*-hexane at 4 °C for 12 h.

2.7. Electroantennogram (EAG) Recording

The EAG procedure used in this study was similar to that reported elsewhere [34]. The tested compounds were released from Pasteur pipettes containing a piece of filter paper (0.4 cm × 5 cm) impregnated with 10 μL of each freshly prepared solution of the test compounds (1 μg/μL of each in hexane) and passed over the antennae. In addition to the pipettes containing the test preparations, one Pasteur pipette containing a filter paper impregnated with 10 μL of the hexane solvent was used as a control. The puff containing the test substance was delivered into a continuously humidified and purified air stream moving at 800 mL/min and passing for 0.1 s through the impregnated filter paper in the pipettes. A control stimulation was made at the beginning and end of every four tested compounds. The test compounds were then applied randomly at intervals of 40 s. The EAG amplitudes in response to the synthetic compounds were expressed in relation to the responses to the control because of the large differences in the overall sensitivity between the individual antennae and to compensate for the decline in antennal sensitivity during each measurement session. In this normalization procedure, the responses to the control were defined as 100%. The values obtained between the two controls were corrected, according to the values of the references by linear interpolation. The compounds were tested using 15 antennae of males and 15 antennae of females.

2.8. Olfactometer Bioassays

The response of beetles to odor sources was tested using a modified open-arena olfactometer [9,35]. The synthetic compound was placed on a piece of filter paper (1.2 cm × 1.2 cm) in the odor region. *n*-hexane was dropped on the same filter papers placed on either side of the odor region. One beetle of each sex was released at the center of the larger circle for each trial. The number of times that the beetle entered the odor region within 5 min after release was recorded. Males and females of *T. brevipilosus* were tested for their behavioral attraction response to six relevant synthetic compounds at five concentrations, which ranged from 0.01 ng/μL to 100 ng/μL. For each concentration, *n*-hexane was used as the control, and 40 biological replicates were performed. Mean attraction values were calculated as follows: Mean numbers = Total numbers of attraction/Total replicates.

2.9. Field Trapping

According to the results of chemical analyses, electrophysiology experiments, and laboratory olfactometer bioassays, different blends of *R*-(+)-*α*-pinene, *S*-(−)-*α*-pinene, *S*-(−)-*β*-pinene, *cis*-verbenol, *trans*-verbenol, and myrtenol were formulated for field trapping tests inside a plantation of *P. kesiya* from November to December 2016. Then, 10 mL of the blend was placed into a series of polyethylene bottles. Black cross-barrier traps (Pherobio Technology Co., Ltd., Beijing, China) were suspended at a 1.5-m distance from the ground. Randomized block experimental designs were used to evaluate the results of the traps. The traps were separated by at least 15 m from each other within a block. Each treatment block contained seven treatments, and five blocks were spaced at a distance of 1 km from each other. A blank control with an un-baited trap was placed in each treatment block. Seven kinds of lures used for field trapping with a corresponding ratio were listed in Table 1.

Table 1. Seven kinds of lures used for field trapping (*v*/*v*).

Compounds	A	B	C	D	E	F	K
S-(−)-*α*-pinene	10	10	-	-	5	5	-
R-(+)-*α*-pinene	-	-	10	10	5	5	-
S-(−)-*β*-pinene	1	1	1	1	1	1	-
trans-verbenol	0.025	0.0125	0.025	0.0125	0.0125	0.025	-
cis-verbenol	-	0.0125	-	0.0125	0.0125	-	-
myrtenol	0.5	0.5	0.5	0.5	0.5	0.5	-

A–F: The lures of A–F; K: The lure of the control.

2.10. Statistical Analyses

The number of times that walking beetles entered the odor region was used to evaluate the attraction level and inhibition for each treatment group. The SPSS statistical analysis software (version 16.0; IBM Inc., Chicago, IL, USA) was used to process data, and all the figures were drawn using Excel 2010 software. A heat map population distribution of *T. brevipilosus* on *P. kesiya* was constructed based on the mean numbers of beetles found in different stages and the mother gallery lengths, while one-way ANOVA was used to compare the amounts of the chemicals (nanogram) identified from the hindgut extracts of females and males during different trunk-breeding phases. The data from these procedures were analyzed with Tukey's multiple range test ($p < 0.05$) as well as the EAG response data from female and male beetles with respect to each stimulus and different compound concentrations. The mean numbers of the walking beetles versus the different concentrations of the chemicals and field trapping data were then $\log_{10}(x + 1)$ transformed. This resulted in homogeneous variances and data were also analyzed using one-way ANOVA followed by Tukey's multiple range test ($p < 0.05$).

3. Results

3.1. Population Dispersal on Trunk Surfaces

Individual *T. brevipilosus* breeding attacks on *P. kesiya* were observed over three consecutive months during the three years of the study. The results showed that both the timing and duration of the breeding attacks were similar, usually beginning in November and extending through January. The pioneer beetles preferentially attacked the upper crown of the individual hosts first at the early stage of the breeding attacks in November (Figure 1). The subsequent beetles gradually moved downward in the trunk over the course of the next two months (December and January), which exhibits a top-down attack pattern (Figure 1A,B).

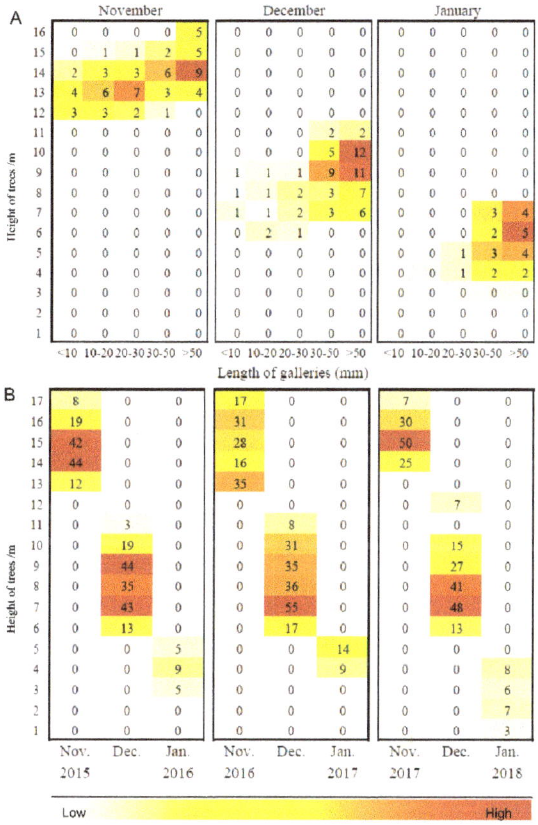

Figure 1. Population distribution heat map for *T. brevipilosus* on *P. kesiya*. (**A**) Length of galleries of *T. brevipilosus* from November to January 2015–2018. (**B**) Distribution of *T. brevipilosus* from November through January 2015–2018.

3.2. Chemical Analysis of Hindgut Extracts

Ten components were identified from the hindgut extracts of the female *T. brevipilosus*, including four terpenes and six oxygenated monoterpenes (Table 2). The amount of myrtenol at stage I was greater and significantly different compared to that of the other stages ($F_{(4,15)} = 4.402$, $p = 0.015$) and declined rapidly from stage I to III. The amount of *cis*-verbenol, *trans*-verbenol, and verbenone decreased from stage I to III, but these were not significant differences ($F_{(2,10)} = 2.121$, $p = 0.160$, $F_{(2,10)} = 0.587$, $p = 0.574$, and $F_{(2,10)} = 4.057$, $p = 0.051$, respectively). The data showed that the amount of *cis*-verbenol and verbenone increased suddenly at stage V and exhibited significant differences at stages III and IV ($F_{(4,15)} = 3.324$, $p = 0.039$, $F_{(4,15)} = 5.367$, $p = 0.007$, respectively). However, this was not the case for stages I and II (p >0.05). In addition, the amounts of terpene and styrene at stage V were significantly lower than those at stage I, but terpinen-4-ol exhibited the opposite pattern. There were no significant differences for β-phellandrene and linalool in the five stages ($p > 0.05$).

For males (Table 2), the data showed that the amounts of *cis*-verbenol, myrtenol, and verbenone significantly decreased from stage I to stage III ($F_{(2,6)} = 5.283$, $p = 0.048$, $F_{(2,6)} = 29.530$, $p = 0.001$, and $F_{(2,6)} = 10.051$, $p = 0.012$, respectively). The amount of *trans*-verbenol decreased from stage I to III, but there were no remarkable differences ($F_{(2,6)} = 3.021$, $p = 0.124$).

Table 2. Chemical constituents identified from the extract of female and male hindguts of *T. brevipilosus* collected from galleries of different lengths.

Retention Time	Component	Gender	Amount of Chemicals from Beetles Collected from Galleries of Different Length (in Nanograms)				
			<10 mm †	10–20 mm ‡	20–30 mm §	30–50 mm ¶	>50 mm ⊥
7:53	styrene	female	2.73 ± 0.49 a	2.45 ± 0.62 a	0.18 ± 0.06 b	0.13 ± 0.04 b	0.52 ± 0.18 b
		male	5.10 ± 0.18 a	1.86 ± 0.53 a	0.20 ± 0.13 a	0.40	0.27
9:36	α-pinene	female	2.47 ± 1.25 b	2.12 ± 0.50 b	20.78 ± 5.58 a	23.17 ± 4.90 a	17.77 ± 6.03 a
		male	6.68 ± 4.46 a	8.37 ± 7.39 a	14.10 ± 4.94 a	12.10	9.54
11:42	*S*-(−)-β-pinene	female	0.00 ± 0.00 b	0.00 ± 0.00 b	1.71 ± 1.01 ab	2.49 ± 1.09 a	0.00 ± 0.00 b
		male	0.00 ± 0.00 a	0.00 ± 0.00 a	1.53 ± 0.99 a	0.00	0.00
14:33	β-phellandrene	female	0.94 ± 0.77 a	0.21 ± 0.07 a	3.46 ± 2.03 a	5.77 ± 2.29 a	3.69 ± 1.53 a
		male	2.24 ± 2.07 a	0.50 ± 0.41 a	3.23 ± 2.20 a	0.68	0.45
18:94	linalool	female	0.76 ± 0.57 a	0.17 ± 0.08 a	0.23 ± 0.13 a	0.56 ± 0.28 a	1.09 ± 0.37 a
		male	0.26 ± 0.22 a	1.08 ± 0.83 a	0.88 ± 0.16 a	0.95	0.90
21:63	*cis*-verbenol	female	6.82 ± 3.69 ab	2.23 ± 0.52 ab	0.65 ± 0.29 b	0.69 ± 0.26 b	19.77 ± 10.99 a
		male	4.61 ± 2.46 a	3.30 ± 1.09 a	0.43 ± 0.25 b	1.27	2.24
21:87	*trans*-verbenol	female	0.87 ± 0.61 a	0.66 ± 0.18 a	0.27 ± 0.05 a	1.39 ± 0.95 a	0.55 ± 0.23 a
		male	0.76 ± 0.24 a	0.39 ± 0.15 a	0.20 ± 0.09 a	0.12	0.20
24:29	terpinen-4-ol	female	0.35 ± 0.35 b	0.13 ± 0.13 b	0.37 ± 0.37 b	0.31 ± 0.31 b	11.61 ± 5.62 a
		male	0.07 ± 0.07 a	0.05 ± 0.05 a	0.18 ± 0.18 a	0.20	0.56
24:99	myrtenol	female	102.24 ± 59.32 a	40.29 ± 15.55 ab	4.11 ± 1.40 b	12.81 ± 5.23 b	21.87±2.64 ab
		male	88.56 ± 40.22 a	30.27 ± 10.84 a	1.39 ± 0.34 b	4.55	7.32
25:75	verbenone	female	4.63 ± 2.21 ab	3.35 ± 0.94 ab	0.49 ± 0.03 b	1.33 ± 0.33 b	10.41 ± 4.43 a
		male	9.35 ± 1.89 a	5.61 ± 1.88 a	1.12 ± 0.37 b	0.52	2.59

† Four replicates for females and three replicates for males. ‡ Six replicates for females and three replicates for males. § Three replicates for females and three replicates for males. ¶ Four replicates for females and one replicate for males. ⊥ Three replicates for females and one replicate for males. Lowercase letters in the same row indicate significant differences ($p < 0.05$) in mean quantities of potential semiochemicals.

Similarly, there were no significant differences for oxygenated monoterpenes (*cis*-verbenol, *trans*-verbenol, myrtenol, and verbenone) between the females and males of *T. brevipilosus* collected from galleries of the same length ($p > 0.05$).

3.3. GC-EAD Analyses

In analyses of solvent extracts of female *T. brevipilosus* by GC-EAD, up to five compounds elicited responses from the antennae of female *T. brevipilosus*, with component 1 (Retention time: 4.72 min), component 2 (Rt: 5:60 min), component 3 (Rt: 9:69 min), component 4 (Rt: 10:75 min), and component 5 (Rt: 10:87 min). By comparing the GC-EAD retention times with those synthetic standards, antennae responded to *S*-(−)-α-pinene (1), *S*-(−)-β-pinene (2), *trans*-verbenol (3), myrtenol (4), and verbenone (5) (Figure 2). Among these chemicals, *trans*-verbenol and verbenone (peaks 3 and 5) elicited a strong response from both sexes. *S*-(−)-β-pinene (peak 2) only elicited a response from females of *T. brevipilosus*.

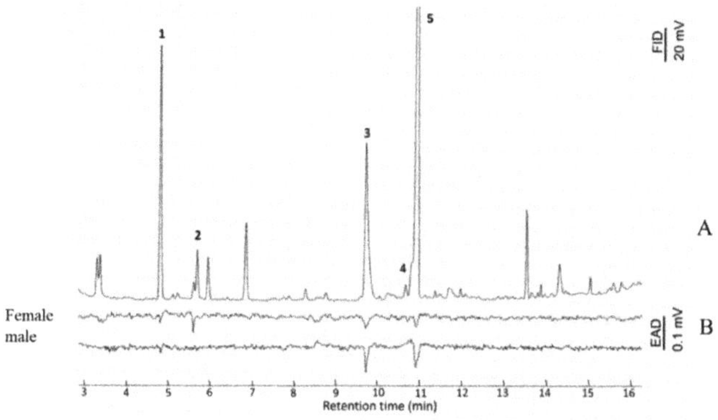

Figure 2. Simultaneously recorded responses from flame ionization detector (FID, **A**) and electroantennographic detection (EAD, **B**, sensing element: female and male *T. brevipilosus* antenna) gas chromatography-electroantennographic detection (GC-EAD) to the volatiles from hindgut of female *T. brevipilosus* ($n = 20$). Peak 1: *S*-(−)-α-pinene, 2: *S*-(−)-β-pinene, 3: *trans*-verbenol, 4: myrtenol, 5: verbenone.

3.4. Electroantennography (EAG)

Six compounds identified from the hindguts and another six host volatiles were tested in the EAG experiment. The results revealed that five compounds elicited strong stimulation of the antennae of the females and males, specifically *cis*-verbenol, *trans*-verbenol, myrtenol, *S*-(−)-α-pinene, and *R*-(+)-α-pinene (Figure 3). However, there were no significant differences between females and males for the same substance.

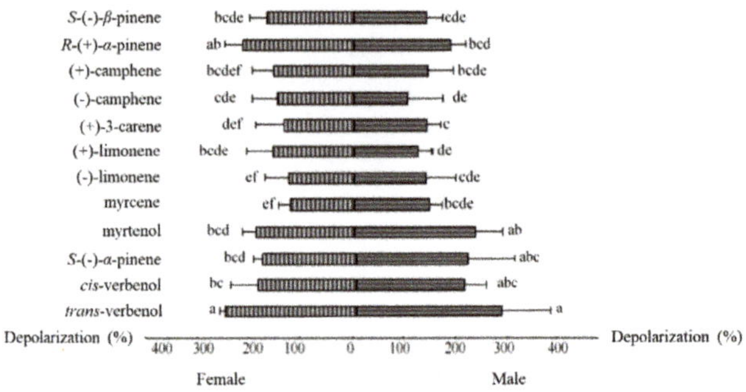

Figure 3. Mean values (± Standard deviation) of the electroantennography response of *T. brevipilosus* females and males to a series of compounds. Mean values marked with the same letter are not significantly different at $p < 0.05$ based on the least significant difference test. Lowercase letters indicate the differences in EAG responses of the same sex bark beetles to different compounds. No significant differences between females and males for the same substance were found.

3.5. Olfactometer Bioassays

Five compounds were tested at five concentrations using a modified open-arena olfactometer (Figure 4). These results revealed that, compared with *n*-hexane, both sexes of *T. brevipilosus* were significantly attracted by *cis*-verbenol and *trans*-verbenol at a 1 ng/μL concentration (Figure 4A,B). Furthermore, a 0.1 ng/μL concentration of *S*-(−)-α-pinene was strongly attractive for both genders (Figure 4E). The mean numbers of the females attracted to myrtenol were significantly higher at 1 ng/μL and 10 ng/μL concentrations than at the other concentrations, which was the case at a 1 ng/μL concentration for males (Figure 4C). Both sexes of *T. brevipilosus* displayed weak attractive tendencies to verbenone at the five concentrations, but there were no significant differences between them (Figure 4D).

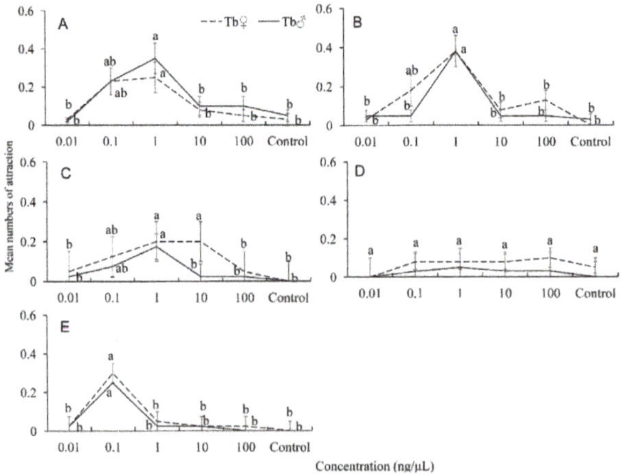

Figure 4. The mean numbers of attracted *T. brevipilosus* to five synthetic compounds at five concentrations (i.e., 0.01 ng/μL, 0.1 ng/μL, 1 ng/μL, 10 ng/μL, and 100 ng/μL) with an *n*-hexane control. Mean numbers = Total numbers of attraction/Total replicates. Lowercase letters indicate significant differences ($p < 0.05$) between the same treatments. The error bars were a standard error. (**A**) *cis*-verbenol. (**B**) *trans*-verbenol. (**C**) myrtenol. (**D**) verbenone. (**E**) *S*-(−)-α-pinene.

3.6. Field Trapping

In the field trapping experiment, traps baited with blends of *S*-(−)-α-pinene, *S*-(−)-β-pinene, *trans*-verbenol, and myrtenol led to the capture of more *T. brevipilosus* individuals within each replicate. The capture by treatment B, baited with blends of *S*-(−)-α-pinene, *S*-(−)-β-pinene, *cis*-verbenol, *trans*-verbenol, and myrtenol, was less than that by treatment A (Figure 5). However, there was no significant difference between the treatments. Replacing *S*-(−)-α-pinene with its enantiomer *R*-(+)-α-pinene significantly reduced the catch numbers. While a mix of the three monoterpenes (i.e., *S*-(−)-α-pinene, *R*-(+)-α-pinene, and *S*-(−)-β-pinene, in a 1:1:0.2 ratio, *v*/*v*) along with oxygenated monoterpenes could enhance the catch numbers, the differences were not significant. Fewer numbers of beetles were caught than expected. One possible reason for this is that the mean height of the host, *P. kesiya*, can be more than 10 meters since our traps were suspended 1.5 m above the ground. Another reason is that the top-down attack pattern during the breeding period may have possibly reduced the number of flights and decreased the probability of trap captures.

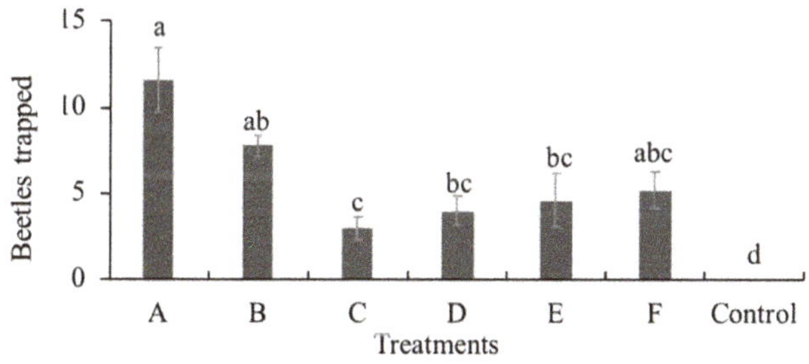

Figure 5. Mean numbers of *T. brevipilosus* caught per trap with seven different baits (between 20 November 2016 and 3 December 2016). Lowercase letters indicate significant differences between trap baits at $p < 0.05$ (one-way ANOVA followed by Tukey's tests). The error bars were standard error. The sample contains *R*-(+)-α-pinene (Rαp), *S*-(−)-α-pinene (Sαp), *S*-(−)-β-pinene (Sβp), *cis*-verbenol (cV), *trans*-verbenol (tV), and myrtenol (Mol). (**A**) Sαp:Sβp:tV:Mol, (**B**) Sαp:Sβp:cV:tV:Mol, (**C**) Rαp:Sβp:tV:Mol, (**D**) Rαp:Sβp:cV:tV:Mol, (**E**) Sαp:Rαp:Sβp:cV:tV:Mol, (**F**) Sαp:Rαp:Sβp:tV:Mol, and control: unbaited.

4. Discussion

Aggressive bark beetle species are able to attack and kill living and sometimes quite healthy trees [36]. Epidemics caused by these "aggressive" species may dramatically alter the state and function of forest ecosystems over large areas [37]. Typical examples of aggressive bark beetle species are *Ips typographus* (Linnaeus, 1758) in Europe [38], and *Dendroctonus frontalis* Zimmermann, 1868 and *D. ponderosae* in North America [39,40]. In China, *T. yunnanensis* is regarded as one of the most aggressive species of *Tomicus* [8,9,16]. Chen et al. regarded the aggressiveness of *T. brevipilosus*, which is a sympatric species, as similar to that of *T. yunnanensis* [9]. In our study, *T. brevipilosus* had the ability to breed in healthy *P. kesiya* and killed trees on its own by mass attack, which could explain why *P. kesiya* has suffered high levels of tree mortality in recent years. Accordingly, *T. brevipilosus* should be regarded as one of the most aggressive species of the *Tomicus* genus.

Typically, the location preferences of individual species are related to the place where the pioneer beetles initiate the attack, and these locations vary among *Tomicus* species [12,41]. Earlier research showed that *T. piniperda* tends to initiate attacks along the lower trunks of *P. sylvestris*, and *T. minor*, which flies later, tends to colonize the upper portions of the trunks [2]. In contrast, the initial attacks by *T. yunnanensis* tend to be along the upper parts of *P. yunnanensis*, and the *T. minor* attacks are more concentrated along the lower trunk [9,12]. Chen et al. revealed that *T. brevipilosus*, *T. yunnanensis*, and *T. minor* coexist together on *P. yunnanensis*, and *T. brevipilosus* had to adjust its infestation location in response to the other two *Tomicus* species, which likely decreased interspecific competition and resulted in two general patterns of infestation [9]. In the first pattern, *T. brevipilosus* breeds preferentially within the middle and lower trunk regions were not already occupied by the other two *Tomicus* species. In the second pattern, *T. brevipilosus* attacks alone, infests the lower parts of the trunk first, and then progressively moves upwards along the trunk into the crown, which exhibits a bottom-up attack pattern. However, our study showed that the pioneer beetles preferentially attack the crown of *P. kesiya* first and gradually move down the trunk, which shows a top-down attack pattern. One possible explanation for *T. brevipilosus* attacking the crowns first before attacking the trunks of *P. kesiya* is that it is the dominant species on *P. kesiya*, and the location of the initial attack was not affected by the presence of other *Tomicus* species. Without competition, it is easy for this beetle to preferentially infest the crown of hosts first during the spring flight.

Aggregation pheromones are used to signal a mass attack by the beetles on pines and allows the insects to coordinate feeding and mating in time and space [42,43]. Two oxygenated monoterpenes, *cis*-verbenol and *trans*-verbenol, have been suggested to be major candidate aggregation pheromone compounds of *T. minor* [26]. In earlier work, Poland et al. concluded that *trans*-verbenol acts as an aggregation pheromone component that influences migrant North American *T. piniperda* populations [22]. In our study, both *cis*-verbenol and *trans*-verbenol were detected in the hindguts of *T. brevipilosus* initially breeding on trunks. In addition, the emergence time for verbenol agreed with previous reports that suggested this aggregation pheromone is generally produced in bark beetles [42,44]. One hypothesis to explain this shared signal is that *Tomicus* species individuals utilize the same pheromones to take full advantage of available resources [45]. Accordingly, the results of the GC-EAD, EAG, and dose-concentration experiments clearly indicated that both *cis*-verbenol and *trans*-verbenol are candidate components of the aggregation pheromone.

Lanne et al. revealed that myrtenol is the major component of the volatile compounds in the hindguts of both sexes of *T. piniperda*, which gave an intermediate response in electroantennography tests [26]. However, the compound demonstrated low entrapment. Our research found that a larger amount of myrtenol was present in the hindguts of the females and males of *T. brevipilosus* at stages I and II when the beetles were in contact with the resin-containing phloem and xylem. Moreover, the antennae from beetles of these stages were characterized by a very strong EAG response. At concentrations of 1 ng/μL and 10 ng/μL, females showed a higher walking response than that observed at other concentrations, and this was the case at the 1 ng/μL concentration for males. Further work is needed to prove that this component is a key constituent of aggregation pheromones of this species.

S-(−)-α-pinene and *S*-(−)-β-pinene can be considered to be an active kairomone or a coattractant aggregation pheromone for *T. brevipilosus*, since the beetles displayed strong attraction in the EAD analysis and behavioral tests. In early research, *S*-(−)-α-pinene mixed with *trans*-verbenol resulted in more trap catches for *T. piniperda* [22,23,46]. In addition, *T. destruens* was clearly attracted by the blend of the host volatiles and ethanol, which acted synergistically. The maximum response was observed at release rates of 300 and 900 mg/day, respectively [47]. Lastly, some bark beetle aggregation pheromones arise through oxygenation of α-pinene by oxidases from beetles or their associated microorganisms [48–51].

Clearly, the amounts of the oxygenated monoterpenes, *cis*-verbenol, *trans*-verbenol, and myrtenol, reached a maximum at stage I when the pioneer beetles bored through the outer bark and contacted the resin-containing phloem and xylem of the host. These amounts subsequently declined from stage I to III. The variation found in this study is in agreement with previous reports that the production of major pheromones by beetles exhibits large quantitative variation, i.e., increasing during gallery initiation and then terminating or declining rapidly afterwards [42]. An abnormal phenomenon was that the amounts of oxygenated monoterpenes, *cis*-verbenol, and verbenone rose suddenly at stage V. One possible reason was that the higher concentration of oxygenated monoterpenes inhibited the attraction between the *Tomicus* at the later stage of trunk breeding. Additionally, our earlier study suggested that *trans*-verbenol and verbenone at higher concentrations inhibited the attraction of *T. minor* and *T. yunnanensis*. Consequently, further studies need to be conducted to clarify this phenomenon.

5. Conclusions

In summary, it is reasonable to assume that the initial location preference along with the aggregation pheromones regulate the top-down attack pattern. *T. brevipilosus* pioneer beetles preferentially attacked the crown of the hosts first at the early stage of spring flight and release aggregation pheromones (i.e., myrtenol, *cis*-verbenol, and *trans*-verbenol) immediately to initiate a mass attack. Then, in the later phase of gallery excavation, the amount of *cis*-verbenol and verbenone increases suddenly, and the high concentration of oxygenated monoterpenes inhibits the attraction of conspecifics. To avoid intraspecific competition and to better utilize limited resources, the subsequent beetles selectively attack the lower area of the trunk to avoid overcrowding within the attack zones. Therefore, the preference of the initial attack zones together with the aggregation pheromones play an important role in mediating the top-down attack pattern of *T. brevipilosus* on *P. kesiya*. However, considering that aggregation pheromones are one of the major factors that regulate the aggressive behavior of *T. brevipilosus* during the reproduction phase, it is still unclear which of these components play an important role in guiding this beetle to infest the host trees. Thus, further research to provide critical information for monitoring and/or mass trapping of this economically important forest pest insect still needs to be conducted.

Author Contributions: F.L. and C.W. contributed equally to the work. F.L. and C.W. contributed equally to the conceptualization. F.L. Writing—review & editing: C.W. Writing—original draft: S.Z. Data curation: X.K. Visualization: Z.Z. Validation: F.L., C.W., and P.W. Investigation: all the authors approved the version to be published.

Funding: This work was supported by the Special Fund for Basic Scientific Research of Central Public Research Institutes (CAFYBB2016QA008), the Institute Special Fund for Basic Research, Institute of Forest Ecology, Environment, and Protection, Chinese Academy of Forestry (CAFYBB2016SY017), and The National Natural Science Foundation of China, (31370655) and (31770693).

Acknowledgments: We thank the staff of Mengxian Forest Station in Yunnan Province for collecting *T. brevipilosus* individuals.

Conflicts of Interest: The authors declare no conflicts of interest.

References

1. Bakke, A. Ecological studies on bark beetles (Coleoptera: Scolytidae) associated with Scots pine (*Pinus sylvestris* L.) in Norway, with particular reference to the influence of temperature. *Medd. Nor. Skogforsøksves.* **1968**, *21*, 441–602.
2. Långström, B. Life cycles and shoot feeding of the pine shoot beetles. *Stud. For. Suec.* **1983**, *163*, 1–29.
3. Långström, B.; Hellqvist, C. Induced and spontaneous attacks by pine shoot beetles on young Scots pine trees: Tree mortality and beetle performance. *J. Appl. Entomol.* **1993**, *115*, 25–36. [CrossRef]
4. Peverieri, G.S.; Faggi, M.; Marziali, L.; Panzavolta, T.; Bonuomo, L.; Tiberi, R. Use of attractant and repellent substances to control *Tomicus destruens* (Coleoptera: Scolytidae) in *Pinus pinea* and *P. pinaster* pine forests of Tuscany. *Entomologica* **2016**, *38*, 91–102.
5. Ye, H. Approach to the reasons of *Tomicus piniperda* (L.) population epidemics. *J. Yunnan Univ. (Nat. Sci.)* **1992**, *14*, 211–216.
6. Ye, H.; Lieutier, F. Shoot aggregation by *Tomicus piniperda* L. (Col: Scolytidae) in Yunnan, southwestern China. *Ann. For. Sci.* **1997**, *54*, 635–641.
7. Långström, B.; Li, L.S.; Hongpin, L.; Peng, C.; Haoran, L.; Hellqvist, C.; Lieutier, F. Shoot feeding ecology of *Tomicus piniperda* and *T. minor* (Col., Scolytidae) in southern China. *J. Appl. Entomol.* **2002**, *126*, 333–342. [CrossRef]
8. Lu, J.; Zhao, T.; Ye, H. The shoot-feeding ecology of three *Tomicus* species in Yunnan province, southwestern China. *J. Insect Sci.* **2014**, *14*, 37. [CrossRef]

9. Wang, J.H.; Zhang, Z.; Kong, X.B.; Wang, H.B.; Zhang, S.F. Intraspecific and interspecific attraction of three *Tomicus* beetle species during the shoot-feeding phase. *Bull. Entomol. Res.* **2015**, *105*, 225–233. [CrossRef]
10. Byers, J.A. Attraction of bark beetles, *Tomicus piniperda*, *Hylurgops palliates*, and *Trypodendron domesticum* and other insects to short-chain alcohols and monoterpenes. *J. Chem. Ecol.* **1992**, *18*, 2385–2402. [CrossRef]
11. Ye, H.; Li, L.S. The distribution of *Tomicus piniperda* (L) population in the crown of Yunnan pine during the shoot feeding period. *Acta Entomol. Sin.* **1994**, *37*, 311–316.
12. Ye, H.; Ding, X.S. Impacts of *Tomicus minor* on distribution and reproduction of *Tomicus piniperda* (Col., Scolytidae) on the trunk of the living *Pinus yunnanensis* trees. *J. Appl. Entomol.* **1999**, *123*, 329–333.
13. Lieutier, F.; Ye, H.; Yart, A. Shoot damage by *Tomicus* sp. (Coleoptera: Scolytidae) and effect on *Pinus yunnanensis* resistance to subsequent reproductive attacks in the stem. *Agric. For. Entomol.* **2003**, *5*, 227–233. [CrossRef]
14. Ji, M.; Dong, X.Q.; Liu, H.P.; Li, L.S.; Xu, H.; Yang, X.P.; Li, H.R.; Ze, S.Z. Preliminary study on remote sensing detection of Yunnan pine forest damaged by *Tomicus piniperda*. *J. West China For. Sci.* **2007**, *36*, 87–90.
15. Browne, F.G. *Pests and Diseases of Forest Plantation Trees. An Annotated List of the Principal Species Occurring in the British Commonwealth*; The Clarendon: Oxford, UK, 1968.
16. Kirkendall, L.R.; Faccoli, M.; Ye, H. Description of the yunnan shoot borer, *Tomicus yunnanensis* Kirkendall & Faccoli sp. n. (Curculionidae, Scolytinae), an unusually aggressive pine shoot beetle from southern China, with a key to the species of *Tomicus*. *Zootaxa* **2008**, *1819*, 25–39.
17. Chen, P.; Lu, J.; Haack, R.A.; Ye, H. Attack pattern and reproductive ecology of *Tomicus brevipilosus* (Coleoptera: Curculionidae) on *Pinus yunnanensis* in southwestern china. *J. Insect Sci.* **2015**. [CrossRef] [PubMed]
18. Jia, P.; Li, J.Z.; Tong, Q.; Dai, Y.J.; Wan, L.M.; Liu, Y. The occurrence and distribution patterns of *Tomicus brevipilosus* adult in treetops (in Chinese). *J. Southwest For. Univ.* **2014**, *6*, 70–74.
19. Wan, L.M.; Luo, F.M.; Tong, Q.; Dai, Y.J.; Jia, P.; Liu, Y. The spatial distribution patterns of *Tomicus brevipilosus* (Eggers) adults in trunk. *J. Sichuan For. Sci. Tech.* **2014**, *35*, 49–53.
20. Byers, J.A.; Lanne, B.S.; Löfqvist, J.; Schlyter, F.; Bergström, G. Olfactory recognition of host-tree susceptibility. *Naturwissenschaften* **1985**, *72*, 324–326. [CrossRef]
21. Byers, J.A. Chemical ecology of bark beetle. *Experientia* **1989**, *45*, 271–283. [CrossRef]
22. Poland, T.M.; Groot, P.D.; Burke, S.; Wakarchuk, D.; Haack, R.A.; Nott, R.; Scarr, T. Development of an improved attractive lure for the pine shoot beetle, *Tomicus piniperda* (Coleoptera: Scolytidae). *Agric. For. Entomol.* **2003**, *5*, 293–300. [CrossRef]
23. Schroeder, L.M. Attraction of the bark beetle *Tomicus piniperda* and some other bark- and wood-living beetles to the host volatiles a-pinene and ethanol. *Entomol. Exp. Appl.* **1988**, *46*, 203–210. [CrossRef]
24. Poland, T.M.; De Groot, P.; Haack, R.A.; Czokajlo, D. Evaluation of semiochemicals potentially synergistic to a-pinene for trapping the larger European pine shoot beetle, *Tomicus piniperda* (Col., Scolytidae). *J. Appl. Entomol.* **2004**, *128*, 639–644. [CrossRef]
25. Faccoli, M.; Anfora, G.; Tasin, M. Responses of the mediterranean pine shoot beetle *Tomicus destruens* (Wollaston) to pine shoot and bark volatiles. *J. Chem. Ecol.* **2008**, *34*, 1162–1169. [CrossRef] [PubMed]
26. Lanne, B.S.; Schlyter, F.; Byers, J.A.; Löfqvist, J.; Leufvén, A.; Bergström, G.; Van Der Pers, J.N.C.; Unelius, R.; Baeckstrom, P.; Norin, T. Differences in attraction to semiochemicals present in sympatric pine shoot beetles, *Tomicus minor*, and *T. piniperda*. *J. Chem. Ecol.* **1987**, *13*, 1045–1067. [CrossRef] [PubMed]
27. Zhao, T.; Li, L.S.; Zhou, N. The attraction of Yunnan pine to pine shoot beetle and tree volatile compositions. *J. Northeast For. Univ.* **2002**, *30*, 47–49.
28. Sun, J.H.; Clarke, S.R.; Kang, L.; Wang, H.B. Field trials of potential attractants and inhibitors for pine shoot beetles in the Yunnan province, China. *Ann. For. Sci.* **2005**, *62*, 9–12. [CrossRef]
29. Borden, J.H. Disruption of Semiochemical-Mediated Aggregation in Bark Beetles. In *Insect Pheromone Resea: New Direction*; Carde, R.T., Minks, A.K., Eds.; Chapman and Hall: New York, NY, USA, 1997; pp. 421–438.
30. Schlyter, F. Semiochemical diversity in practice: Antiattractant semiochemicals reduce bark beetle attacks on standing trees-a first meta-analysis. *Psyche J. Entomol.* **2012**. [CrossRef]
31. Lindgren, B.S.; Borden, J.H. Displacement and aggregation of mountain pine beetles, *Dendroctonus ponderosae* (Coleoptera: Scolytidae), in response to their antiaggregation and aggregation pheromones. *Can. J. For. Res.* **1993**, *23*, 286–290. [CrossRef]

32. Shea, P.J.; Neustein, M. Protection of a Rare Stand of Torrey Pine from *Ips paraconfusus*. In *Application of Semiochemicals for Management of Bark Beetle Infestations*; Salom, S.M., Hobson, K.R., Eds.; Plenum Press: Ogden, UT, USA, 1995; pp. 39–43.
33. Zhang, Z.; Bian, L.; Sun, X.; Luo, Z.; Xin, Z.; Luo, F.; Chen, Z. Electrophysiological and behavioural responses of the tea geometrid *Ectropis obliqua* (Lepidoptera: Geometridae) to volatiles from a non-host plant, rosemary, *Rosmarinus officinalis* (Lamiaceae). *Pest Manag. Sci.* **2015**, *71*, 96–104. [CrossRef]
34. Batista-Pereira, L.G.; Santangelo, E.M.; Stein, K.; Unelius, C.R.; Eiras, A.E.; Corrêa, A.G. Electrophysiological studies and identification of possible sex pheromone components of Brazilian populations of the sugarcane borer, *Diatraea saccharalis*. *Z. Für Nat. C* **2002**, *57*, 753–758. [CrossRef]
35. Byers, J.A.; Birgersson, G.; Francke, W. Aggregation pheromones of bark beetles, *Pityogenes quadridens* and *P. bidentatus*, colonizing scotch pine: Olfactory avoidance of interspecific mating and competition. *Chemoecology* **2013**, *23*, 251–261. [CrossRef]
36. Christiansen, E.; Waring, R.H.; Berryman, A.A. Resistance of conifers to bark beetle attack: Searching for general relationships. *For. Ecol. Manag.* **1987**, *22*, 89–106. [CrossRef]
37. Cedervind, J.; Pettersson, M.; Långström, B. Attack dynamics of the pine shoot beetle, *Tomicus piniperda* (Col.; Scolytinae) in Scots pine stands defoliated by *Bupalus piniaria* (Lep.; Geometridae). *Agric. For. Entomol.* **2003**, *5*, 253–261. [CrossRef]
38. Christiansen, E.; Bakke, A.; Christiansen, E.; Bakke, A. The Spruce Bark Beetle in Eurasia. In *Dynamics of Forest Insects Populations*; Berryman, A.A., Ed.; Plenum Press: New York, NY, USA, 1988; pp. 479–503.
39. Flamm, R.O.; Coulson, R.N.; Payne, T.L. The Southern Pine Beetle. In *Dynamics of Forest Insects Populations*; Berryman, A.A., Ed.; Plenum Press: New York, NY, USA, 1988; pp. 531–553.
40. Raffa, K.F.; Berryman, A.A. Interacting selective pressures in conifer-bark beetle systems: A basis for reciprocal adaptations? *Am. Nat.* **1987**, *129*, 234–262. [CrossRef]
41. Rudinsky, J.A. Ecology of Scolytidae. *Annu. Rev. Entomol.* **1962**, *7*, 327–348. [CrossRef]
42. Wood, D.L. The role of pheromones, kairomones, and allomones in the host selection and colonization behavior of bark beetles. *Annu. Rev. Entomol.* **1982**, *27*, 411–446. [CrossRef]
43. Seybold, S.J.; Bohlmann, J.; Raffa, K.F. Biosynthesis of coniferophagous bark beetle pheromones and conifer isoprenoids: Evolutionary perspective and synthesis. *Can. Entomol.* **2000**, *132*, 697–753. [CrossRef]
44. Cano-Ramírez, C.; Armendáriz-Toledano, F.; Macías-Sámano, J.E.; Sullivan, B.T.; Zúñiga, G. Electrophysiological and behavioral responses of the bark beetle *Dendroctonus rhizophagus* to volatiles from host pines and conspecifics. *J. Chem. Ecol.* **2012**, *38*, 512–524. [CrossRef]
45. Ayres, B.D.; Ayres, M.P.; Abrahamson, M.D.; Teale, S.A. Resource partitioning and overlap in three sympatric species of *Ips* bark beetles (Coleoptera: Scolytidae). *Oecologia* **2001**, *128*, 443–453. [CrossRef]
46. Song, L.W.; Ren, B.Z.; Sun, S.H.; Zhang, X.J.; Zhang, K.P.; Gao, C.Q. Field trapping test on semiochemicals of pine shoot beetle *Tomicus piniperda* L. *J. Northeast For. Univ.* **2005**, *33*, 38–40.
47. Gallego, D.; Galián, J.; Diez, J.J.; Pajares, J.A. Kairomonal responses of *Tomicus destruens* (Col., Scolytidae) to host volatiles α-pinene and ethanol. *J. Appl. Entomol.* **2008**, *132*, 654–662. [CrossRef]
48. Farooq, A.; Tahara, S.; Choudhary, M.I.; Attaur, R.; Ahmed, Z.; Başer, K.H.C.; Demirci, F. Biotransformation of (−)-α-pinene by *Botrytis cinerea*. *Z. Für Nat. C* **2002**, *57*, 303–306. [CrossRef]
49. Divyashree, M.S.; George, J.; Agrawal, R. Biotransformation of terpenic substrates by resting cells of *Aspergillus niger* and *Pseudomonas* putida isolates. *J. Food Sci. Tech.* **2006**, *43*, 73–76.
50. Seybold, S.J.; Huber, D.P.W.; Lee, J.C.; Graves, A.D.; Bohlmann, J. Pine monoterpenes and pine bark beetles: A marriage of convenience for defense and chemical communication. *Phytochem. Rev.* **2006**, *5*, 143–178. [CrossRef]
51. Shi, Z.H.; Sun, J.H. Quantitative variation and biosynthesis of hindgut volatiles associated with the red turpentine beetle, *Dendroctonus valens* LeConte, at different attack phases. *Bull. Entomol. Res.* **2010**, *100*, 273–277. [CrossRef]

forests

MDPI

Article

The Use of qPCR Reveals a High Frequency of *Phytophthora quercina* in Two Spanish Holm Oak Areas

Beatriz Mora-Sala *, Mónica Berbegal and Paloma Abad-Campos

Instituto Agroforestal Mediterráneo, Universitat Politècnica de València, Camino de Vera s/n, 46022 Valencia, Spain; mobermar@etsia.upv.es (M.B.); pabadcam@eaf.upv.es (P.A.-C.)
* Correspondence: beamosa@upvnet.upv.es

Received: 17 October 2018; Accepted: 8 November 2018; Published: 10 November 2018

Abstract: The struggling Spanish holm oak woodland situation associated with *Phytophthora* root rot has been studied for a long time. *Phytophthora cinnamomi* is considered the main, but not the only species responsible for the decline scenario. This study verifies the presence and/or detection of *Phytophthora* species in two holm oak areas of Spain (southwestern "dehesas" and northeastern woodland) using different isolation and detection approaches. Direct isolation and baiting methods in declining and non-declining holm oak trees revealed *Phytophthora cambivora*, *Phytophthora cinnamomi*, *Phytophthora gonapodyides*, *Phytophthora megasperma*, and *Phytophthora pseudocryptogea* in the dehesas, while in the northeastern woodland, no *Phytophthora* spp. were recovered. Statistical analyses indicated that there was not a significant relationship between the *Phytophthora* spp. isolation frequency and the disease expression of the holm oak stands in the dehesas. *Phytophthora quercina* and *P. cinnamomi* TaqMan real-time PCR probes showed that both *P. cinnamomi* and *P. quercina* are involved in the holm oak decline in Spain, but *P. quercina* was detected in a higher frequency than *P. cinnamomi* in both studied areas. Thus, this study demonstrates that molecular approaches complement direct isolation techniques in natural and seminatural ecosystem surveys to determine the presence and distribution of *Phytophthora* spp. This is the first report of *P. pseudocryptogea* in Europe and its role in the holm oak decline should be further studied.

Keywords: *Quercus ilex* L.; *Phytophthora cinnamomi*; *Phytophthora quercina*; *Phytophthora pseudocryptogea*; qPCR

---

1. Introduction

Holm oak (*Quercus ilex* L.) grows spontaneously throughout the Mediterranean basin, from the Iberian Peninsula to Turkey in the North and from Morocco to Tunisia in the South, having its optimum growing conditions in the Western Mediterranean regions [1]. Holm oak is a low nutrient demanding species, which prefers dry soils situated in Spain from sea level up to 2000 m high, although the most dense holm oak forests' altitude ranges from 200 to 800 m. This species is well adapted to Mediterranean xeric conditions, with an early active taproot development and little branching at the expense of shoot development [2].

In Spain, holm oak is the most abundant evergreen *Fagaceae* tree species, covering almost all Spanish provinces except the Canary Islands and Galicia regions, where it is scarce [1]. About 2.8 M ha of the Spanish forestry surface are holm oak woodlands and 2.4 M ha are oak rangelands (henceforth called dehesas) (which consist mainly of holm oaks mixed with cork oaks (*Quercus suber* L.), and even a deciduous oak (*Quercus faginea* Lam.)) [3].

Holm oak constitutes a fundamental pillar of the Spanish dehesa, an agro–silvo–pastoral system, benefiting from the use of its fruit mainly for livestock during the autumn season and the grass growing underneath the canopy for grazing. Its wood it is also a valuable asset. In addition, it hosts migrant birds from Central and Northern Europe during the winter season. This complex system is suffering a significant decline due to biotic and abiotic factors [4,5]. The Spanish dehesas' decline associated with *Phytophthora* root rot has been studied since the end of the 20th century [6–11]. *Phytophthora cinnamomi* Rands. is considered the main pathogen responsible for the decline of this ecosystem [4,7–10,12,13], but it is not the only *Phytophthora* species infecting holm oaks [14–16].

On the other hand, Spanish natural oak woodlands are also undergoing this decline caused by *Phytophthora* spp. [17,18]. Several studies across Europe demonstrate the association of declining oak woodlands with *Phytophthora quercina* T. Jung, among other species, causing root infections [19–24]. In addition, abiotic factors, such as increasing temperatures and water stress, are being enhanced by climate changing conditions, which have a negative impact on the tree health status, weakening the stands and making holm oaks more susceptible to *Phytophthora* and *Pythium* infection [9,24–27]. Moreover, in view of the lack of regeneration of the stands, reforestations and afforestations are conducted with nursery material, with the consequent risk of introducing alien *Phytophthora* species to natural ecosystems [14,28–31].

Some *Phytophthora* species infect plants without causing external symptoms and this plant material is transported worldwide, allowing pathogens to be disseminated without generating any alert at the inspection points [30,32]. Denman et al. [33] reported that leaves from holm oak and rhododendron saplings remained asymptomatic when they were infected with *Phytophthora ramorum* Werres, De Cock, and Man in't Veld and *Phytophthora kernoviae* Brasier, Beales, and S.A. Kirk, two invasive species affecting ornamental and natural ecosystems. In 2006, imported ornamental *Grevillea* plants, which were asymptomatic, were found to be infected with *Phytophthora niederhauserii* Z.G. Abad and J.A. Abad [30]. Thus, visual screening for monitoring *Phytophthora* without complementary tests is not an appropriate management tool. The direct isolation of *Phytophthora* species on semiselective media from affected tissue or baiting techniques do not always generate quick and sensitive results, making it difficult to accurately monitor forest areas [5,20]. Economic and environmental losses caused by *Phytophthora* worldwide [31,34,35] require the use of all available techniques to detect and identify invasive species as quickly as possible. Combining direct isolation and baiting techniques with molecular tools, such as quantitative real-time PCR, increases the specificity, reproducibility, and sensitivity of the assessments, adding efficiency and accuracy to the diagnosis, an essential part of forest management strategies.

The aim of this study was to verify the presence of *Phytophthora* species in the holm oak rhizosphere in southwestern Spanish dehesas and in a northeastern Spanish holm oak woodland. In addition, the association between the *Phytophthora* species and the symptomatology of the holm oaks was studied in the dehesas by taking samples from declining and non-declining stands. For this purpose, different *Phytophthora* spp. isolation and detection approaches were performed: Direct isolation on semiselective media and apple and soil baiting using leaf material. Moreover, as *P. cinnamomi* and *P. quercina* are considered among the main pathogens associated with holm oak decline, their presence and relative abundance were studied in the samples using specific TaqMan real-time PCR probes.

2. Materials and Methods

2.1. Study Sites and Sampling

Studies were conducted in autumn 2012 and 2013 at 10 and 15 mature dehesas, respectively, located in the Extremadura region (southwestern Spain) (Table 1). This region has siliceous soils with *Pyro bourgaeanae-Querceto rotundifoliae sigmetum* vegetation series, and calcareous soils with *Paeonio coriaceae-Querceto rotundifoliae sigmetum* vegetation series, within an altitude ranging from 300 to 600 m [36]. At each site, two different areas were studied: A declining area where three symptomatic trees were randomly selected and a non-declining area with three randomly selected asymptomatic trees. Trees severely affected by aerial pathogens or insect pests were discarded. In the 2012 survey, one soil sample including fine roots from the rhizosphere around the base of each tree was collected (60 samples in total) by making three 20–30 cm deep holes at approximately 1 m distance from the trunk and bulked, obtaining a representative 0.5 kg sample, as described by Pérez-Sierra et al. [18] (Table 1). In the 2013 survey, sites 1 to 5 were sampled as described above, but in the remaining ten sites, two pooled samples per site were collected (Table 1). In sites 6 to 15 from 2013, one pooled sample from 3 declining trees and one pooled sample from 3 non-declining trees were collected at approximately 1 m distance from the trunk of each tree at 20–30 cm depth. Fifty samples in total were collected in 2013.

Table 1. Description of the survey conducted in 2012 and 2013 in the dehesas of the Extremadura region and in 2013 in the oak woodland of Montseny Biosphere Reserve.

2012 Dehesas			
Site	**Number of Samples**	***X* Coordinate**	***Y* Coordinate**
1	6	748324.99	4428259.51
2	6	248632.54	4460613.6
3	6	752500	4418487
4	6	694464.02	4431470.91
5	6	752500	4418487
6	6	750948.57	4437972.39
7	6	742685	4456109
8	6	753940.25	4450439.88
9	6	248428	4459568
10	6	749280	4457282
2013 Dehesas			
Site	**Number of Samples**	***X* Coordinate**	***Y* Coordinate**
1	6	748324.99	4428259.51
2	6	248632.54	4460613.6
3	6	752500	4418487
4	6	694464.02	4431470.91
5	6	761398,91	4425067.28
6	2	750948.57	4437972.39
7	2	742685	4456109
8	2	753940.25	4450439.88
9	2	248428	4459568
10	2	749580	4457274
11	2	279799	4430500
12	2	285614.18	4435261.32
13	2	281973	4432507
14	2	724766.4	4438845.56
15	2	246007.77	4396525.56
2013 Montseny Biosphere Reserve (Oak Woodland)			
Site	**Number of Samples**	***X* Coordinate**	***Y* Coordinate**
MS 2	1	450134	4625428
MS 6	1	458610	4621206
MS 12	1	457172	4620252
MS 13	1	457197	4620078
MS 14	1	455346	4619895
MS 16	1	454763	4621083
MS 18	1	455161	4621632
MS 22	1	455266	4618911
MS 23	1	454086	4619117
MS 24	1	453979	4619403
MS 25	1	452961	4620152
MS 26	1	452734	4619947
MS 27	1	451398	4622040
MS 28	1	450537	4622715
MS 29	1	449829	4622703

As differences in the *Phytophthora* spp. have been identified in eastern Spanish holm oak surveys [18], a study area located in northeastern Spain was included in the study. Montseny mountains is a 31,063 ha area located in Catalonia that since 1978 has been a biosphere reserve. It is made up of primarily siliceous rocks, with limestone rocks located on the western slopes of the mountains [37,38]. *Quercus ilex* is located in the lower altitudes among a *Quercetum-ilicis-galloprovinciale* vegetation series [38,39]. Fifteen holm oak declining stands showing defoliation, dead branches, and dieback symptoms and whose altitude ranged from 293 to 868 m were sampled as described above for 2012 just after a precipitation period in autumn 2013 (Table 1).

All the samples from the different surveys were transported to the laboratory, where roots were separated from soil for processing and soil was conserved at 5 °C until processing.

2.2. *Phytophthora spp. Isolation*

Roots from each sample were carefully washed under tap water and blotted on filter paper and direct isolation was performed on CMA-PARPB, as described by Jeffers and Martin [40], with and without the addition of hymexazol. Green apple baits were used for soil isolation. Granny Smith apples were surface disinfested with 95% ethanol. Four perpendicular 1 cm^2 holes were cut, filled with soil and remains of fine roots, and moistened with sterile water. These filled holes were sealed with tape and incubated in covered trays at 20 °C. The apples were examined daily until lesions developed. Small tissue fragments from the edge of the lesions were plated on CMA-PARPB with and without hymexazol and incubated at 20 °C in the dark. Oomycete-like colonies grown both from root and soil samples were transferred to potato dextrose agar (PDA) (Biokar-Diagnostics, Beauvais, France) and incubated at 20 °C in the dark for 7 days for further identification. Pure cultures of all putative *Phytophthora* isolates were obtained by transferring single hyphal tips to PDA plates.

Additionally, in the 2013 surveys, soils were also baited using leaflets of *Camellia* sp., *Rhododendron* sp., and *Viburnum* sp., following the methods described by Jung et al. [41,42]. Isolations were made using CMA-PARBPH as the selective agar medium [40] and processed as described above.

2.3. *Culture DNA Extraction, Sequencing, and Statistical Analyses*

DNA was extracted from pure cultures of putative *Phytophthora* grown on PDA by scraping the mycelium and grinding to a fine powder under liquid nitrogen, using the commercial kit EZNA Plant Miniprep Kit (Omega Bio-Tek, Doraville, GA, USA) following the manufacturer's instructions. Ribosomal DNA ITS amplifications were carried out using the universal primers ITS6 and ITS4 [43,44]. The PCR reaction final volume was 25 µL: PCR buffer 1×, 2.5 mM $MgCl_2$, 200 µM each dNTP, 0.4 µM of each primer, 1 U of DNA Taq polymerase (Dominion MBL, Córdoba, Spain), and 1 µL of template DNA. All PCR reactions were performed in a PTC 200 thermocycler (MJ Research Inc., Waltham, MA, USA) with the following parameters: 94 °C for 3 min; 35 cycles of 94 °C for 30 s, 55 °C for 30 s, and 72 °C for 45 s; and 72 °C for 10 min. Amplified products were sequenced at Macrogen Europe (Amsterdam, The Netherlands). The isolates were identified to the species level by conducting Basic Local Alignment Search Tool (BLAST) and comparing with the sequence data on international collection databases (*Phytophthora* Database, https://www.phytophthoradb.org and GenBank, https://www.ncbi.nlm.nih.gov/genbank/).

The total number of *Phytophthora* spp. isolates (*Phytophthora* pool) obtained and the number of isolates from each *Phytophthora* species were converted into frequencies relative to the total number of *Phytophthora* isolates recovered in the dehesas surveys. An analysis of variance (ANOVA) was performed with the 2012 and 2013 dehesas' data using a general linear model (GLM) in SAS version 9.0 (SAS Institute, Cary, NC, USA), in order to study the relationship between the frequency and diversity of *Phytophthora* spp. and the symptomatology shown by the trees in the dehesas. Mean values were compared using the Fischer's least significant difference (LSD) procedure at p-value = 0.05.

2.4. Environmental Samples: DNA Extraction and P. cinnamomi and P. quercina qPCRs

Roots and soil from both types of holm oak stands were tested with specific TaqMan probes for the main two oak *Phytophthora* pathogens, *P. cinnamomi* and *P. quercina* [16,45]. Each soil sample was passed through a 2 mm sieve to remove the organic matter and gravel. Once it was homogenized, 50–80 g per sample was lyophilized overnight and pulverized using FRITSCH Variable Speed Rotor Mill-PULVERISETTE 14 (ROSH, Oberstein, Germany). DNA was extracted in duplicate with the Power Soil DNA Isolation Kit (MO BIO Laboratories, Carlsbad, CA, USA) following the manufacturer's protocol. The root samples were first ground using a mortar and pestle under liquid nitrogen and then extraction was performed from 60 to 80 mg using the Power Plant Pro DNA Isolation Kit (MO BIO Laboratories, Carlsbad, CA, USA).

Real-time PCR was performed on a Rotor-Gene Q 5plex HRM (QIAGEN, Hilden, Germany) and data were analyzed with the Software Version 2.0.2. (QIAGEN) following the MIQE guidelines [46]. The primers quercina_F (GGTCTTGTCTGGCGTATGG), quercina_R (AGCTACTTGTTCAGACCGAAG), and the hydrolysis probe (6-FAM/GCTGTAAAA/ZEN/GGCGG CGGCTGTTGC/IaBlk-FQ/) designed by Català et al. [16] were used to detect *P. quercina* in DNA from all the soil and root samples collected in the study. In addition, *P. cinnamomi* was also tested with the primers P cin FF (CAATTAGTTGGGGGCCTGCT), P cin RF (GCAGCAGCAGCCGTCG), and the P cin hydrolysis probe (TTGACATCGACAGCCGCCGC) [45]. The qPCRs were performed in a total volume of 25 μL using Premix Ex TAQ (Probe qPCR; Takara Biotechnology (Dalian), Co., Ltd., China). Reactions consisted of 12.5 μL Premix Ex Taq (2×), 2.5 μL of primers–probe mix (500 nM of each primer and 250 nM probe), 1 μL of BSA (5 mg/mL) and 2 μL of template DNA. Two-step PCR was performed with the following cycling conditions: 95 °C for 1 min; 45 cycles of 95 °C for 5 s and 60 °C for 45 s for *P. quercina*, while for *P. cinnamomi*, 45 cycles of 95 °C for 5 s and 60 °C for 60 s. Two replicates were performed alongside standard dilution curves of *P. quercina* (isolate Ps-982 from Mediterranean Agroforestry Institute–UPV collection) and *P. cinnamomi* (isolate Ps-727 from Mediterranean Agroforestry Institute–UPV collection). Probe sensitivity was tested with serial dilution of each DNA ranging from 0.2 ng/μL to 2 fg/μL for *P. quercina* DNA; 2 ng/μL *P. cinnamomi* DNA (2 ng/μL) was serially diluted (1:10, 1:10^2, 1:10^3, 1:10^4, 1:10^5, 2:10^6). Negative samples were diluted and tested again to avoid false negatives.

3. Results

3.1. Phytophthora spp. Isolation

In the 2012 survey, *Phytophthora* spp. were detected in three dehesas, which represented 30% of the sampled sites. Three isolates of *Phytophthora* were recovered through the apple baiting method, one of each of: *Phytophthora cambivora* (Petri) Buisman (from a non-declining site), *P. cinnamomi*, and *Phytophthora gonapodyides* (H.E. Petersen) Buisman (from declining sites) (Table 2).

In 2013, the dehesas were surveyed and the soils baited in addition to the other methods already described for 2012. *Phytophthora* spp. were recovered in the 2013 dehesas survey from 21 holm oak samples, which represented 42% of the total samples, with 20% from declining samples and the remaining 22% from non-declining samples. A total of 165 Oomycetes isolates were obtained in 2013: 59 *Phytophthora* spp. isolates (clustered into four species) and 107 *Pythium* spp. isolates. In 2013, 39% of the *Phytophthora* spp. isolates were recovered from declining sites, and 61% were recovered from non-declining sites (Table 3). Regarding the isolation method, 13.4% of the *Phytophthora* spp. isolates were isolated directly from roots, 5.1% from apple baits, and 81.3% from leaf baits. As for the diversity of species obtained in 2013, *P. cinnamomi*, *P. gonapodyides*, *Phytophthora megasperma* Drechsler and *Phytophthora pseudocryptogea* Safaiefarahani, Mostowfizadeh, G.E. Hardy, and T.I. Burgess were isolated (Table 3). The range of abundance according to isolation was 39% *P. cinnamomi*, 35.6% *P. gonapodyides*, 20.3% *P. megasperma*, and 5.1% *P. pseudocryptogea*.

Table 2. Number of isolates of *Phytophthora* spp. obtained from *Quercus ilex* roots and soil in the 2012 survey in the dehesas of the Extremadura region according to the symptomatology of the sampled trees and results of the TaqMan real-time PCR assays. Results obtained from samples in each site are grouped according to whether samples were from declining or non-declining trees.

Site	Symptomatology	Isolates			qPCR			
					CIN		QUE	
		CAM	CIN	GON	Roots	Soil *	Roots	Soil *
1	d	0	0	1 [a]	ndt	0/3	ndt	3/3
1	nd	0	0	0	ndt	2/3	ndt	3/3
2	d	0	0	0	ndt	0/3	ndt	1/3
2	nd	0	0	0	ndt	0/3	ndt	1/3
3	d	0	0	0	ndt	0/3	ndt	2/3
3	nd	0	0	0	ndt	1/3	ndt	0/3
4	d	0	0	0	ndt	0/3	ndt	1/3
4	nd	0	0	0	ndt	0/3	ndt	3/3
5	d	0	0	0	ndt	0/3	ndt	2/3
5	nd	0	0	0	ndt	1/3	ndt	1/3
6	d	0	0	0	ndt	1/3	ndt	3/3
6	nd	0	0	0	ndt	0/3	ndt	2/3
7	d	0	0	0	ndt	0/3	ndt	2/3
7	nd	0	0	0	ndt	0/3	ndt	3/3
8	d	0	0	0	ndt	2/3	ndt	2/3
8	nd	0	0	0	ndt	0/3	ndt	3/3
9	d	0	0	0	ndt	2/3	ndt	0/3
9	nd	1 [a]	0	0	ndt	1/3	ndt	2/3
10	d	0	1 [a]	0	ndt	1/3	ndt	3/3
10	nd	0	0	0	ndt	0/3	ndt	3/3

d = declining; nd = non-declining; CAM = *Phytophthora cambivora*; CIN = *Phytophthora cinnamomi*; GON = *Phytophthora gonapodyides*; QUE = *Phytophthora quercina*; [a] = isolated from soil with apple baiting; ndt = not determined; * = number of positive samples detected out of the total number of samples.

Table 3. Number of isolates of *Phytophthora* spp. obtained from *Q. ilex* roots and soil in the 2013 survey in the dehesas of the Extremadura region according to whether samples were from declining or non-declining trees and results of the TaqMan real-time PCR assays. PCR results obtained from samples in sites 1 to 5 are grouped according to whether samples were from declining or non-declining trees.

Site	Symptomatology	Isolates				qPCR			
						CIN		QUE	
		CIN	GON	PSC	MEG	Roots *	Soil *	Roots *	Soil *
1	d	0	1 [b]	0	0	0/3	2/3	2/3	3/3
1	nd	4 [r,b]	8 [r,b]	0	4 [b]	2/3	2/3	3/3	3/3
2	d	0	0	0	0	0/3	1/3	2/3	1/3
2	nd	0	3 [b]	0	3 [b]	1/3	2/3	1/3	3/3
3	d	4 [b]	0	0	0	3/3	2/3	0/3	0/3
3	nd	0	0	0	0	2/3	1/3	1/3	1/3
4	d	0	0	0	2 [b]	0/3	2/3	3/3	3/3
4	nd	0	3 [b]	0	0	0/3	1/3	2/3	3/3
5	d	3 [b]	0	0	0	2/3	0/3	1/3	2/3
5	nd	2 [b]	6 [b]	0	0	1/3	1/3	3/3	3/3
6	d	0	0	0	0	−	−	+	−
6	nd	0	0	0	0	−	−	−	+
7	d	0	0	0	0	+	+	+	+
7	nd	0	0	0	0	−	−	+	+
8	d	0	0	0	0	+	−	−	+
8	nd	0	0	0	0	+	−	+	+

Table 3. *Cont.*

Site	Symptomatology	Isolates				qPCR			
						CIN		QUE	
		CIN	GON	PSC	MEG	Roots *	Soil *	Roots *	Soil *
9	d	0	0	0	0	−	−	−	+
9	nd	0	0	0	1 [b]	+	+	+	−
10	d	3 [r,b]	0	0	0	+	+	+	−
10	nd	0	0	0	0	−	+	−	−
11	d	1 [b]	0	0	0	−	+	+	−
11	nd	0	0	0	0	−	−	+	+
12	d	0	0	0	0	+	−	−	+
12	nd	0	0	0	0	+	−	−	+
13	d	0	0	0	0	+	+	−	+
13	nd	0	0	0	0	−	−	+	−
14	d	0	0	0	1 [b]	−	−	−	−
14	nd	0	0	0	0	−	−	+	+
15	d	5 [a,b]	0	2 [b]	0	+	+	+	+
15	nd	1 [r]	0	1 [a]	0	+	−	−	+

d = declining; nd = non-declining; CIN = *P. cinnamomi*; GON = *P. gonapodyides*; MEG = *Phytophthora megasperma*; PSC = *Phytophthora pseudocryptogea*; QUE = *P. quercina*; [r] = isolated from roots; [a] = isolated from soil with apple baiting; [b] = isolated from baiting soil with leaves; * = number of positive detected samples out of the total number of samples; + = positive; − = negative.

Twenty-three isolates of *P. cinnamomi* were isolated from eight samples in the dehesas in 2013 (Table 3); percentages from declining and non-declining samples are shown in Figure 1. Twenty-one cultures of *P. gonapodyides* were isolated from five samples, with most of the samples from non-declining sites (Table 3, Figure 1). Twelve *P. megasperma* isolates were isolated from five samples (Table 3), and most of these samples were from non-declining trees (Figure 1). Three *P. pseudocryptogea* cultures were isolated from two samples (Table 3, Figure 1).

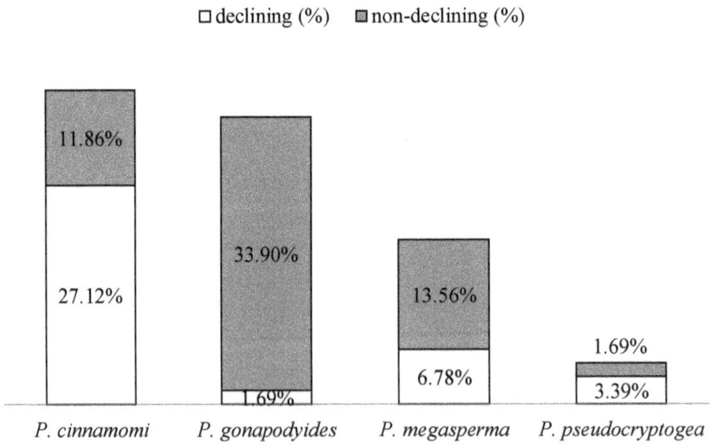

Figure 1. Percentage of each *Phytophthora* species cultures isolated in the dehesas 2013 survey according to whether the holm oaks were declining or non-declining.

The statistical analysis showed that the factors' symptomatology (p-value = 0.3626) and dehesa (p-value = 0.3087) were not significant for the frequency of the different *Phytophthora* species present in the dehesas in 2013. Considering the different species isolated separately, only the presence of *P. gonapodyides* was significantly higher in non-declining samples (p-value = 0.0366). The presence of either one species or another was not significantly associated with the dehesa factor (*P. cinnamomi p*-value = 0.2277, *P. gonapodyides p*-value = 0.9176 and *P. megasperma p*-value = 0.7029). *P. pseudocryptogea* was only isolated in one dehesa, but from both declining and non-declining sites.

No *Phytophthora* isolates were recovered from the 15 samples of the Montseny Biosphere Reserve by direct isolation on semiselective media from affected tissues and/or the baiting.

3.2. Environmental Samples: Hydrolysis Probes—P. cinnamomi and P. quercina qPCRs

The *Phytophthora quercina* standard curve plot showed that the correlation between the Cq-value and the DNA concentration was high (r^2 = 0.99966), with an efficiency of 0.90389. For *P. quercina*, the limit of detection (LOD) was established at a DNA concentration of 2 fg/µL.

Phytophthora quercina was detected in all the surveyed dehesas in 2012 (65.1% of the samples). Of these, 31.8% came from declining holm oak trees and 33.3% from non-declining trees (Table 2). In 2013, *P. quercina* was detected in all the surveyed dehesas (79.6% of the samples) (Table 3). A total of 66.7% of the soil samples were positive for *P. quercina*: 27.8% were from declining soil samples, while 38.9% were from non-declining soil samples. A total of 55.6% of the root samples were positive for *P. quercina*: 24.1% were from declining holm oak fine roots, while 31.5% were from non-declining holm oak roots. In the survey conducted in Montseny Biosphere Reserve in 2013, 66.7% of the samples were positive for *P. quercina*, of which 40% were from root samples and 53.3% from soil samples (Table 4).

Table 4. Results of the TaqMan real-time PCR assays obtained from *Q. ilex* roots and soil in the 2013 survey in the oak woodland of Montseny Biosphere Reserve.

Site	qPCR			
	CIN		QUE	
	Roots	Soil	Roots	Soil
MS 2	−	+	+	+
MS 6	+	+	−	−
MS 12	+	+	−	−
MS 13	−	−	+	+
MS 14	−	−	−	+
MS 16	−	−	−	+
MS 18	−	−	−	−
MS 22	−	−	+	+
MS 23	+	−	+	+
MS 24	+	+	+	−
MS 25	−	−	−	−
MS 26	+	+	+	−
MS 27	−	−	−	+
MS 28	+	−	−	−
MS 29	+	−	−	+

CIN = *P. cinnamomi*. QUE = *P. quercina*. + = positive detection; − = negative detection.

The *Phytophthora cinnamomi* standard curve revealed a high correlation between the Cq-value and the DNA concentration (r^2 = 0.99731), with a reaction efficiency of 0.92014. The LOD was established at 4 fg/µL.

Phytophthora cinnamomi was detected in seven out of the ten surveyed dehesas in 2012 (19.7% of the samples) (Table 2). A total of 12.1% of the soil detections came from declining holm oak trees and 7.6% from non-declining. In 2013, *P. cinnamomi* was detected in 11 dehesas (57.4% of the samples) (Table 3). A total of 38.9% of the soil samples were positive for *P. cinnamomi*; 22.2% were from declining trees, while 16.7% were from non-declining trees. A total of 33.9% of the root samples were positive for *P. cinnamomi*, with 22.2% from declining oak fine roots and 14.8% from non-declining oak roots. In Montseny Biosphere Reserve, 53.3% of the samples were positive for *P. cinnamomi*, of which 46.7% were from root samples and 33.3% from soil samples (Table 4).

4. Discussion

This study provides evidence that molecular approaches complement direct isolation methods of *Phytophthora* species from fine roots from holm oak in natural (Montseny Biosphere Reserve) and seminatural (dehesas) ecosystems, confirming that it is not only *P. cinnamomi* that is involved in the holm oak decline in Spain, but *P. quercina* is also present. Moreover, this is the first report of *P. pseudocryptogea* in Europe.

Regarding traditional isolation methods, an increase in *Phytophthora* isolation was observed in the dehesas in 2013 compared with the sampling conducted in the dehesas in 2012. This was probably not only due to the implementation of the leaf baiting technique, but also because in 2013, isolation of *Phytophthora* spp. from fine roots was more successful. This could be explained by the fact that under favorable environmental conditions, *Phytophthora* spp. infected the tree root systems and rotted fine roots containing the pathogens detached from the plant, so the pathogens can establish again in the soil [47]. *Phytophthora* spp. dispersion requires warm temperatures and free water to produce infective zoospores; if not, they remain as resistant structures in the soil [34]. Furthermore, the efficiency of *Phytophthora* isolation techniques can be compromised by the climatic conditions suffered during the period previous to the survey and by the presence of other microorganisms [8,34]. In fact, the dehesa regions where the surveys were conducted received less precipitation in 2012 than in 2013 [48]. Thus, according to this, the environmental conditions for *Phytophthora* spp. isolation were more favorable in 2013 than in 2012 in the southwestern Spanish dehesas, as they were recovered in 2013 from fresh lesions [49]. Another possible explanation for the low efficiency of *Phytophthora* recovery in Montseny Biosphere Reserve is the presence of other fast-growing species in the samples, such as *Pythium* spp., making the isolation difficult. *Pythium* spp. were recorded in very low numbers in dehesas in 2012 (explained by the absence of favorable environmental conditions), but their presence was very relevant in the 2013 dehesas and in the Montseny Biosphere Reserve surveys. The genus *Pythium* is present in almost all soils and, as the isolation medium used for *Phytophthora* isolation is semiselective [34,40], *Pythium* spp. were also isolated with a high frequency in our study and were able to mask *Phytophthora* spp. presence.

Oak decline, associated with abiotic and biotic factors, has been occurring across Europe during the past decades [4,11,22,28,42,50]. Among the several *Phytophthora* species that have been associated with this decline, *P. cinnamomi* has been considered the main biotic factor responsible for oak mortality in Spain since the 1990s [8,9,51]. In 2013, *P. cinnamomi* was the most frequent species isolated in the infested dehesas, as was expected according to previous studies [7,52]. In addition to *P. cinnamomi*, Corcobado et al. [15] reported that *P. gonapodyides* was also involved in oak decline in this region. In our surveys, *P. gonapodyides*, *P. megasperma*, and *P. pseudocryptogea* were recovered at low frequencies, and these species may play an important role as causal agents of the disease, as reported in other studies, where they were also recovered at low frequencies [14,25,28,52]. There is no statistical evidence to support a differential distribution of *Phytophthora* species among the dehesas in 2013. Moreover, statistical results indicated that there was not a significant relationship between the *Phytophthora* spp. isolation frequency and the symptomatology of the holm oak stands. Our results showed a higher percentage of *Phytophthora* spp. recovery in 2013 from non-declining sites than from declining sites. *Phytophthora cinnamomi*, which was found in six dehesas, either from declining or from non-declining trees, is a primary root pathogen of woody species, considered a hemibiotrophic organism with life strategies which can change from biotrophic to necrotrophic, according to the environmental conditions [53–55]. This species is also present in plant reservoirs and, depending on its behavior, will determine if the plant remains asymptomatic or not [54–57]. Furthermore, *P. cinnamomi* is highly aggressive to holm oaks, as demonstrated previously [11,12,25,58–60]. Tsao [61] stated that a certain percentage of lost roots is required for symptoms to emerge and our results in the 2013 survey provide evidence that a tree symptomatology is not always an indication about the conditions of its root system. Statistical analyses showed that *P. gonapodyides* is more frequent in non-declining stands, in agreement with the results of Vettraino et al. [28], while the other species found did not show any statistical pattern. *Phytophthora gonapodyides* is known to attack the small or fine feeder roots [62] and to produce a wilting toxin [41]. Nevertheless, Brasier et al. [62] stated that *P. gonapodyides* is often in balance with the unstressed oak root system, but this can change under stress conditions, contributing to a rapid decline. Hansen et al. [63] suggested that some *Phytophthora* spp. from clade 6 ecologically related to *Phytophthora chlamydospora* could cause limited root damage with no above ground disease symptoms contributing to the oak decline. *Phytophthora megasperma* isolated in the present study had been previously associated with oak decline [28,42]. Although it is considered a pathogen of herbaceous plants and agricultural trees, it can become a serious problem when the oak balance is broken due to other factors, such as droughts or waterlogging [34]. A similar behavior has been indicated for other *Phytophthora* spp. such as *P. gonapodyides* [10]. *Phytophthora megasperma, P. quercina, P. psychrophila, Phytophthora drechsleri*, and *Phytophthora syringae* have also been associated with oak decline [18,19,25], although these species were not found in our samples. Nevertheless, *P. pseudocryptogea* in the present study was isolated for the first time in Europe and from a holm oak-rangeland in Spain. This species was described by Safaiefarahani et al., who re-evaluated the *P. cryptogea* complex [64]. Although it was isolated from three soil samples in the present study, from three soil samples, the role of *P. pseudocryptogea* in holm oak decline remains unknown. The pathogenicity of this species in holm oak should be further studied.

The results obtained with the *P. quercina* probe are relevant, since it has always been thought that the holm oak decline in acidic soils in Spain is caused primarily by *P. cinnamomi. Phytophthora cinnamomi* was present in a high number of samples in both study locations, as was expected, but surprisingly it was not the most frequent species detected. *Phytophthora quercina* was shown as the most frequent species in this study, and the number of positive samples was higher in both studied areas. Molecular diagnoses provide faster and more sensitive detection of *Phytophthora* spp. [16,45,65–72].

5. Conclusions

Different *Phytophthora* species were detected and identified in the study areas, regardless of whether they cause symptoms of decline or not. Further research is needed to clarify the effect of these pathogens in combination and abiotic factors in the oak stands. The implementation of the different direct and baiting isolation techniques for the isolation of *Phytophthora* spp., along with the available molecular detection techniques, allows a better diagnosis and understanding of the role of *Phytophthora* spp. in the holm oak forest areas.

Author Contributions: B.M.-S. conducted field sampling, performed experimental work and data analysis in lab experiments, discussed the results, and wrote the paper. M.B. participated in the data analyses, discussed the results, and revised the manuscript. P.A.-C. designed the study and revised the manuscript.

Funding: This research was supported by funding from the project AGL2011-30438-C02-01 (Ministerio de Economía y Competitividad, Spain).

Acknowledgments: We would like to thank M. León from the Instituto Agroforestal Mediterráneo–UPV (Spain) for technical assistance. The authors are grateful to A. Solla and his team from the Centro Universitario de Plasencia–Universidad de Extremadura (Spain) for helping in the southwestern sample collection.

Conflicts of Interest: The authors declare no conflict of interest.

References

1. Romane, F.; Terradas, J. *Quercus ilex L. Ecosystems: Function, Dynamics and Management*; Kluwer Academic Publishers: Dordrecht, The Netherlands, 1992; p. 377, ISBN 0-79231-764-5.
2. Moro, R. *Guía de los árboles de España*; Omega: Barcelona, Spain, 2007; p. 407, ISBN 8-42821-043-8.
3. MAGRAMA. Diagnóstico del Sector Forestal Español. Análisis y Prospectiva. Serie Agrinfo/Medioambiente nº 8. Ed. Ministerio de Agricultura, Alimentación y Medio Ambiente. Madrid, Spain, 2014. Available online: https://www.mapa.gob.es/ministerio/pags/Biblioteca/Revistas/pdf_AAYPP%2FAPMA_2014_8.pdf (accessed on 30 August 2016).

4. Camilo-Alves, C.S.P.; da Clara, M.I.E.; de Almeida Ribeiro, N.M.C. Decline of Mediterranean oak trees and its association with *Phytophthora cinnamomi*: A review. *Eur. J. For. Res.* **2013**, *132*, 411–432. [CrossRef]
5. Sena, K.; Crocker, E.; Vincelli, P.; Barton, C. *Phytophthora cinnamomi* as a driver of forest change: Implications for conservation and management. *For. Ecol. Manag.* **2018**, *409*, 799–807. [CrossRef]
6. Cobos, J.M.; Montoya, R.; Tuset, J.J. New damages to the oak woodlands in Spain—Preliminary evaluation of the possible implication of *Phytophthora cinnamomi*. In *Recent Advances in Studies on Oak Decline, Proceedings of the International Congress Recent Advances in Studies on Oak Decline, Selva di Fasano, Brindisi, Italy, 13–18 Septemper 1992*; Luisi, N., Lerario, P., Vannini, A., Eds.; Dipartamento di Patología Vegetóle, Universitá degli Studi: Bari, Italy, 1992; pp. 163–169.
7. Tuset, J.J.; Hinarejos, C.; Mira, J.L.; Cobos, M. Implicación de *Phytophthora cinnamomi* Rands en la enfermedad de la seca de encinas y alcornoques. *Bol. Sanid. Veg. Plagas* **1996**, *22*, 491–499.
8. Brasier, C.M. Oak tree mortality in Iberia. *Nature* **1992**, *360*, 539. [CrossRef]
9. Brasier, C.M. *Phytophthora cinnamomi* and oak decline in southern Europe—Environmental constraints including climate change. *Ann. Sci. For.* **1996**, *53*, 347–358. [CrossRef]
10. Brasier, C.M.; Robredo, F.; Ferraz, J.F.P. Evidence for *Phytophthora cinnamomi* involvement in Iberian oak decline. *Plant Pathol.* **1993**, *42*, 140–145. [CrossRef]
11. Gallego, F.J.; Perez de Algaba, A.; Fernandez-Escobar, R. Etiology of oak decline in Spain. *Eur. J. For. Pathol.* **1999**, *29*, 17–27. [CrossRef]
12. Sánchez, M.E.; Caetano, P.; Ferraz, J.; Trapero, A. *Phytophtora* disease of *Quercus ilex* in south-western Spain. *For. Pathol.* **2002**, *32*, 5–18. [CrossRef]
13. Sánchez, M.E.; Sánchez, J.E.; Navarro, R.M.; Fernández, P.; Trapero, A. Incidencia de la podredumbre radical causada por *Phytophthora cinnamomi* en masas de *Quercus* en Andalucía. *Bol. Sanid. Veg. Plagas* **2003**, *29*, 87–108.
14. Sanchez, M.E.; Andicoberry, S.; Trapero, A. Pathogenicity of three *Phytophthora* spp. causing late seedling rot of *Quercus ilex* ssp. *ballota*. *For. Pathol.* **2005**, *35*, 115–125. [CrossRef]
15. Corcobado, T.; Cubera, E.; Pérez-Sierra, A.; Jung, T.; Solla, A. First report of *Phytophthora gonapodyides* involved in the decline of *Quercus ilex* in xeric conditions in Spain. *New Dis. Rep.* **2010**, *22*, 33. [CrossRef]
16. Català, S.; Berbegal, M.; Pérez-Sierra, A.; Abad-Campos, P. Metabarcoding and development of new real-time specific assays reveal *Phytophthora* species diversity in Holm Oak forests in eastern Spain. *Plant Pathol.* **2017**, *66*, 115–123. [CrossRef]
17. Luque, J.; Parladé, J.; Pera, J. Pathogenicity of fungi isolated from *Quercus suber* in Catalonia (NE Spain). *For. Pathol.* **2000**, *30*, 247–263. [CrossRef]
18. Pérez-Sierra, A.; López-García, C.; León, M.; García-Jiménez, J.; Abad-Campos, P.; Jung, T. Previously unrecorded low-temperature *Phytophthora* species associated with *Quercus* decline in a Mediterranean forest in eastern Spain. *For. Pathol.* **2013**, *43*, 331–339. [CrossRef]
19. Jung, T.; Cooke, D.E.L.; Blaschke, H.; Duncan, J.M.; Oßwald, W. *Phytophthora quercina* sp. nov., causing root rot of European oaks. *Mycol. Res.* **1999**, *103*, 785–798. [CrossRef]
20. Nechwatal, J.; Schlenzig, A.; Jung, T.; Cooke, D.E.L.; Duncan, J.M.; Oßwald, W.F. A combination of baiting and PCR techniques for the detection of *Phytophthora quercina* and *P. citricola* in soil samples from oak stands. *For. Pathol.* **2001**, *31*, 85–97. [CrossRef]
21. Balci, Y.; Halmschlager, E. *Phytophthora* species in oak ecosystems in Turkey and their association with declining oak trees. *Plant Pathol.* **2003**, *52*, 694–702. [CrossRef]
22. Balci, Y.; Balci, S.; MacDonald, W.L.; Gottschalk, K.W. Relative susceptibility of oaks to seven species of *Phytophthora* isolated from oak forest soils. *For. Pathol.* **2008**, *38*, 394–409. [CrossRef]
23. Jönsson, U.; Jung, T.; Rosengren, U.; Nihlgard, B.; Sonesson, K. Pathogenicity of Swedish isolates of *Phytophthora quercina* to *Quercus robur* in two different soils. *New Phytol.* **2003**, *158*, 355–364. [CrossRef]
24. Martín-García, J.; Solla, A.; Corcobado, T.; Siasou, E.; Woodward, S. Influence of temperature on germination of *Quercus ilex* in *Phytophthora cinnamomi*, *P. gonapodyides*, *P. quercina* and *P. psychrophila* infested soils. *For. Pathol.* **2015**, *45*, 215–223. [CrossRef]
25. Sánchez, M.E.; Caetano, P.; Romero, M.A.; Navarro, R.M.; Trapero, A. *Phytophthora* root rot as the main factor of oak decline in southern Spain. In *Progress in Research on Phytophthora Diseases of Forest Trees, Proceedings of the Third International IUFRO Working Party S07.02.09, Freising, Germany, 11–18 September 2004*; Brasier, C.M., Jung, T., Oßwald, W., Eds.; Forest Research: Farnham, UK, 2006; pp. 149–154, ISBN 0-85538-721-1.

26. Romero, M.A.; Sánchez, J.E.; Jiménez, J.J.; Belbahri, L.; Trapero, A.; Lefort, F.; Sánchez, M.E. New *Pythium* taxa causing root rot in Mediterranean *Quercus* species in southwest Spain and Portugal. *J. Phytopathol.* **2007**, *115*, 289–295. [CrossRef]
27. Jiménez, A.J.; Sánchez, E.J.; Romero, M.A.; Belbahri, L.; Trapero, A.; Lefort, F.; Sánchez, M.E. Pathogenicity of *Pythium spiculum* and *P. sterilum* on feeder roots of *Quercus rotundifolia*. *Plant Pathol.* **2008**, *57*, 369. [CrossRef]
28. Vettraino, A.M.; Barzanti, G.P.; Bianco, M.C.; Ragazzi, A.; Capretti, P.; Paoletti, E.; Vannini, A. Occurrence of *Phytophthora* species in oak stands in Italy and their association with declining oak trees. *For. Pathol.* **2002**, *32*, 19–28. [CrossRef]
29. Rizzo, D.M.; Garbelotto, M.; Hansen, E.M. *Phytophthora ramorum*: Integrative research and management of an emerging pathogen in California and Oregon forests. *Ann. Rev. Phytopathol.* **2005**, *43*, 309–335. [CrossRef] [PubMed]
30. Brasier, C.M. The biosecurity threat to the UK and global environment from international trade in plants. *Plant Pathol.* **2008**, *57*, 792–808. [CrossRef]
31. Jung, T.; Orlikowski, L.; Henricot, B.; Abad-Campos, P.; Aday, A.G.; Aguín Casal, O.; Bakonyi, J.; Cacciola, S.O.; Cech, T.; Chavarriaga, D.; et al. Widespread *Phytophthora* infestations in European nurseries put forest, semi-natural and horticultural ecosystems at high risk of *Phytophthora* diseases. *For. Pathol.* **2016**, *46*, 134–163. [CrossRef]
32. Hullbert, J.M.; Agne, M.C.; Burgess, T.I.; Roets, F.; Wingfield, M.J. Urban environments provide opportunities for early detections of *Phytophthora* invasions. *Biol. Invasions* **2017**, *19*, 3629–3644. [CrossRef]
33. Denman, S.; Kirk, S.A.; Moralejo, E.; Webber, J.F. Phytophthora ramorum and Phytophthora kernoviae on naturally infected asymptomatic foliage. *Bull. OEPP* **2009**, *39*, 105–111. [CrossRef]
34. Erwin, D.C.; Ribeiro, O.K. *Phytophthora Diseases Worldwide*; APS Press: St. Paul, MN, USA, 1996; p. 562, ISBN 0-89054-212-0.
35. Hernández-Lambraño, R.E.; González-Moreno, P.; Sánchez-Agudo, J.A. Environmental factors associated with the spatial distribution of invasive plant pathogens in the Iberian Peninsula: The case of *Phytophthora cinnamomi* Rands. *For. Ecol. Manag.* **2018**, *419–420*, 101–109. [CrossRef]
36. Jiménez, M.P.; Díaz-Fernández, P.M.; Iglesias, S.; De Tuero, M.; Gil, L. *Regiones de procedencia Quercus ilex L.*; Instituto Nacional para la Conservación de la Naturaleza: Madrid, Spain, 1996; p. 100, ISBN 84-8014-143-3.
37. Boada, M.; Puig, J.; Barriocanal, C. The effects of isolation and Natural Park coverage for landrace in situ conservation: And approach from the Montseny mountains (NE Spain). *Sustainability* **2013**, *5*, 654–663. [CrossRef]
38. García, C. Estudio faunístico y ecológico de la familia Phoridae en el P. N. del Montseny. Ph.D Thesis, Universidad de Barcelona, Barcelona, Spain, 2013.
39. Peñuelas, J.; Boada, M. A global change-induced biome shift in the Montseny mountains (NE Spain). *Glob. Chang. Biol.* **2003**, *9*, 131–140. [CrossRef]
40. Jeffers, S.N.; Martin, S.B. Comparison of two media selective for *Phytophthora* and *Pythium* species. *Plant Dis.* **1986**, *70*, 1038–1043. [CrossRef]
41. Jung, T.; Blaschke, H.; Neumann, P. Isolation, identification and pathogenicity of *Phytophthora* species from declining oak stands. *Eur. J. For. Pathol.* **1996**, *26*, 253–272. [CrossRef]
42. Jung, T.; Blaschke, H.; Oßwald, W. Involvement of soilborne *Phytophthora* species in Central European oak decline and the effect of site factors on the disease. *Plant Pathol.* **2000**, *49*, 706–718. [CrossRef]
43. White, T.J.; Bruns, S.; Lee, S.; Taylor, J. Amplification and direct sequencing of fungal ribosomal RNA genes for phylogenetics. In *PCR Protocols: A Guide to Methods and Applications*; Academic Press: San Diego, CA, USA, 1990; pp. 315–322.
44. Cooke, D.E.L.; Drenth, A.; Duncan, J.M.; Wagels, G.; Brasier, C.M. A Molecular Phylogeny of *Phytophthora* and Related Oomycetes. *Fungal Genet. Biol.* **2000**, *30*, 17–32. [CrossRef] [PubMed]
45. Kunadiya, M.; White, D.; Dunstan, W.A.; Hardy, G.E.St.J.; Andjic, V.; Burgess, T.I. Pathways to false-positive diagnoses using molecular detection methods; *Phytophthora cinnamomi* a case study. *FEMS Microbiol. Lett.* **2017**, *364*, fnx009. [CrossRef] [PubMed]
46. Bustin, S.A.; Benes, V.; Garson, J.A.; Hellemans, J.; Huggett, J.; Kubista, M.; Mueller, R.; Nolan, T.; Pfaffl, M.W.; Shipley, G.L.; et al. The MIQE guidelines: Minimum information for publication of quantitative real-time PCR experiments. *Clin. Chem.* **2009**, *55*, 611–622. [CrossRef] [PubMed]

47. Shearer, B.L.; Tippett, J.T. *Jarrah Dieback: The Dynamics and Management of Phytophthora cinnamomi in the Jarrah (Eucalyptus marginata) Forests of South-Western Australia*; Lewis, M., Ed.; Department of Conservation and Land Management: Como, Australia, 1989; ISSN 1032-8106.
48. AEMET. Available online: http://www.aemet.es/es/ (accessed on 20 May 2016).
49. Shearer, B.L.; Shea, S.R. Variation in seasonal population fluctuations of *Phytophthora cinnamomi* within and between infected *Eucalyptus marginata* sites of southwestern Australia. *For. Ecol. Manag.* **1987**, *21*, 209–230. [CrossRef]
50. Balci, Y.; Halmschlager, E. Incidence of *Phytophthora* species in oak forests in Austria and their possible involvement in oak decline. *For. Pathol.* **2003**, *33*, 157–174. [CrossRef]
51. Brasier, C.M.; Hamm, P.B.; Hansen, E.M. Cultural characters, protein patterns and unusual mating behaviour of *P. gonapodyides* isolates from Britain and North America. *Mycol. Res.* **1993**, *97*, 1287–1298. [CrossRef]
52. Corcobado, T.; Cubera, E.; Moreno, G.; Solla, A. *Quercus ilex* forests are influenced by annual variations in water table, soil water deficit and fine root loss caused by *Phytophthora cinnamomi*. *Agric. For. Meteorol.* **2013**, *169*, 92–99. [CrossRef]
53. Hardham, A.R.; Blackman, L.M. Molecular cytology of *Phytophthora*–plant interactions. *Australas. Plant Path.* **2010**, *39*, 29. [CrossRef]
54. Crone, M.; McComb, J.; O'Brien, P.A.; Hardy, G.E. Survival of *Phytophthora cinnamomi* as oospores, stromata, and thick-walled chlamydospores in roots of symptomatic and asymptomatic annual and herbaceous perennial plant species. *Fungal Biol.* **2013**, *117*, 112–123. [CrossRef] [PubMed]
55. Crone, M.; McComb, J.; O'Brien, P.A.; Hardy, G.E. Assessment of Australian native annual/herbaceous perennial plant species as asymptomatic or symptomatic hosts of *Phytophthora cinnamomi* under controlled conditions. *For. Pathol.* **2013**, *43*, 245–251. [CrossRef]
56. Gómez-Aparicio, L.; Ibáñez, B.; Serrano, M.S.; De Vita, P.; Ávila, J.M.; Pérez-Ramos, I.M.; García, L.V.; Sánchez, M.E.; Marañón, T. Spatial patterns of soil pathogens in declining Mediterranean forests: Implications for trees regeneration. *New Phytol.* **2012**, *194*, 1014–1024. [CrossRef] [PubMed]
57. Serrano, M.S.; de Vita, P.; Fernández-Rebollo, P.; Sánchez, M.E. Calcium fertilizers induce soil suppressiveness to *Phytophthora cinnamomi* root rot of *Quercus ilex*. *Eur. J. Plant Pathol.* **2012**, *132*, 271–279. [CrossRef]
58. Robin, C.; Desprez-Loustau, M.L. Testing variability in pathogenicity of *Phytophthora cinnamomi*. *Eur. J. Plant Pathol.* **1998**, *104*, 465–475. [CrossRef]
59. Robin, C.; Desprez-Loustau, M.L.; Capron, G.; Delatour, C. First record of *Phytophthora cinnamomi* on cork and holm oaks in France and evidence of pathogenicity. *Ann. Sci. For.* **1998**, *55*, 869–883. [CrossRef]
60. Robin, C.; Capron, G.; Desprez-Loustau, M.L. Root infection by *Phytophthora cinnamomi* in seedlings of three oak species. *Plant Pathol.* **2001**, *50*, 708–716. [CrossRef]
61. Tsao, P.H. Why many *Phytophthora* root rots and crown rots of tree and horticultural crops remain undetected. *Bull. OEPP* **1990**, *20*, 11–17. [CrossRef]
62. Brasier, C.M.; Cooke, D.E.L.; Duncan, J.M.; Hansen, E.M. Multiple new phenotypic taxa from trees and riparian ecosystems in *Phytophthora gonapodyides-P. megasperma* ITS Clade 6, which tend to be high-temperature tolerant and either inbreeding or sterile. *Mycol. Res.* **2003**, *107*, 277–290. [CrossRef] [PubMed]
63. Hansen, E.M.; Reeser, P.; Sutton, W.; Brasier, C.M. Redesignation of *Phytophthora* taxon *Pgchlamydo* as *Phytophthora chlamydospora* sp. nov. *N. Am. Fungi* **2015**, *10*, 1–14.
64. Safaiefarahani, B.; Mostowfizadeh-Ghalamfarsa, R.; Hardy, G.E.; Burgess, T.I. Re-evaluation of *Phytophthora cryptogea* species complex and the description of a new species, *Phytophthtora pseudocryptogea* sp. nov. *Mycol. Prog.* **2015**, *14*, 108. [CrossRef]
65. Ippolito, A.; Schena, L.; Nigro, F.; Soleti, V.L.; Yaseen, T. Real-time detection of *Phytophthora nicotianae* and *P. citrophthora* in citrus roots and soil. *Eur. J. Plant Pathol.* **2004**, *110*, 833–843. [CrossRef]
66. Bonants, P.J.; van Gent-Pelzer, M.P.; Hooftman, R.; Cooke, D.E.L.; Guy, D.C.; Duncan, J.M. A combination of baiting and different PCR formats, including measurement of real-time quantitative fluorescence, for the detection of *Phytophthora fragariae* in strawberry plants. *Eur. J. Plant Pathol.* **2004**, *110*, 689–702. [CrossRef]
67. Tooley, P.W.; Martin, F.N.M.; Carras, M.M.; Frederick, R.D. Real-time fluorescent Polymerase Chain Reaction detection of *Phytophthora ramorum* and *Phytophthora pseudosyringae* using mitochondrial gene regions. *Am. Phytopathol. Soc.* **2006**, *96*, 336–345. [CrossRef] [PubMed]

68. Schena, L.; Hughes, K.J.D.; Cooke, D.E.L. Detection and quantification of *Phytophthora ramorum*, *P. kernoviae*, *P. citricola* and *P. quercina* in symptomatic leaves by multiplex real-time PCR. *Mol. Plant Pathol.* **2006**, *7*, 365–379. [CrossRef] [PubMed]
69. Schena, L.; Duncan, J.M.; Cooke, D.E.L. Development and application of a PCR-based 'molecular tool box' for the identification of *Phytophthora* species damaging forests and natural ecosystems. *Plant Pathol.* **2008**, *57*, 64–75. [CrossRef]
70. Hughes, K.J.D.; Tomlinson, J.A.; Giltrap, P.M.; Barton, V.; Hobden, E.; Boonham, N.; Lane, C.R. Development of a real-time PCR assay for detection of *Phytophthora kernoviae* and comparison of this method with a conventional culturing technique. *Eur. J. Plant Pathol.* **2011**, *131*, 695–703. [CrossRef]
71. Than, D.J.; Hughes, K.J.D.; Boonhan, N.; Tomlinson, J.A.; Woodhall, J.W.; Bellgard, S.E. A TaqMan real-time PCR assay for the detection of *Phytophthora* 'taxon Agathis' in soil, pathogen of Kauri in New Zealand. *For. Pathol.* **2013**, *43*, 324–330. [CrossRef]
72. Touseef, H.; Bir, P.S.; Firoz, A. A quantitative real-time PCR based method for the detection of *Phytophthora infestans* causing Late blight of potato, in infested soil. *Saud. J. Biol. Sci.* **2014**, *21*, 380–386.

forests

MDPI

Article

Calcium and Potassium Imbalance Favours Leaf Blight and Defoliation Caused by *Calonectria pteridis* in Eucalyptus Plants

Thaissa P. F. Soares [1], Edson A. Pozza [1], Adélia A. A. Pozza [2], Reginaldo Gonçalves Mafia [3] and Maria A. Ferreira [1,*]

1. Department of Plant Pathology/DFP, Universidade Federal de Lavras/UFLA, Lavras, MG 37200-000, Brazil; thaissapfs@gmail.com (T.P.F.S.); edsonpozza@gmail.com (E.A.P.)
2. Department of Plant Pathology/DCS, Universidade Federal de Lavras/UFLA, Lavras, MG 37200-000, Brazil; adeliapozza@gmail.com
3. Centro de Tecnologia, Fibria Celulose S.A., Rod. Aracruz Barra do Riacho, Km 25., Aracruz-ES 29197-900, Brazil; rgoncalves@fibria.com.br

* Correspondence: mariaferreira@dfp.ufla.br; Tel.: +55-(35)-3829-1799

Received: 26 October 2018; Accepted: 13 December 2018; Published: 18 December 2018

Abstract: The supply of nutrients in balanced proportions leads to greater crop yields and represents an alternative practice for the management of plant diseases. Accordingly, we investigated the effect of the doses of and the nutritional balance between calcium (Ca) and potassium (K) on the severity of leaf spot and defoliation caused by the fungus *Calonectria pteridis*. Moreover, the effect of the treatments on the growth of interspecific hybrid eucalyptus clone seedlings (*Eucalyptus grandis* Hill ex Maiden × *E. urophylla* S.T. Blake), which are highly susceptible to the disease, was evaluated. The 25 treatments comprised combinations of one of five doses of Ca (1.2, 3.0, 6.0, 9.0 and 12.0 mmol L^{-1}) with one of five doses of K (0.8, 2.0, 4.0, 8.0 and 12.0 mmol L^{-1}) and five replicates of each treatment were included in the study. The supply of high concentrations of K favoured *C. pteridis* infection and resulted in high disease severity, although defoliation was not observed. However, the supply of both nutrients in excess (12.0 mmol L^{-1} Ca × 9.0 mmol L^{-1} K) resulted in a higher disease severity and an increased defoliation percentage (82 and 64%, respectively). Defoliation not associated with Calonectria leaf blight disease was observed with the imbalanced treatments, that is, the treatments combining a low concentration of one nutrient and an excess concentration of the other nutrient. The supply of K at a level near the standard dose (6 mmol L^{-1}) and of Ca at a dose above 4 mmol L^{-1} (standard dose) ensured high mean values for the morphological variables root and shoot biomass, plant height and chlorophyll a and b contents. These treatments also resulted in low disease severity and defoliation percentages, indicating that a balanced supply of Ca and K ensures reductions in disease severity and defoliation and contributes to higher growth.

Keywords: disease resistance; plant nutrition-disease relationship; eucalyptus disease; leaf spot; control

1. Introduction

Eucalyptus (*Eucalyptus* sp.) farming has grown rapidly to meet the growing world demand for pulp, paper and coal for the steel industry. In addition, the use of eucalyptus biomass and pulp as raw materials for the production of biofuels and biomaterials has been evaluated [1–3]. Brazil is the world's largest producer of eucalyptus, with a planted area of 5.67 million hectares [1]. More recently, eucalyptus plantations have undergone rapid expansion toward the northern and north-eastern regions of the country and the warm-temperature and high-humidity conditions of these regions are favourable for several diseases, particularly leaf blight caused by *Calonectria* species [4].

Calonectria species (anamorph: *Cylindrocladium*) are known to cause the disease Calonectria leaf blight (CLB) [5], which is mainly observed in eucalyptus trees and other diseases, including damping-off, cutting rot and root rot, have also been associated with *Calonectria* sp. [4]. The symptoms of CLB in most *Eucalyptus* species include spots that are initially small, circular or elongated and grey or light brown in colour and these spots progress and extend throughout the leaf limb to result in leaf drop and sometimes severe defoliation [6,7]. Defoliation caused by the pathogen is mainly observed in stages of plant development and interferes with physiological processes, such as photosynthesis, to induce a volumetric increase in the plant [8].

The leaf blight and severe defoliation caused by CLB in the northern and north-eastern regions of the country are mainly the result of infection with *Calonectria pteridis* Crous, M.J. Wingf. & Alfenas and can occur during the seedling production phase in the field, particularly in forests aged up to three years. In nurseries, the environmental conditions implemented for the production of cuttings, which include high humidity frequent irrigation, favour the growth of this pathogen and the plant's spatial density also favours the development of CLB. The disease is currently controlled by integrated cultivation and chemical methods as well as by the selection and multiplication of resistant genotypes, which is a much more effective approach. These integrated practices reduce the initial inoculum and render the environmental conditions unfavourable to the pathogen [9]. In the field, the selection and use of resistant genetic materials is the only commonly used control method [4].

However, the planting of clones that are considered resistant in regions other than those where the selective breeding was performed constitutes a major problem. In many cases, high disease severity has been observed in clones that are considered resistant but are planted in regions with a distinct climate that favours fungal growth. Thus, the fact that the breeding programs do not consider a variety of regions and climatic factors has resulted in the inappropriate use of these clones.

Therefore, the integrated management of CLB through an adequate nutritional balance might represent an alternative method for the control of leaf blight disease in eucalyptus in both nurseries and the field. Plant nutrition is known to exert direct and indirect effects on the activation of defence mechanisms against infection by phytopathogens [10,11]. Mineral nutrients are critical for the formation of physical and/or chemical barriers involved in plant resistance. Thus, balanced fertilization represents an alternative control method but this approach requires knowledge of the interactions between nutrients and their influence on the intensity of a given disease. Such data are important because of the synergism and antagonism between mineral elements, which mainly occur due to competition for the same absorption sites [12,13].

The balance between calcium and potassium has been well studied from a nutritional perspective because these nutrients play a key role in plant defence, growth and metabolism [11,14]. Potassium is involved in enzyme activation, protein synthesis, stomatal opening and plant water flow and interacts at various levels in processes associated with hormone defence, such as in the jasmonic acid and ethylene pathways [15]. In turn, calcium influx through the plasma membrane signals plant defence responses, which demonstrates that calcium plays a fundamental role in pathogen recognition [16]; moreover, calcium participates in cell wall synthesis and might represent one of the more important nutrients that can potentially aid the management of plant diseases [17].

In eucalyptus, the influence of nutrients on the severity of leaf spot caused by CLB is unknown. In addition, there is scarce knowledge regarding the effects of an unbalanced supply of nutrients on both the resistance and the morphological characteristics of the plant. The balanced mineral fertilization constitutes a strategy to restrict the conditions that favour pathogen attack and symptom expression. Then, these strategies can be easily integrated into management programs for eucalyptus diseases. The objectives of this study were to investigate the effect of the calcium/potassium (Ca/K) ratio on the severity of leaf blight and the degree of defoliation caused by *Calonectria pteridis* and to evaluate the growth responses of eucalyptus under conditions of nutrient imbalance.

2. Materials and Methods

2.1. Plant Preparation and Implantation

The experiment was conducted in a greenhouse with temperature and relative humidity conditions of 25 ± 2 °C and 80%, respectively, which were monitored by a datalogger HT-500 data logger (Instrutherm®, São Paulo, SP, Brazil) located inside a meteorological shelter located 2 m from the experimental site. Seedlings of the hybrid clone of *Eucalyptus grandis* × *E. urophylla*, which are susceptible to *C. pteridis*, were removed from 53-cm^3 tubes at 50 days of age. The cultivation substrate was then removed under running water. The exposed roots were carefully washed and disinfested with 1% sodium hypochlorite for 1 min and the seedlings were transplanted to 6-L trays containing Clark's nutrient solution [18] modified with the respective treatments for acclimatization.

In the acclimation phase, the seedlings remained in trays with nutrient solution diluted to 25% ionic strength for 7 days. After new leaf pairs and/or secondary roots emerged, the solutions were replaced with 60% ionic strength solution. The seedlings remained under this condition for another 7 days.

After the adaptation phase, uniform seedlings within each treatment were transplanted into 3-L plastic pots containing Clark's nutrient solution with 100% ionic strength according to the respective treatments and the seedlings were maintained under these conditions for another 25 days. The total experimental period from the acclimation phase to the end of the evaluations was 39 days.

The 25 treatments consisted of combinations of one of five doses of K (1.2, 3.0, 6.0, 9.0 and 12.0 mmol L^{-1}), which was provided as KNO_3 and KCl_2, with one of five doses of Ca (0.8, 2.0, 4.0, 8.0 and 12.0 mmol L^{-1}), which was provided as $Ca(NO_3)_2$ and $CaCl_2$, in a 5 × 5 factorial scheme. The experiment was performed based on a randomized block design with five replications and each experimental unit (pot) included two eucalyptus seedlings.

The micronutrient solutions were prepared with Fe-EDTA (89.53 µmol L^{-1}), manganese sulphate (91.01 µmol L^{-1}), zinc sulphate (0.76 µmol L^{-1}), copper sulphate (0.3153 µmol L^{-1}), boric acid (46.25 µmol L^{-1}) and ammonium molybdate (0.10 µmol L^{-1}). All the solutions were prepared using pure analysis-grade reagents. The concentrations of K and Ca used in the experiment were established based on the standard doses in Clark's solution [18] for these nutrients: thus, the five concentrations included the control doses of 6 mmol L^{-1} K and 4 mmol L^{-1} Ca and two doses higher than and two doses lower the standard doses.

The roots were maintained under constant aeration in the nutrient solutions. The pH of the solutions was checked at three-day intervals with a digital pH meter (HANNA® Instruments, Ann Arbor, MI, USA) and corrected when necessary to 5.5–6.0 through the addition of 0.1 mol L^{-1} KOH or HCl. The volume of the pots was refreshed as needed with deionized water. The electrical conductivity (EC) in each treatment was monitored weekly with a conductivity meter (HANNA® Instruments, Ann Arbor, MI, USA) at all stages to determine whether the solution needed to be replaced and the solution was not replaced if the decrease in EC did not reach 50% of the initial K^+ value.

The outside of the pots was painted with reflective silver paint to prevent the entrance of light and algal proliferation. In addition, the plant stems were attached to Styrofoam blocks in the upper face of the pots to prevent the entrance of light and support the seedlings.

2.2. Inoculum Preparation and Inoculation

Cultures of the *C. pteridis* isolate A2 were prepared in potato-dextrose-agar (PDA) medium. The fungus isolate was obtained from eucalyptus plantations in the state of Maranhão and maintained in the fungal collection of the Laboratory of Pathology, Federal University of Lavras (Universidade Federal de Lavras—UFLA). For inoculum production, the mycelium was cultured for 7 days for subsequent induction of sporulation by scraping the aerial mycelium and submersing the cultures in water [19]. The inoculum suspension was prepared by adding 20 mL of distilled water + 0.05% Tween 20 to the culture surface and then removing the conidia by scraping with a brush. The suspension was filtered through a double layer of sterilized gauze and the inoculum concentration was adjusted to 1.0×10^4 conidia/mL^{-1} using a Neubauer chamber. The suspension was sprayed on all the leaves of the seedlings until runoff. The seedlings were then covered with dark plastic bags to ensure constant leaf wetting in this environment and deionized water was frequently sprayed to ensure maintenance of the humidity levels. The seedlings remained under these conditions, which were ideal for germination and pathogen penetration, for 48 h.

2.3. Experimental Evaluation

2.3.1. Disease and Defoliation Assessment

Severity of CLB were evaluated when the first symptoms were observed in the leaves, that is, at 24, 48, 72, 96 and 120 h after inoculation (hai).

Disease severity was assessed by photographing the second fully expanded leaf pair at a distance of 10 cm after each evaluation period. The images were analysed using Quant® software [20] to obtain the percentage of the injured area as a function of the total leaf area.

The area under the disease progress curve (AUDPC) was calculated according to Shaner and Finney [21]:

$$AUDPC = \sum_{i=l}^{n-1} \left(\frac{y_i + y_{i+l}}{2} \right) (t_{i+l} - t_i) \tag{1}$$

where y_i is the proportion of the disease at the i-th observation; t_i is the time (in days) of the i-th observation; and n is the total number of observations.

The severity values obtained at 120 h were analysed separately as the final severity (FS). The total number of leaves prior to inoculation and that remaining at the end of the evaluation period were counted to determine the defoliation percentage according to the following equation:

$$FS\ (\%) = 100 - \frac{100 \times Qf}{Qi} \tag{2}$$

where D represents the percentage of defoliation, Qi is the number of leaves prior to inoculation and Qf refers to the final quantity of leaves.

2.3.2. Plant Growth and Chlorophyll Content

Four morphological variables were evaluated: shoot height (H), which was measured from the collar to the apical bud; collar diameter (CD), which was measured at the base of the collar; dry shoot weight (DSW); and dry root weight (DRW). At the end of the experiment, the roots were separated from the shoots and all the parts were washed in distilled water, packed in paper bags and dried in an oven at 60 °C to a constant weight.

The Dickson quality index (DQI) [21] was obtained using the following formula:

$$DQI = \frac{TDW}{H/CD} \tag{3}$$

where TDW (total dry weight) corresponds to the sum of DSW and DRW, H represents the height and CD is the collar diameter.

The chlorophyll content of the leaves was determined using a portable chlorophyll meter (PAD-502®, Minolta, Osaka, Japan) [22]. Four readings were performed on the second fully expanded leaf pair (two readings on each leaf) one day prior to inoculation and five days after inoculation. The mean of the four readings was used for the analysis. The chlorophyll a and b contents (mg gmf^{-1}) were determined using the following equations in accordance to a previous study [23]:

$$hlorophyll\ a = \frac{(12.7 \times A_{663} - 2.69 \times A_{645})V}{1000\ FMW} \tag{4}$$

$$Chlorophyll\ b = \frac{(22.9 \times A_{645} - 4.68 \times A_{663})V}{1000\ FMW} \tag{5}$$

where A is the absorbance at the indicated wavelength, V refers to the final volume of the chlorophyll-acetone extract and FMW represents the fresh weight in grams of plant material used (mg (g MF)$^{-1}$).

The following equations were used to determine the chlorophyll *a* and *b* concentrations:

$$Ya = (0.008 \times X) - 0.0053 \tag{6}$$

$$Yb = (0.0004 \times X) - 0.0032 \tag{7}$$

where X corresponds to the reading of the chlorophyll meter, *Ya* refers to chlorophyll *a*; and *Yb* corresponds to chlorophyll *b*.

2.4. Determination of the Leaf Nutrient Contents

Leaves were collected prior of the implementation of the treatments and after the end of the experiment to determine the nutritional status of the cuttings. The Ca and K contents were determined according to the method described by Malavolta et al. [24]. The reference values used to interpret the results of the leaf analysis were based on the work conducted by Martinez et al. [25] for eucalyptus.

2.5. Statistical Analysis

The data were subjected to the Shapiro-Wilk [26] and Bartlett tests to evaluate the normality of the distribution of the residuals and the homogeneity of the variances, respectively. The data did not need to be transformed because the values met the assumptions of the analysis of variance. The means of the studied variables were compared using the F test ($p < 0.05$) and significant values were subjected to a regression analysis and fitted to the linear models that are most representative of the studied pathosystem. The significant interactions between the studied quantitative factors were plotted on response surface diagrams. The statistical analyses were performed using the statistical software R and graphs were produced using SigmaPlot v. 16.0 (SigmaPlot, SYSTAT Software, Inc., San Jose, CA, USA).

3. Results

3.1. Disease Severity

The first symptoms were observed at 48 hai and an interaction was observed between the K and Ca doses for the AUDPC (p = 0.0000), final severity (p = 0.0000) and defoliation (p = 0.0000). At all doses supplied, the disease showed progression over time and mean severity values greater than 30% were observed at the last evaluation (Figure 1).

The supply of 12 mmol L^{-1} K with all tested doses of Ca resulted in a high final severity (FS) (59 to 78%) (Figure 2A). However, the highest FS (mean of 82%) was obtained with 9 mmol L^{-1} K and 12 mmol L^{-1} Ca. A moderate severity response was observed with the standard doses of the nutrients (6 mmol L^{-1} K × 4 mmol L^{-1} Ca) and with the treatments with low concentrations of K. The interaction between 9 mmol L^{-1} K and 8 mmol L^{-1} Ca yielded the lowest FS of 24.19%.

The combinations of 3 mmol L^{-1} K with 9 mmol L^{-1} Ca and of 4 mmol L^{-1} K with 8 mmol L^{-1} Ca resulted in the lowest AUDPC values (1598.9 and 1955.8, respectively) (Figure 2B), whereas the supply of 9 mmol L^{-1} K and 12 mmol L^{-1} Ca yielded higher AUDPC values. The highest defoliation was also obtained with the treatments consisting of 9 mmol L^{-1} K with 12 mmol L^{-1} Ca and 12 mmol L^{-1} K with 12 mmol L^{-1} Ca, which yielded defoliation percentages of 64% and 54%, respectively (Figure 2C). High defoliation percentages ranging from 39% to 63% were also observed after the treatments consisting of low doses of K and varying doses of Ca (except for values close to the standard dose). In addition, low doses of Ca combined with variable doses of K resulted in substantial defoliation (30%–63%). The supply of Ca at a dose higher than the established standard dose (4 mmol L^{-1})

combined with the standard dose of K (6 mmol L^{-1}) resulted in low defoliation and the interaction between 6 mmol L^{-1} K and 8 mmol L^{-1} Ca yielded a low defoliation percentage of 3.9% (Figure 2C).

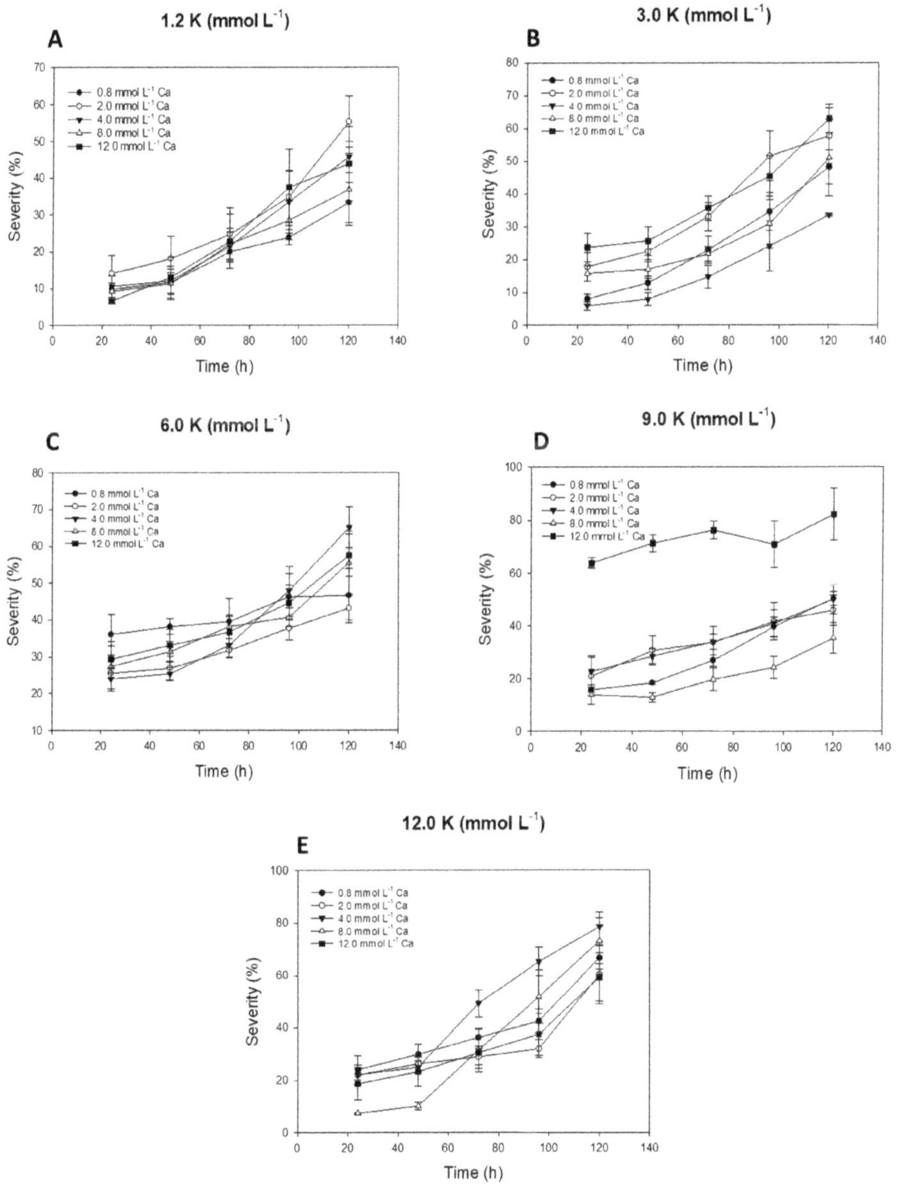

Figure 1. Severity (%) of *Calonectria* leaf blight in eucalyptus over time in the presence of different doses of Ca (mmol L^{-1}) combined with (**A**) 1.2 mmol L^{-1}, (**B**) 3.0 mmol L^{-1}, (**C**) 6.0 mmol L^{-1}, (**D**) 9.0 mmol L^{-1} and (**E**) 12.0 mmol L^{-1} K. The bars represent the standard error.

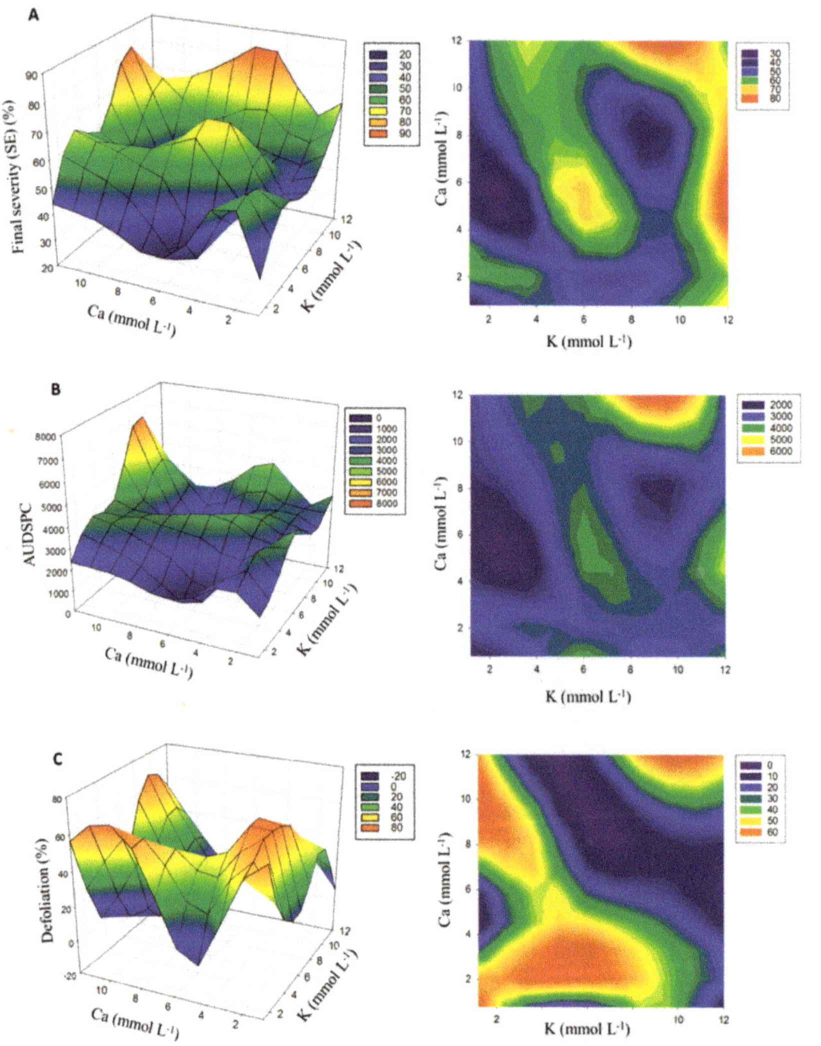

Figure 2. Influence of the supply of Ca and K (mmol L^{-1}) on the final severity (FS) (**A**), area under the disease progression curve (AUDPC) (**B**) and defoliation (**C**) of eucalyptus seedlings.

3.2. *Plant Growth and Chlorophyll Content*

The interaction of the Ca and K doses affected the DRW ($p = 0.0000$) and DSW ($p = 0.0000$) as well as the DQI ($p = 0.0008$). In turn, the supply of Ca, regardless of the supply of K, was significant for the H ($p = 0.0048$) and chlorophyll ($p = 0.0001$) content prior to inoculation. No treatment effect on the CD ($p = 0.89950$) was observed.

The highest DRW was obtained with 6 mmol L^{-1} K combined with 8 mmol L^{-1} Ca, which resulted in a mean DRW of 3.69 g $plant^{-1}$ (Figure 3A). This treatment also led to a high DSW, with a mean value of 10.9 g $plant^{-1}$ (Figure 3B). The other doses did not result in significant weight gains. The supply of any dose of K combined with a high dose of Ca significantly increased the DQI (Figure 3C). However, the treatment consisting of 8 mmol L^{-1} K with 12 mmol L^{-1} Ca exerted a notably distinct effect, with a mean of 0.23 g $plant^{-1}$.

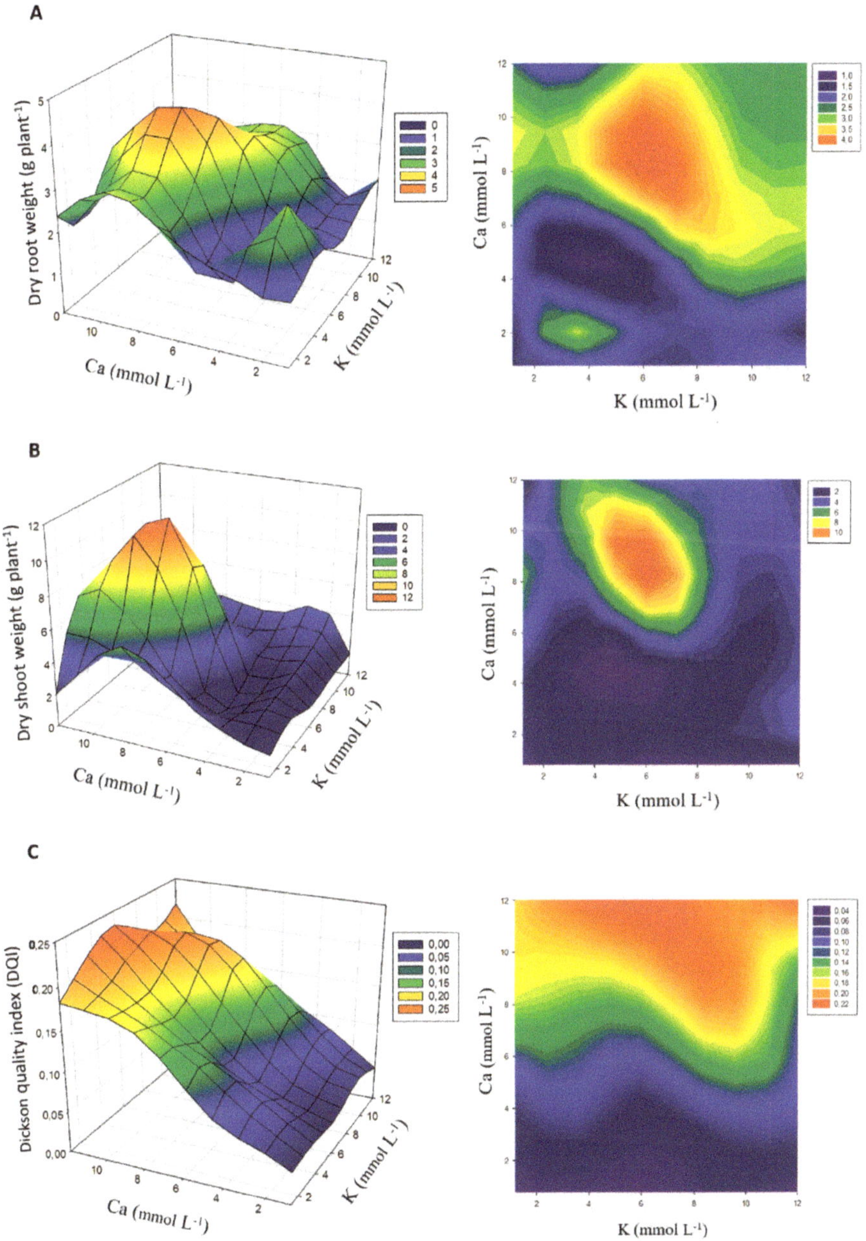

Figure 3. Influence of the supply of Ca and K (mmol L^{-1}) on the dry root weight (g $plant^{-1}$) (**A**), dry shoot weight (g $plant^{-1}$) (**B**) and Dickson quality index (DQI) (**C**) of eucalyptus.

An increase in the chlorophyll *a* and *b* contents (Figure 4) and H (Figure 5) was obtained with the supply of Ca at a dose up to 8 mmol L^{-1} and the supply of higher doses decreased the values of these variables.

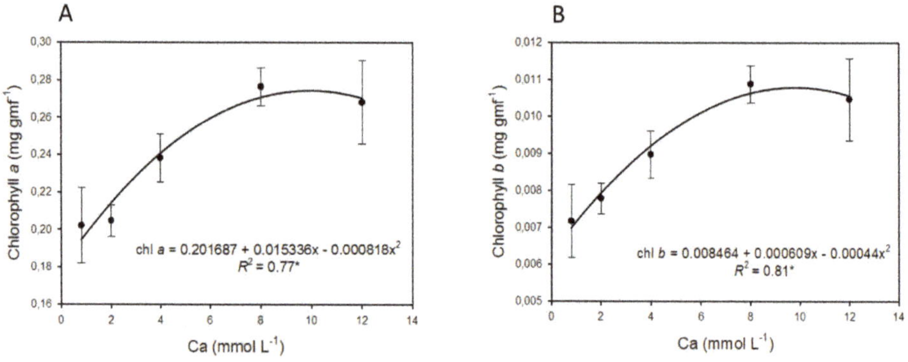

Figure 4. Contents of chlorophyll *a* (**A**) and *b* (**B**) of eucalyptus seedlings as a function of the calcium supply (mmol L^{-1}). Chl: Chlorophyll. * Significant at $p \leq 0.05$. The bars represent the standard error.

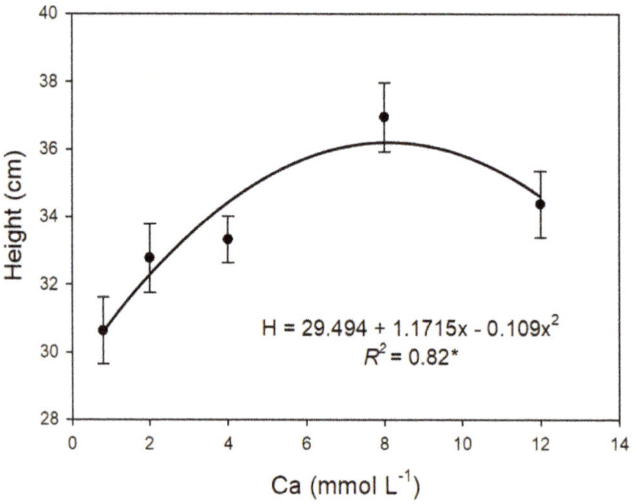

Figure 5. Effect of the calcium supply (mmol L^{-1}) on the height of eucalyptus seedlings. * Significant at $p \leq 0.05$. The bars represent the standard error.

3.3. Determination of Leaf Nutrient Contents

The interaction between Ca and K exerted an effect on the contents of these same nutrients in the leaves ($p = 0.0302$ and $p = 0.0040$, respectively). The leaf content of K relative to the initial content of 8.6 g kg^{-1} was increased by all the treatments. Specifically, all the treatments resulted in leaf contents of K above the range considered adequate for this nutrient (10–12 g kg^{-1}) according to Martinez et al. [25]. However, the highest K contents in the leaves were obtained with the treatments consisting of 12 mmol L^{-1} K combined with 0.8 to 8 mmol L^{-1} Ca (Figure 6A).

The leaf content of Ca decreased in almost all the treatments relative to the initial content of this nutrient in the leaves, 19.1 g kg^{-1}, except in the 1.2 and 12 mmol L^{-1} K and Ca treatment, with a value of 23.7 g kg^{-1} (Figure 6B). The leaf Ca contents were only below the adequate range according to Martinez et al. [25] (i.e., 8–12 g kg^{-1}) in the treatment with 12 mmol L^{-1} K combined with doses of Ca below 8 mmol L^{-1}.

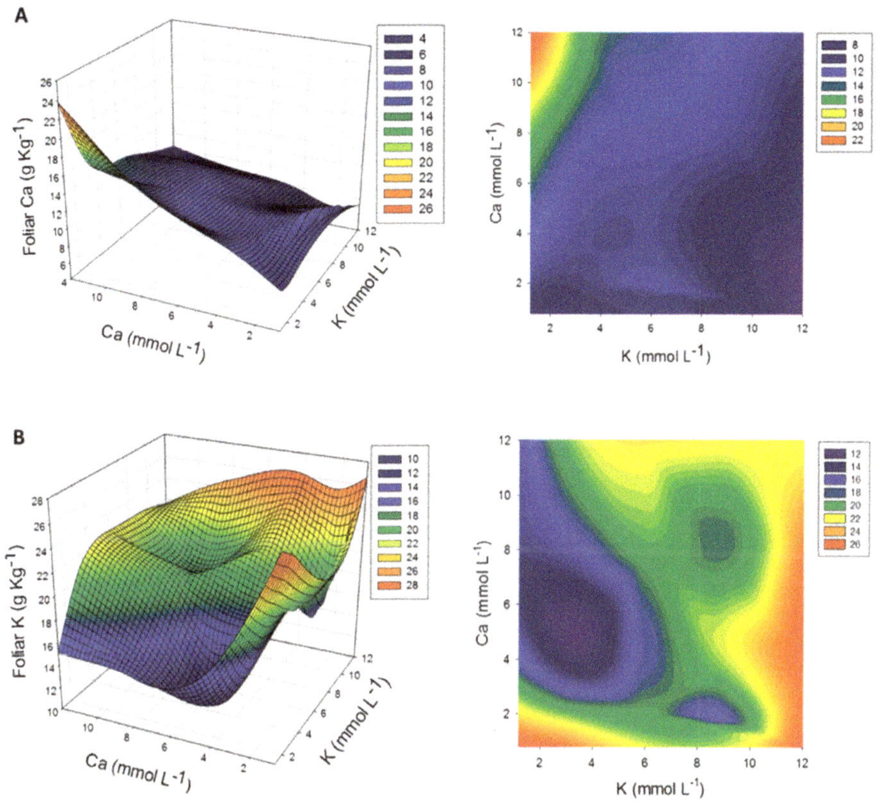

Figure 6. Leaf contents of calcium (**A**) and potassium (**B**) (g kg^{-1}) as a function of the supply of different doses of Ca and K (mmol L^{-1}) in the nutrient solution.

4. Discussion

In this study, eucalyptus defoliation caused by *C. pteridis* varied according to the Ca and K doses. An excessive supply of both nutrients led to high disease severity combined with a high percentage of defoliation. In eucalyptus, disease-induced defoliation is the main damage resulting from a reduction in the photosynthetically active area, which can reduce the plant diameter and H, as observed by Pulrolnik [27] in *E. grandis* trees.

In addition, deficiency in either Ca or K resulted in high defoliation in eucalyptus seedlings, even if the treatments resulted in a low disease severity, indicating that defoliation was not always associated with the disease. Treatment with K at a dose below 6 mmol L^{-1} (standard), regardless of the Ca dose, resulted in a high percentage of defoliation. Hillocks and Chinodya [28] investigated the relationship between K deficiency and leaf spot caused by *Alternaria macrospora*, which leads to defoliation in cotton and showed that K-deficient plants show early defoliation, even in the absence of the leaf pathogen. Prior to these results, premature defoliation had been attributed only to leaf spot caused by *Alternaria*. Thus, the results of the present study regarding the effect of K deficiency on defoliation are consistent with those obtained by Hillocks and Chinodya [28]. Early defoliation as a consequence of K deficiency is characterized by chlorophyll degradation and negative chlorophyll fluorescence, as observed by Hu [29], who found that K deficiency causes premature senescence of cotton leaves, decreases the leaf area and increases the rate of yellow leaves.

As observed for K, the treatments with doses of Ca below 4 mmol L^{-1} (standard dose) combined with varying doses of K resulted in significant defoliation occurred that was not associated with CLB.

This behaviour can be explained by the influence of Ca on the K contents in the leaves. According to Tsialtas [30], high concentrations of Ca^{2+} in cotton leaves inhibit the exit of K^+ from cells under stress conditions. This conclusion was also reached in other studies [31,32], which demonstrated that an increase in the apoplastic Ca concentration reduces the stress caused by K^+ efflux from the mesophyll of leaves and thus ensures higher K retention under K-deficiency conditions. The results of the present study suggest that Ca deficiency directly influences the occurrence of defoliation due to an imbalance in the Ca/K ratio. High Ca concentrations allow greater K retention and indirectly contribute to the maintenance of cellular turgor [30].

The eucalyptus seedlings treated with 12 mmol L^{-1} K combined with any of the tested doses of Ca except 12 mmol L^{-1} presented high disease severity without defoliation. Several factors, including stomatal movement, might have contributed to the high disease severity obtained with these treatments. K exerts an important influence on stomatal opening and closing and induces rapid stomatal closure when provided at a sufficient dose; thus, pathogen infection via the stomata is directly affected by the supply of K [33]. In this study, an excessive supply of K might have reduced the stomatal closure responses and increased the stomatal opening movements, thereby leading to infection by *C. pteridis*, which penetrates exclusively via the stomata [34]. This stomatal response in the presence of high concentrations of K was detailed by Smith and Stewart [35], who subjected *Striga hermonthica* (Del.) Benth. plants to high concentrations of K (>200 millimolar) and observed atypical behaviour regarding the maintenance of stomatal opening and closing movements.

Although high concentrations of K favoured *C. pteridis* infection, they also contributed to the non-occurrence of defoliation, even though the seedlings showed a high necrotic leaf percentage due to the disease. This finding is again due to the notable influence of K on defoliation.

Compared with the FS value, the AUDPC was not representative in explaining the behaviour of CLB disease in eucalyptus and only a high disease percentage due to the interaction between 9 mmol L^{-1} K and 12 mmol L^{-1} Ca was observed based on the AUDPC, whereas other interactions, including 9 mmol L^{-1} K and 12 mmol L^{-1} Ca, affected the FS value. The AUDPC did not indicate the actual behaviour of CLB disease as observed at the final severity.

The Ca and K doses that yielded low disease severity and defoliation percentages also contributed positively to the evaluated morphological characteristics, leading to a high crop yield. Therefore, the overlapping of all the analysed variables would show that the regions in the graphs that favour all the desirable characteristics correspond to the interactions of K at doses near the standard dose (6 mmol L^{-1}) with Ca at doses higher than 4 mmol L^{-1}.

An increase in DRW was primarily obtained by the supply of Ca at a high dose and the treatments with Ca doses greater than 4 mmol L^{-1}, regardless of the amount of K supplied, yielded the higher mean DRW values. The importance of Ca in root development was also demonstrated by Rodrigues [36], who found that the application of limestone to eucalyptus plants promoted an increase in the density of fine roots up to 40 cm in the soil. Furthermore, according Ivashuta et al. [37], the protein required for cell root cell expansion, which determines the plant's development, is dependent on Ca and called Ca-dependent protein kinase 1 (CDPK1). Therefore, Ca plays a key role in root development.

The treatment with 6 mmol L^{-1} K $\times$ 8 mmol L^{-1} Ca resulted in good root development and great shoot development in the eucalyptus seedlings; moreover, similar doses of Ca and K resulted in greater increases in the DSW compared with the other treatment doses. Both nutrients participate in physiological processes related to plant growth and influence cambium activity and xylem development, which are important in wood formation [38]. K participates in the regulation of the osmotic potential and cell expansion and Ca is involved in a number of processes, such as cell division, cell wall chemistry and cell differentiation; therefore, both nutrients promote increases in the cambium width [39]. Moreover, according to Laclau et al. [40], fertilization with K significantly increases the leaf biomass of *E. grandis* as well as its leaf life span.

The significant increase in the dry organ weights (i.e., DRW and DSW) obtained with 6 mmol L^{-1} K and 8 mmol L^{-1} Ca indicated that the amount of Ca supplied to eucalyptus plants in the standard solution was insufficient to achieve the desired morphological characteristics. The DQI results validate this conclusion because the treatments that yielded the highest indices included Ca at doses above the standard dose (4 mmol L^{-1}) and the findings were independent of the dose of K. In addition, the highest mean chlorophyll *a* and *b* contents and H values were observed in the seedlings supplied 8 mmol L^{-1} Ca.

According to the results of the leaf analysis, the treatments with a high K supply resulted in lower leaf contents of Ca compared with those obtained with the other treatments, indicating that these nutrients compete for the same absorption site in the plant [40]. In general, reducing the supply of one of the nutrients increased the leaf content of the other nutrient. Only the treatments with low doses of K yielded a leaf content of Ca above the range considered adequate by Martinez et al. [25]. Thus, an excess supply of K reduces the uptake of Ca and leads to the accumulation of K in the leaves. In addition, the high leaf content of K observed with the treatments consisting of 12 mmol L^{-1} K confirmed that imbalances in this nutrient cause issues in stomatal movement and thereby lead to high disease severity.

5. Conclusions

In practice, the use of mineral fertilization regimes in the production of eucalyptus seedlings that increase the supply of Ca while maintaining the dose of K at a level close to the standard dose used in this study can ensure higher crop growth and reduce losses related to defoliation due to increases in disease and defoliation susceptibility attributed to unbalanced fertilization.

Author Contributions: Conceptualization, T.P.F.S. and M.A.F.; Methodology, T.P.F.S., E.A.P., and A.A.A.P.; Software, T.P.F.S.; Validation, T.P.F.S., E.A.P., A.A.A.P., R.G.M. and M.A.F.; Formal Analysis, T.P.F.S.; Investigation, T.P.F.S.; Resources, E.A.P. and M.A.F.; Data Curation, T.P.F.S., E.A.P., A.A.A.P., R.G.M. and M.A.F.; Writing—Original Draft Preparation, T.P.F.S.; Writing—Review & Editing, T.P.F.S., E.A.P., A.A.A.P., R.G.M. and M.A.F.; Visualization, T.P.F.S., E.A.P., A.A.A.P., R.G.M. and M.A.F.; Supervision, M.A.F.; Project Administration, M.A.F.; Funding Acquisition, M.A.F.

Funding: This research was funded by Coordenação de Aperfeiçoamento de Pessoal de Nível Superior grant number (scholarship) and The APC was funded by Maria A. Ferreira, Fundação de Desenvolvimento Científico e Cultural.

Acknowledgments: We would like to acknowledge the funding and scholarships from the Coordenação de Aperfeiçoamento de Pessoal de Nível Superior Coordination for the Improvement of Higher Education Personnel (CAPES).

Conflicts of Interest: The authors declare no conflict of interest.

References

1. Ibá—Indústria Brasileira de Árvores. Available online: http://iba.org/images/shared/Biblioteca/IBA_RelatorioAnual2017.pdf (accessed on 3 October 2017).
2. Hinchee, M.; Rottmann, W.; Mullinax, L.; Zhang, C.; Chang, S.; Cunningham, M.; Pearson, L.; Nehra, N. Short-rotation woody crops for bioenergy and biofuels applications. In *Biofuels*; Springer: New York, NY, USA, 2009; pp. 139–156.
3. Shepherd, M.; Bartle, J.; Lee, D.J.; Brawner, J.; Bush, D.; Turnbull, P. *Eucalypts* as a biofuel feedstock. *Biogeosciences* **2011**, *2*, 639–657. [CrossRef]
4. Alfenas, A.C.; Zauza, E.A.V.; Mafia, R.G.; Assis, T.F. *Clonagem e Doenças do Eucalipto*, 2nd ed.; Editora, U.F.V., Viçosa, M.G., Eds.; Editora UFV: Viçosa, Brazil, 2009; p. 500.
5. Rodas, C.A.; Lombard, L.; Gryzenhout, M.; Slippers, B.; Wingfield, M.J. *Cylindrocladium* blight of. *Eucalyptus grandis* in Colombia. *Australas. Plant Pathol.* **2005**, *34*, 143–149. [CrossRef]
6. Alfenas, A.C.; Ferreira, F.A. A mancha de folha do eucalipto no Brasil causada por três espécies de *Cylindrocladium*—Uma revisão da descrição da doença. *Revista Árvore* **1979**, *3*, 47–56.

7. Ferreira, F.A.; Alfenas, A.C.; Moreira, A.M.; Demuner, N.L. Mancha-depteridis doença foliar de eucalipto em areas tropicais brasileiras. *Fitopatol. Bras.* **1995**, *20*, 107–110.
8. Alfenas, A.C.; Zauza, E.A.V.; Mafia, R.G.; Assis, T.F. *Clonagem e Doenças do Eucalipto*; Editora UFV: Viçosa, Brazil, 2004; p. 442.
9. Ferreira, E.M.; Alfenas, A.C.; Mafia, L.A.; Mafia, R.G. Eficiência de fungicidas sistêmicos para o controle de *Cylindrocladium candelabrum* em eucalipto. *Fitopatol. Bras.* **2006**, *31*, 468–475. [CrossRef]
10. Datnoff, L.E.; Rodrigues, F.A.; Seebold, K.W. Silicon and Plant Nutrition. In *Mineral Nutrition and Plant Disease*; Datnoff, L.E., Elmer, W.H., Huber, D.M., Eds.; APS Press: Saint Paul, MN, USA, 2007; pp. 233–246.
11. Wang, M.; Zheng, Q.; Shen, Q.; Guo, S. The critical role of potassium in plant stress response. *Int. J. Mol. Sci.* **2013**, *14*, 7370–7390. [CrossRef] [PubMed]
12. Jakobsen, T.J. Interaction between Plant Nutrients: 1. Theory and Analytical Procedures. *Acta Agric. Scand. Section B–Soil Plant Sci.* **1992**, *42*, 208–212. [CrossRef]
13. Jakobsen, T.J. Interaction between Plant Nutrients: III. Antagonism between Potassium, Magnesium and Calcium. *Acta Agric. Scand. Section B–Soil Plant Sci.* **1993**, *43*, 1–5. [CrossRef]
14. Zhang, L.; Du, L.; Poovaiah, B.W. Calcium signaling and biotic defense responses in plants. *Plant Signal. Behav.* **2014**, *9*, e973818. [CrossRef]
15. Armengaud, P.; Breitling, R.; Amtmann, A. The potassium-dependent transcriptome of *Arabidopsis* reveals a prominent role of jasmonic acid in nutrient signaling. *Plant Physiol.* **2004**, *136*, 2556–2576. [CrossRef]
16. Ma, W.; Qi, Z.; Smigel, A.; Walker, R.K.; Verma, R.; Berkowitz, G.A. Ca^{2+}, cAMP, and transduction of nonself perception during plant immune responses. *Proc. Nat. Acad. Sci. USA* **2009**, *106*, 20995–21000. [CrossRef] [PubMed]
17. Rahman, M.; Punja, Z.K. Calcium and plant disease. In *Mineral Nutrition and Plant Disease*; Datnoff, L.E., Elmer, W.H., Huber, D.M., Eds.; American Phytopathological Society Press: Saint Paul, MN, USA, 2007; Volume 1, pp. 79–93.
18. Clark, R.B. Characterization of phosphates in intact maize roots. *J. Agric. Food Chem.* **1975**, *23*, 458–460. [CrossRef] [PubMed]
19. Alfenas, R.F. Produção de Inóculo de *Cylindrocladium pteridis* em Condições Controladas. Master's Thesis, Federal University of Viçosa, Viçosa, Brazil, October 2009.
20. Vale, F.X.R.; Fernandes, F.E.I.F.; Liberato, J.R. QUANT—A software for plant disease severity assessment. In Proceedings of the Anais International Congress of Plant Pathology, Christchurch, New Zealand, 2–7 February 2003.
21. Dickson, A.; Leaf, A.L.; Hosner, J.F. Quality appraisal of white spruce and white pine seedling stock in nurseries. *For. Chron.* **1960**, *36*, 10–13. [CrossRef]
22. Shaner, G.; Finney, R.E. The effect of nitrogen fertilization on the expression of slow-mildew resistance in Knox wheat. *Phytopathology* **1977**, *67*, 1051–1056. [CrossRef]
23. Matsumoto, S.N.; Carvalho, F.M.; Viana, A.E.S.; Malta, M.R.; Castro, L.G. Initial growth of coffee plants (*Coffea Arabica* L.) submitted to different phosphate doses in nutritive solution. *Coffee Sci.* **2008**, *31*, 58–67.
24. Malavolta, E.; Vitti, G.C.; Oliveira, S.A. *Avaliação do Estado Nutricional das Plantas: Princípios e Aplicações*, 2nd ed.; POTAFOS: Piracicaba, Brazil, 1997; p. 319.
25. Martinez, H.E.P.; Carvalho, J.G.; Souza, R.B. Diagnose foliar. In *Recomendações para Uso de Corretivos e Fertilizantes em Minas Gerais: 5 Aproximação*; Ribeiro, A.C., Guimarães, P.T.G., Alvarez, V.V.H., Eds.; CFSEMG: Viçosa, Brazil, 1999; pp. 143–168.
26. Shapiro, S.S.; Wilk, M.B. An analysis of variance test for normality (complete samples). *Biometrika* **1965**, *52*, 591–611. [CrossRef]
27. Pulrolnik, K.; Reis, G.G.; Reis, M.G.F.; Monte, M.A.; Fontan, I.C.I. Crescimento de plantas de clone de *Eucalyptus grandis* [hillexmaiden] submetidas a diferentes tratamentos de desrama artificial, na região de cerrado. *Revista Árvore* **2005**, *29*, 495–505. [CrossRef]
28. Hillocks, R.J.; Chindoya, R. The relationship between *Alternaria* leaf spot and potassium deficiency causing premature defoliation of cotton. *Plant Pathol.* **1989**, *38*, 502–508. [CrossRef]
29. Hu, W.; Lv, X.; Yang, J.; Chen, B.; Zhao, W.; Meng, Y.; Oosterhuis, D.M. Effects of potassium deficiency on antioxidant metabolism related to leaf senescence in cotton (*Gossypium hirsutum* L.). *Field Crop. Res.* **2016**, *191*, 139–149. [CrossRef]

30. Tsialtas, J.T.; Shabala, S.; Matsi, T. A prominent role for leaf calcium as a yield and quality determinan in upland cotton (*Gossypium hirsutum* L.) varieties grown under irrigated Mediterranean conditions. *J. Agron. Crop Sci.* **2016**, *202*, 161–173. [CrossRef]
31. Shabala, S.; Pottosin, I. Regulation of potassium transport in plants under hostile conditions: Implications for abiotic and biotic stress tolerance. *Physiol. Plant.* **2014**, *151*, 257–279. [CrossRef]
32. Reddy, K.R.; Hodges, H.F.; Varco, J. Potassium nutrition of cotton. *Miss. Agric. For. Exp. Stn.* **2000**, *1094*, 1–10.
33. Pervez, H.; Ashraf, M.; Makhdum, M.I.; Mahmood, T. Potassium nutrition of cotton (*Gossypium hirsutum* L.) in relation to cotton leaf curl virus disease in aridisols. *Pak. J. Bot.* **2007**, *39*, 529–539.
34. Graça, R.N.; Alfenas, A.C.; Maffia, L.A.; Titon, M.; Alfenas, R.F.; Lau, D.; Rocabado, J.M.A. Factors influencing infection of eucalypts by *Cylindrocladium pteridis. Plant Pathol.* **2009**, *58*, 971–981. [CrossRef]
35. Smith, S.; Stewart, G.R. Effects of potassium levels on the stomatal behavior of the hemi-parasite *Striga hermonthica. Plant Physiol.* **1990**, *94*, 1472–1476. [CrossRef] [PubMed]
36. Rodrigues, F.A.V. Crescimento de Eucalipto em Idade Jovem e Movimentação de Cálcio e Magnésio no Solo em Resposta à Aplicação de Calcário e Gesso Agrícola. Ph.D. Thesis, Universidade Federal de Viçosa, Viçosa, Brazil, April 2013.
37. Ivashuta, S.; Liu, J.; Lohar, D.P.; Haridas, S.; Bucciarelli, B.; VandenBosch, K.A.; Vance, C.P.; Harrison, M.J.; Gantt, J.S. RNA interference identifies a calcium-dependent protein kinase involved in *Medicago truncatula* root development. *Plant Cell* **2005**, *17*, 2911–2921. [CrossRef] [PubMed]
38. Fromm, J. Wood formation of trees in relation to potassium and calcium nutrition. *Tree Physiol.* **2010**, *30*, 1140–1147. [CrossRef] [PubMed]
39. Fageria, V.D. Nutrient interactions in crop plants. *J. Plant Nutr.* **2001**, *24*, 1269–1290. [CrossRef]
40. Laclau, J.P.; Almeida, J.C.R.; Gonçalves, J.L.M.; Saint-Andre, L.; Ventura, M.; Ranger, J.; Moreira, R.M.; Nouvellon, Y. Influence of nitrogen and potassium fertilization on leaf life span and allocation of above-ground growth in Eucalyptus plantations. *Tree Physiol.* **2009**, *29*, 111–124. [CrossRef]

forests

MDPI

Article

Chemosensory Characteristics of Two *Semanotus bifasciatus* Populations

Sufang Zhang, Sifan Shen, Shiyu Zhang, Hongbin Wang, Xiangbo Kong, Fu Liu and Zhen Zhang *

Key Laboratory of Forest Protection, Research Institute of Forest Ecology, Environment and Protection, Chinese Academy of Forestry, Beijing 100091, China

* Correspondence: zhangzhen@caf.ac.cn; Tel.: +86-10-62889567; Fax: +86-10-62884972

Received: 9 July 2019; Accepted: 1 August 2019; Published: 2 August 2019

Abstract: *Semanotus bifasciatus* (Motschulsky) (Cerambycidae: Coleoptera) is a major forest borer in China, and attractants provide a promising method for the control of this pest. Exploration of the chemosensory mechanisms of *S. bifasciatus* is important for the development of efficient attractants for this pest. However, little information is available about the olfactory mechanisms of *S. bifasciatus*. Previous research has indicated that the trapping effects of the same attractant are different between Beijing and Shandong populations of *S. bifasciatus*. To explore the reasons for this, next-generation sequencing was performed to analyze the antennal transcriptome of both sexes of the two *S. bifasciatus* populations, and the olfactory-related genes were identified. Furthermore, the expression levels and single nucleotide polymorphisms (SNPs) of the olfactory-related genes between the two populations were compared. We found that the expression levels of odorant binding proteins (OBPs), odorant receptors (ORs), and sensory neuron membrane proteins (SNMPs) in male *S. bifasciatus* of the Beijing population were obviously lower than those in the Shandong population, and most of the conserved SNPs in OBPs and ORs of the two populations showed more diversity in the Beijing population. Our work provides a foundation for future research of the molecular olfactory mechanisms and pest management of *S. bifasciatus*, as well as other longhorn beetles.

Keywords: antennal transcriptome; chemosensory genes; expression level; SNPs

1. Introduction

Semanotus bifasciatus (Motschulsky) mainly damages *Platycladus orientalis* (L.) Franco, *Sabina chinensis* (L.) Antoine, and other species, and the damage it causes is very serious in some regions [1], which makes this pest a main forest borer species in China. In recent years, the protection of old cypress trees, forest resources, and the ecological environment has been threatened by *S. bifasciatus* in northern China. For example, 67.3% of planted seedlings were damaged by *S. bifasciatus* in Dalian, Liaoning province in 1988 [2]; about 90% of dead cypresses were infested by *S. bifasciatus* in the Lingyan Forest district of Taian, Shandong province in the period 2002–2006; and 97% of the introduced cypress were damaged by *S. bifasciatus* in Xiangshan Mountain park, Beijing in 1979. Thus, the control of *S. bifasciatus* is urgent and important for forest protection.

However, as a boring insect, *S. bifasciatus* remain concealed during egg and larval stages, so it is hard to control this pest. The adult form of *S. bifasciatus* appears in the forest for about half a month only, and at this stage, we can use treatments on them relatively easily. Thus, developing attractants for *S. bifasciatus* based on volatiles or pheromones has attracted the interest of many forest insect scientists, and some volatile lures, using volatiles from certain host plants, have been used to trap this pest [3,4]. The attracting effect of an attractant was tested at two different places—Beijing and Shandong, China—and the results indicated that the responses of two *S. bifasciatus* populations from these two places differed greatly (unpublished data). This phenomenon indicated that the olfactory sensibilities of the two *S. bifasciatus* populations were different, and this deserves further research. The results of such research may help us to develop accurate attractants for different pest populations.

However, the olfactory response mechanisms of *S. bifasciatus* are unclear, and only fragmented information has been available until now [5]. Thus, our work focuses on the olfactory variety of *S. bifasciatus* from different places.

In insects, environmental odors are mainly detected by olfactory sensory neurons (OSNs) located in sensilla on the antennae [6]. During transformation of chemical volatile signals to electrical nervous impulses, insects rely on at least six olfactory-related gene groups, including two binding protein families, three receptor families, and the sensory neuron membrane proteins (SNMPs) [7,8]. The three receptor families expressed in chemosensory neurons [9–11], include odorant receptors (ORs), ionotropic receptors (IR), and gustatory receptors (GRs) [12,13]. Insect ORs are seven-transmembrane domain proteins, and their topologies are reversed, comparing to vertebrate ORs [14,15]. ORs always appear as dimers, comprised by one specific OR [16] and one conserved gene, called the odorant receptor co-receptor (Orco) [17]. Most GRs are expressed in gustatory receptor neurons, which are involved in contact chemoreception [18]. IRs refer to receptors that are related to ionotropic glutamate receptors (iGluRs) [9]. In addition to the receptor genes, there are two types of binding proteins in olfactory systems that play critical roles in olfaction. Odorant binding proteins (OBP) are small, soluble proteins with six conserved cysteines [19,20]. OBPs (odorant binding proteins) have been proposed to deliver odor molecules to the receptors [19]. Chemosensory proteins (CSPs) are another type of small binding protein, which are more conserved, and they are characterized by two disulfide bridges formed by the four conserved cysteines [21].

Insects display a highly specific and sensitive olfactory sense and complex olfactory behavior, and their olfaction-related genes show large sequence diversity [22–25]. Therefore, identification of olfactory-related genes mainly relies on genomic data [26–28]. Recently, studies on antennal transcriptomes have led to the identification of olfactory-related genes in many insects [29–34], which demonstrated the power of transcriptomic data for olfactory gene identification. However, few studies have compared the olfactory recognition mechanisms between different populations of the same insect species.

In this study, to explore the olfactory variety of different *S. bifasciatus* populations, we sequenced the antennal transcriptomes of both sexes from two places, identified the olfactory-related genes of *S. bifasciatus*, and compared the expression quantity and single nucleotide polymorphisms (SNPs) between the two populations.

2. Materials and Methods

2.1. Insects for Antennal Transcriptome Sequencing

Sections of wood containing overwintering *S. bifasciatus* insects were collected from two sites, the Lingyan Forest district of Taian, Shandong province, and the Mentougou district of Beijing, in March 2018, and were placed in 26 °C ± 2 °C temperature with 50% ± 10% relative humidity and a 16 h light, 8 h dark photo-period. The wood cross-sections were sealed with wax. Eclosion was seen about two weeks later. After eclosion, antennae from 15 female and male *S. bifasciatus* from the two sites were cut off at the base, and the samples were frozen in liquid nitrogen immediately. Antenna from five *S. bifasciatus* were collected as one sample, and a total of three biological replicates were obtained.

2.2. RNA-Seq Library Preparation and Sequencing

We extracted the total RNA of *S. bifasciatus* antenna with TRIzol reagent (Invitrogen, Carlsbad, CA, USA), as previously described [8,35], and then treated them with RNase-free DNase I (TaKaRa, Dalian, Liaoning, China). The RNA purity was checked using the NanoPhotometer® spectrophotometer (IMPLEN, Westlake Village, CA, USA), concentration with Qubit® 2.0 Fluorometer (Life Technologies,

Carlsbad, CA, USA), and integrity on the Bioanalyzer 2100 system (Agilent Technologies, Santa Clara, CA, USA), respectively.

A total amount of 1 μg RNA per sample was used to synthetize complementary DNA (cDNA) [36,37]. Then, sequencing libraries were prepared with an Illumina TruSeq™ RNA Sample Preparation Kit (Illumina, San Diego, CA, USA). An AMPure XP system (Beckman Coulter, Beverly, MA, USA) was used to select cDNA fragments (of 200 bp length, preferentially), and two-end adaptor ligated DNA fragments were enriched with polymerase chain reaction (PCR). Finally, the products were purified and quantified (Agilent Bioanalyzer 2100). The index-coded samples were clustered and sequenced with an Illumina Hiseq 2500, according to the manufacturer's instructions.

2.3. De Novo Assembly of the Sequences

Clean data were obtained from raw sequencing data, using the following treatments: with self-written Perl scripts, we removed reads containing an adapter sequence, uncertain bases (reads with >10% N), and sequences more than 50% bases with error rates >1%. Pollution of the data was ruled out by comparing with the NT database. Assembly was performed with Trinity (version: trinityrnaseq-r20131110) [38], and redundancy was treated with TGICL [39].

2.4. Annotation

All unigenes were aligned with Blast (The Basic Local Alignment Search Tool) against the NR, SWISSPROT, KEGG, and KOG databases (cut-off by 1e-5), and we selected the highest sequence similarity targets as the functional annotation of the transcripts. Then, Gene Ontology (GO) annotation was performed with Blast 2GO [40,41], and the molecular function, biological process, and cellular component of the genes were assigned [42].

We identified the chemosensory genes in *S. bifasciatus* using contig tBLASTx searches with a query sequence of chemosensory genes from other insects, such as *Drosophila melanogaster*, *Bombyx mori*, and other long-horned beetles. Open reading frames (ORFs) of the candidate genes were found, which were further verified using BLAST (http://blast.ncbi.nlm.nih.gov/Blast.cgi).

2.5. Gene Expression Quantification

FPKM (fragments per kilobase of exon per million fragments mapped) values were used to measure the expression level of the genes [43]. FPKM value and standard error were obtained from the three biological replicates. Differentially expressed genes (DEGs) of the two *S. bifasciatus* populations were selected using DESeq (http://bioconductor.org/packages/release/bioc/html/DESeq.html) [44]. The compatible-hits-norm model was used to normalize the Unigene expression levels and DEGs [45], and the false discovery rate (FDR) method ($p \leq 0.05$) was used to identify DEGs [46].

2.6. SNP Calling

We used the Genome Analysis Toolkit (GATK2) software to carry out SNP calling [47]; the GATK standard filter method and other parameters were used to filter the Raw vcffiles (cluster Window Size: 35; MQ0 ≥ 4; and (MQ0/(1.0 × DP)) > 0.1, QUAL < 10, QUAL < 30.0, or QD < 5.0, or HRun > 5) and retain the SNPs with distance >5. The SNPs that showed stable diversity between the two *S. bifasciatus* populations or between sexes were selected as conserved SNPs.

2.7. Data Analysis

The expression levels of the identified chemosensory genes were compared according to the FPKM values and based on the three biological replicates. Significant difference between the two populations were tested using ANOVA test (in SPSS 18, SPSS Inc., Chicago, IL, USA).

3. Results

3.1. Antennal Transcriptomes Assembly and Annotation

The transcriptomic sequence data were generated using an antenna cDNA library and Illumina HiSeq TM2000/MiSeq technology. About 6 Gbp of clean sequence data were obtained for each sample. Assembly resulted in 55,355 contigs, with an N50 of 2813 bp. The number of contigs longer than 1 kb was 29,791. Gene annotation to several databases obtained gene annotation information for 35,188 genes.

3.2. Chemosensory Gene Identification of *S. bifasciatus*

We focused on olfactory gene families to perform a detailed analysis of the transcriptomes. In total, 18 OBPs, 21 CSPs, 66 ORs, 24 GRs, 14 IRs, and 4 SNMPs in *S. bifasciatus* were identified (Supplementary Table S1).

As the trapping effects of the attractant differed obviously between Shandong and Beijing, we further analyzed the possible reasons for this situation. As the expression levels and gene sequence variations were two factors that may have been related to the trapping effects, we further analyzed the differences of these aspects in the two *S. bifasciatus* populations.

3.3. Chemosensory Gene Expression Characteristics of Two *S. bifasciatus* Populations

Chemosensory gene expressions of the two *S. bifasciatus* populations in both sexes were analyzed. Two types of binding proteins (OBPs and CSPs) related to chemosensory were compared between the two populations firstly. Among the 18 identified OBPs, 4 OBPs (OBP3, OBP4, OBP11, and OBP15) were expressed significantly differently between females of the two populations (Figure 1A), while 8 (OBP2, OBP5, OBP7, OBP9, OBP12, OBP13, OBP15, and OBP17) were expressed significantly differently between males of the two populations (Figure 1B). These results indicate that more gene expression varieties were found in males than females of the two populations. A different situation was detected in CSPs, another binding protein of the olfactory system (Figure S1), which means that only a few genes (2 between females and 3 between males of the two populations) were significantly expressed between the two populations.

Among the three type of receptor genes related to chemosensory, we found that the ORs possessed most gene expression differences. Similar to OBPs, varieties of males were more than that of females between the two populations: 10 ORs (OR14, OE16, OR19, OR38, OR41, OR43, OR44, OR47, OR59, and OR63) were expressed significantly differently between females of the two populations (Figure 2A), while 20 ORs (OR2, OR3, OR7, OR9, OR11, OR13, OR14, OR18, OR19, OR22, OR34, OR36, OR40, OR46, OR48, OR49, OR50, OR51, OR56, and OR63) were expressed significantly differently between males of the two populations (Figure 2B). The expression differences of GRs and IRs in the two populations were also compared (Figures S2 and S3), and we found that the differently expressed genes of these two types of receptor were less than that of ORs. Three (in females) and 6 (in males) GRs were differently expressed between the two populations (Figure S2), and 2 (in females) and 4 (in males) IRs were expressed differently between the two populations (Figure S3). It is worth mentioning that SNMPs expressed obviously different in male antenna from the two populations, which means that all the four identified SNMPs expressed significantly differently between males from the two populations (Figure 3). These differently expressed genes may have mediated the olfactory differences of the two *S. bifasciatus* populations.

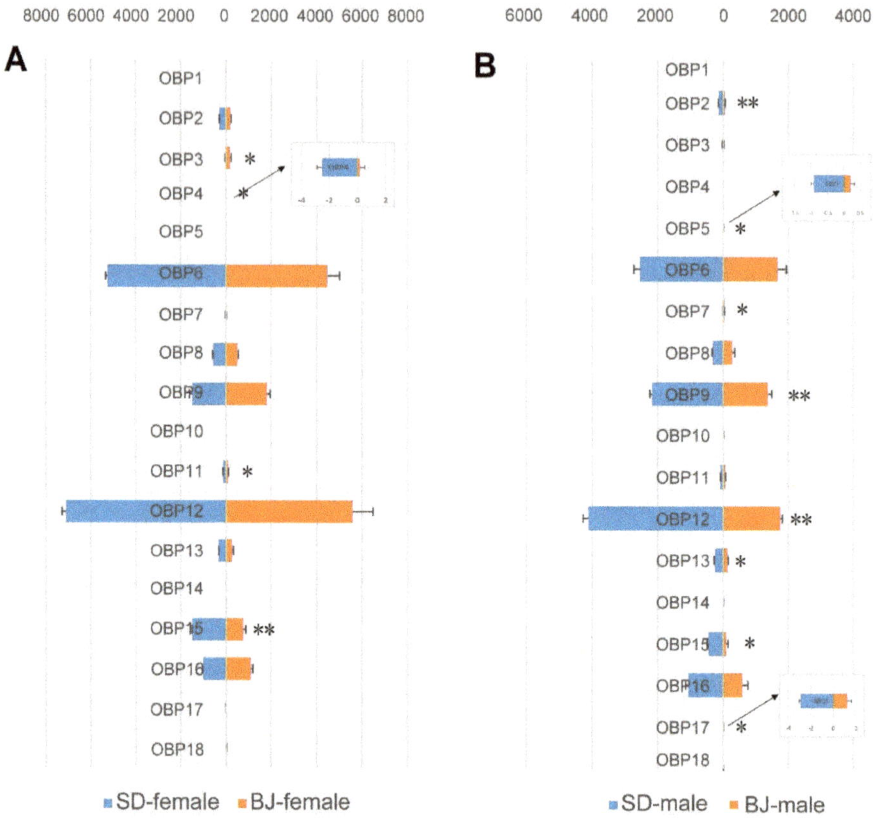

Figure 1. Comparison of odorant binding protein (OBP) genes expression levels in both sexes of the two *S. bifasciatus* populations. The horizontal axis refers to the expression level of the genes (FPKM values, mean ± SE, $n = 3$). (**A**): OBPs expression level comparison between females of the two population; (**B**): OBPs expression level comparison between males of the two population. BJ, Beijing population; SD, Shandong population. * means the difference is significant at the 0.05 level; ** means the difference is significant at the 0.01 level.

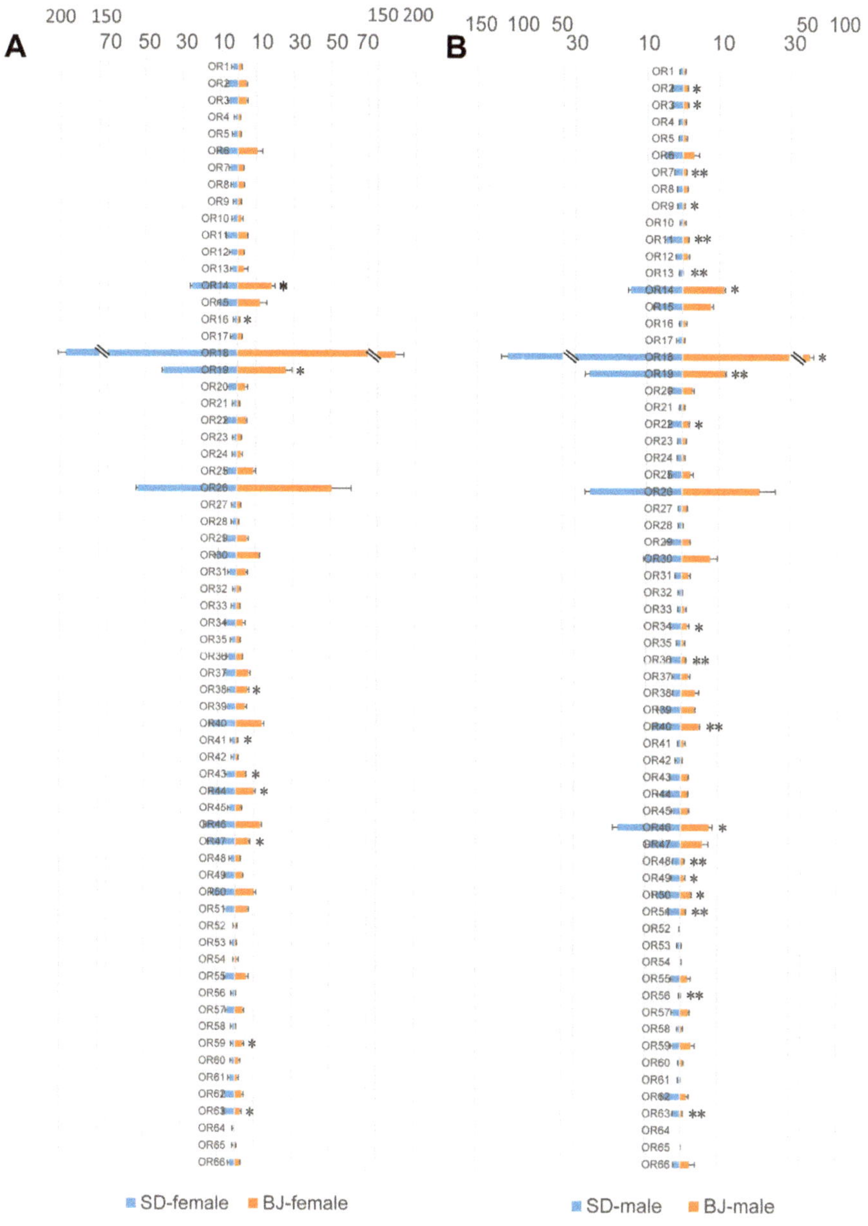

Figure 2. Comparison of odorant receptor (OR) genes expression levels in both sexes of the two *S. bifasciatus* populations. The horizontal axis refers to the expression level of the genes (FPKM values, mean ± SE, n = 3). (**A**): ORs expression level comparison between females of the two population; (**B**): ORs expression level comparison between males of the two population. BJ, Beijing population; SD, Shandong population. * means the difference is significant at the 0.05 level; ** means the difference is significant at the 0.01 level.

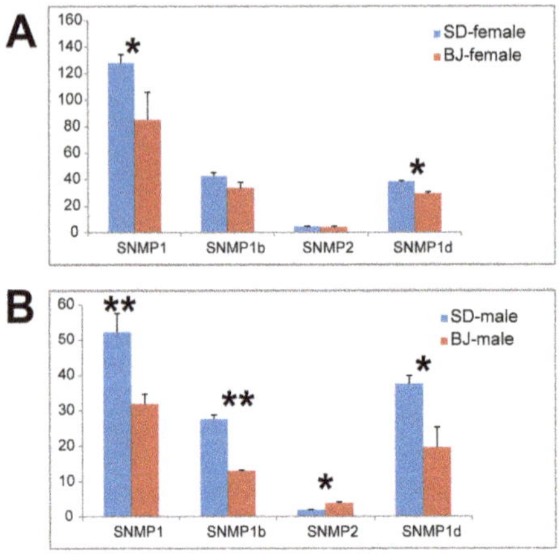

Figure 3. Comparison of sensory neuron membrane protein (SNMP) genes expression levels in both sexes of the two *S. bifasciatus* populations. The vertical axis refers to the expression level of the genes (FPKM values, mean ± SE, $n = 3$). (**A**): SNMPs expression level comparison between females of the two population; (**B**): SNMOs expression level comparison between males of the two population. BJ, Beijing population; SD, Shandong population. * means the difference is significant at the 0.05 level; ** means the difference is significant at the 0.01 level.

3.4. SNP Analysis of Chemosensory Genes in Two S. bifasciatus Populations

We collected the conserved SNP sites of the chemosensory genes of both sexes in the two *S. bifasciatus* populations, where 2 SNPs in OBPs, 2 SNPs in CSPs, 63 SNPs in ORs, 12 SNPs in GR, 16 SNPs in IRs, and 11 SNPs in SNMPs were obtained (Supplementary Table S2). Then we analyzed the SNP sites of each gene. If the acid bases at the SNP site were the same in the three repeats of Shandong population but different in that of the Beijing population, we record this SNP site as Beijing diversity SNP site; reversely, if the acid bases at the SNP site were the same in the three repeats of the Beijing population but different in that of the Shandong population, it was recorded as Shandong diversity SNP site. Similar rules were used in the SNP site recording related to sexes. Figure 4 illustrates the total SNP site numbers that we found as Beijing diversity, Shandong diversity, and diversity between sexes in each chemosensory gene type. Interestingly, most of the SNPs in OBPs and ORs were Beijing diversity ones, while fewer showed diversity in the Shandong population and only several SNPs were correlated to the sexes of *S. bifasciatus*; however, the Beijing diversity and Shandong diversity SNPs number were similar in CSPs, GRs, IRs, and SNMPs.

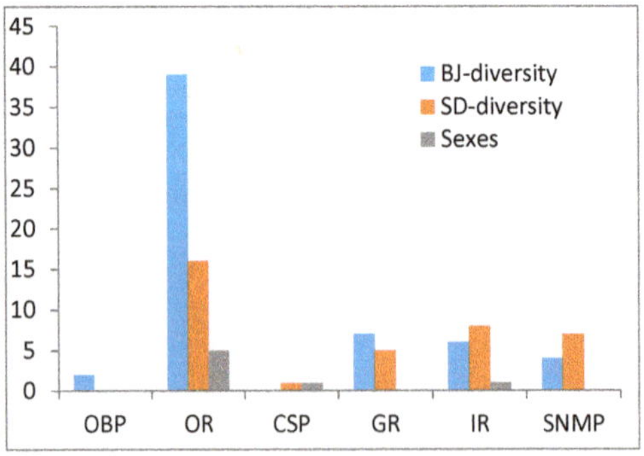

Figure 4. Three categories of collected conserved SNPs. BJ, Beijing; SD, Shandong.

4. Discussion

Little is known about the olfactory mechanisms of *S. bifasciatus*, which are very serious cypress pests [3,4]. At the same time, the olfactory response of *S. bifasciatus* populations from different places to the same attractant are varied (unpublished data). The genetic basis related to physiological differences of insects may be gene expression level differences or sequences variation [48–50]. We found that in males of *S. bifasciatus*, the expression levels of OBPs, ORs, and SNMPs of the Beijing population were obviously lower than those of the Shandong population, and most conserved SNPs of OBPs and ORs from the two populations were the ones that showed more diversity in the Beijing population.

Chemosensory mechanism studies of longhorn beetles have been few, up to now, and chemosensory genes from only a few species have been identified [51]. We identified 18 OBPs, 21 CSPs, 66 ORs, 24 GRs, 14 IRs, and 4 SNMPs in *S. bifasciatus*, and these numbers were compared to those of *Anoplophora glabripennis* (42 OBPs, 12 CSPs, 37 ORs, 11 GRs, 4 IRs, and 2 SNMPs), a longhorn beetle whose genome has been sequenced [52]. Thus, our results applied the chemosensory genes of *S. bifasciatus* relatively completely, and this will facilitate the chemosensory study of *S. bifasciatus* and longhorn beetles.

Previous trapping variation in *S. bifasciatus* from two different sites indicated that the olfactory systems of these two populations were different (unpublished data). The reasons could be quantitative or qualitative diversities of their chemosensory genes [48–50]. Thus, to explore the quantitative differences in the olfactory systems of these two *S. bifasciatus* populations, we analyzed the expression levels of *S. bifasciatus* from the two sites. Interestingly, both of these aspects were different in these two *S. bifasciatus* populations. Many OBP, OR, and SNMP genes expressions were lower in males of *S. bifasciatus* from Beijing than those from Shandong. However, the attracting effect of the lure in the Shandong population was worse than in the Beijing population (unpublished data). We can deduce that either the olfactory levels were not the main reason for the attraction differences, or that the expression level was negatively related to the attraction effects of the lure we used.

To analyze the qualitative differences between the olfactory systems of the two *S. bifasciatus* populations, olfactory gene SNP analysis was performed. For OBPs, ORs, the SNP sites referring to the diversity of the Beijing population were more than that of the Shandong population, which may explain the fact that our lure could attract the Beijing population better, as the diversity of the chemosensory genes may contribute to the recognition of lures [53]. Few studies related to the SNP variations between populations of forest pests have been carried out [53,54], and our work has applied a new sight-to-reason analysis of the population olfactory differences of the same forest pests. However, further analysis of OBP and OR gene functions is necessary to explain the chemosensory recognition diversity of the two *S. bifasciatus* populations.

5. Conclusions

To explore the olfactory variety of *S. bifasciatus*, we used next-generation sequencing to analyze the antennal transcriptome of both sexes of the two populations. The olfactory-related genes were identified. Furthermore, expression levels and SNPs of the olfactory-related genes between the two populations were analyzed. We found that in males, the expression levels of many OBP, OR, and SNMP genes of *S. bifasciatus* from Beijing were obviously lower than those from Shandong, and most conserved SNPs in OBPs and ORs from the two populations showed more diversity in the Beijing population. This work provides a foundation for future research into the molecular olfactory mechanisms of *S. bifasciatus*, as well as other longhorn beetles, and a basis for the development of new and accurate attractants for the management of this forest pest.

Supplementary Materials: The following are available online at http://www.mdpi.com/1999-4907/10/8/655/s1, Table S1: Identified chemosensory genes in *S. bifasciatus* and their unigene ID, Table S2: Collected conserved SNPs in two *S. bifasciatus* populations, Figure S1: Comparison of chemosensory protein (CSP) genes expression levels in both sexes of the two *S. bifasciatus* populations. The horizontal axis refers to the expression level of the genes (FPKM values, mean ± SE, $n = 3$), Figure S2: Comparison of gustatory receptor (GR) genes expression levels in both sexes of the two *S. bifasciatus* populations. The horizontal axis refers to the expression level of the genes (FPKM values, mean ± SE, $n = 3$), Figure S3: Comparison of ionotropic receptor (IR) genes expression levels in both sexes of the two *S. bifasciatus* populations. The vertical axis refers to the expression level of the genes (FPKM values, mean ± SE, $n = 3$).

Author Contributions: S.Z. (Sufang Zhang) designed the experiments and performed the bioinformatic analysis, as well as drafting the manuscript; S.S. assisted in identification of the chemosensory genes; S.Z. (Shiyu Zhang), X.K., H.W., and F.L. collected the insects.; Z.Z. designed the experiments and modified the manuscript.

Funding: This research was funded by The National Key Research and Development Program of China, grant number 2018YFC1200400 and 2017YFD0600102, The Fundamental Research Funds for the Central Non-profit Research Institution of CAF, grant number CAFYBB2018SZ006 and CAFYBB2017QB003, and The National Nature Science Foundation of China, grant number 31670657.

Acknowledgments: We thank the BioMarker Company for transcriptome sequencing.

Conflicts of Interest: The authors declare no conflicts of interest.

References

1. Gao, S.; Xu, Z.; Gong, X. Progress in research on *semanotus bifasciatus*. *For. Pest Dis.* **2007**, *26*, 19–22.
2. Wang, S. *Quarantine Objects of Forest Plant in China*; China Forestry Publishing House: Beijing, China, 1996.
3. Wu, X.; Wang, J.; Liu, H.; Dong, l.; Jin, Y. Chemical analysis and electroantennogram responses in *Semanotus bifasciatus* adults to *Platycladus orientalis*. *Chin. Bull. Entomol.* **2007**, *44*, 671–675.
4. Kong, X.; Zhang, Z.; Wang, H.; Yang, J.; Hu, Y. Analysis of *Platycladus orientalis* volatiles and their elecctroantennogram responses with *Semanotus bifasciatus*. *For. Res.* **2005**, *18*, 260–266.
5. Sun, Y. *Sensilla and Behavioral Responses of Adults of Semanotus bifasciatus (Coleopetera: Cerambycidae) to Volatile Compounds of Platycladus orientalis*; Beijing Forest Universtiy: Beijing, China, 2008.
6. Su, C.Y.; Menuz, K.; Carlson, J.R. Olfactory perception: Receptors, cells, and circuits. *Cell* **2009**, *139*, 45–59. [CrossRef]
7. Vogt, R.G.; Miller, N.E.; Litvack, R.; Fandino, R.A.; Sparks, J.; Staples, J.; Friedman, R.; Dickens, J.C. The insect SNMP gene family. *Insect Biochem. Mol. Biol.* **2009**, *39*, 448–456. [CrossRef]
8. Zhang, S.; Zhang, Z.; Wang, H.; Kong, X. Antennal transcriptome analysis and comparison of olfactory genes in two sympatric defoliators, *Dendrolimus houi* and *Dendrolimus kikuchii* (lepidoptera: Lasiocampidae). *Insect Biochem. Mol. Boil.* **2014**, *52*, 69–81. [CrossRef]
9. Benton, R.; Vannice, K.S.; Gomez-Diaz, C.; Vosshall, L.B. Variant ionotropic glutamate receptors as chemosensory receptors in drosophila. *Cell* **2009**, *136*, 149–162. [CrossRef]
10. Touhara, K.; Vosshall, L.B. Sensing odorants and pheromones with chemosensory receptors. *Annu. Rev. Physiol.* **2009**, *71*, 307–332. [CrossRef]
11. Kaupp, U.B. Olfactory signaling in vertebrates and insects: Differences and commonalities. *Nat. Rev. Neurosci.* **2010**, *11*, 188–200. [CrossRef]
12. Kwon, J.Y.; Dahanukar, A.; Weiss, L.A.; Carlson, J.R. The molecular basis of CO_2 reception in drosophila. *Proc. Natl. Acad. Sci. USA* **2007**, *104*, 3574–3578. [CrossRef]
13. Robertson, H.M.; Kent, L.B. Evolution of the gene lineage encoding the carbon dioxide receptor in insects. *J. Insect Sci.* **2009**, *9*, 19. [CrossRef]
14. Clyne, P.J.; Warr, C.G.; Freeman, M.R.; Lessing, D.; Kim, J.; Carlson, J.R. A novel family of divergent seven-transmembrane proteins: Candidate odorant receptors in *Drosophila*. *Neuron* **1999**, *22*, 327–338. [CrossRef]
15. Vosshall, L.B.; Amrein, H.; Morozov, P.S.; Rzhetsky, A.; Axel, R. A spatial map of olfactory receptor expression in the *Drosophila* antenna. *Cell* **1999**, *96*, 725–736. [CrossRef]
16. Hallem, E.A.; Carlson, J.R. Coding of odors by a receptor repertoire. *Cell* **2006**, *125*, 143–160. [CrossRef]
17. Vosshall, L.B.; Hansson, B.S. A unified nomenclature system for the insect olfactory coreceptor. *Chem. Senses* **2011**, *36*, 497–498. [CrossRef]

18. Vosshall, L.B.; Stocker, R.F. Molecular architecture of smell and taste in *Drosophila*. *Annu. Rev. Neurosci.* **2007**, *30*, 505–533. [CrossRef]
19. Sanchez-Gracia, A.; Vieira, F.G.; Rozas, J. Molecular evolution of the major chemosensory gene families in insects. *Heredity* **2009**, *103*, 208–216. [CrossRef]
20. Vogt, R.G. Biochemical diversity of odor detection: OBPs, ODEs and SNMPs. In *Insect Pheromone Biochemistry and Molecular Biology*; Academic Press: London, UK, 2003; pp. 391–446.
21. Pelosi, P.; Zhou, J.J.; Ban, L.P.; Calvello, M. Soluble proteins in insect chemical communication. *Cell. Mol. Life Sci.* **2006**, *63*, 1658–1676. [CrossRef]
22. Engsontia, P.; Sanderson, A.P.; Cobb, M.; Walden, K.K.O.; Robertson, H.M.; Brown, S. The red flour beetle's large nose: An expanded odorant receptor gene family in *Tribolium castaneum*. *Insect Biochem. Mol. Biol.* **2008**, *38*, 387–397. [CrossRef]
23. Krieger, J.; Grosse-Wilde, E.; Gohl, T.; Dewer, Y.M.E.; Raming, K.; Breer, H. Genes encoding candidate pheromone receptors in a moth (*Heliothis virescens*). *Proc. Natl. Acad. Sci. USA* **2004**, *101*, 11845–11850. [CrossRef]
24. Robertson, H.M.; Wanner, K.W. The chemoreceptor superfamily in the honey bee, *Apis mellifera*: Expansion of the odorant, but not gustatory, receptor family. *Genome Res.* **2006**, *16*, 1395–1403. [CrossRef]
25. Tanaka, K.; Uda, Y.; Ono, Y.; Nakagawa, T.; Suwa, M.; Yamaoka, R.; Touhara, K. Highly selective tuning of a silkworm olfactory receptor to a key mulberry leaf volatile. *Curr. Biol.* **2009**, *19*, 881–890. [CrossRef]
26. Gong, D.-P.; Zhang, H.-J.; Zhao, P.; Xia, Q.-Y.; Xiang, Z.-H. The odorant binding protein gene family from the genome of silkworm, *Bombyx mori*. *BMC Genom.* **2009**, *10*, 332. [CrossRef]
27. Zhou, J.-J.; Kan, Y.; Antoniw, J.; Pickett, J.A.; Field, L.M. Genome and est analyses and expression of a gene family with putative functions in insect chemoreception. *Chem. Senses* **2006**, *31*, 453–465. [CrossRef]
28. Zhou, J.J.; He, X.L.; Pickett, J.A.; Field, L.M. Identification of odorant-binding proteins of the yellow fever mosquito *Aedes aegypti*: Genome annotation and comparative analyses. *Insect Mol. Biol.* **2008**, *17*, 147–163. [CrossRef]
29. Bengtsson, J.M.; Trona, F.; Montagné, N.; Anfora, G.; Ignell, R.; Witzgall, P.; Jacquin-Joly, E. Putative chemosensory receptors of the codling moth, *Cydia pomonella*, identified by antennal transcriptome analysis. *PLoS ONE* **2012**, *7*, e31620. [CrossRef]
30. Grosse-Wilde, E.; Kuebler, L.S.; Bucks, S.; Vogel, H.; Wicher, D.; Hansson, B.S. Antennal transcriptome of *Manduca sexta*. *Proc. Natl. Acad. Sci. USA* **2011**, *108*, 7449–7454. [CrossRef]
31. Legeai, F.; Malpel, S.; Montagne, N.; Monsempes, C.; Cousserans, F.; Merlin, C.; Francois, M.-C.; Maibeche-Coisne, M.; Gavory, F.; Poulain, J.; et al. An expressed sequence tag collection from the male antennae of the noctuid moth *Spodoptera littoralis*: A resource for olfactory and pheromone detection research. *BMC Genom.* **2011**, *12*, 86. [CrossRef]
32. Zhang, Y.-N.; Jin, J.-Y.; Jin, R.; Xia, Y.-H.; Zhou, J.-J.; Deng, J.-Y.; Dong, S.-L. Differential expression patterns in chemosensory and non-chemosensory tissues of putative chemosensory genes identified by transcriptome analysis of insect pest the purple stem borer *Sesamia inferens* (walker). *PLoS ONE* **2013**, *8*, e69715.
33. Mitchell, R.F.; Hughes, D.T.; Luetje, C.W.; Millar, J.G.; Soriano-Agatmn, F.; Hanks, L.M.; Robertson, H.M. Sequencing and characterizing odorant receptors of the cerambycid beetle Megacyllene caryae. *Insect Biochem. Mol. Biol.* **2012**, *42*, 499–505. [CrossRef]
34. Andersson, M.; Grosse-Wilde, E.; Keeling, C.; Bengtsson, J.; Yuen, M.; Li, M.; Hillbur, Y.; Bohlmann, J.; Hansson, B.; Schlyter, F. Antennal transcriptome analysis of the chemosensory gene families in the tree killing bark beetles, *Ips typographus* and *Dendroctonus ponderosae* (coleoptera: Curculionidae: Scolytinae). *BMC Genom.* **2013**, *14*, 1–16. [CrossRef]
35. Zhang, S.-F.; Liu, H.-H.; Kong, X.-B.; Wang, H.-B.; Liu, F.; Zhang, Z. Identification and expression profiling of chemosensory genes in *Dendrolimus punctatus* walker. *Front. Physiol.* **2017**, *8*, 8. [CrossRef]
36. Bogdanova, E.A.; Shagin, D.A.; Lukyanov, S.A. Normalization of full-length enriched cDNA. *Mol. BioSystems* **2008**, *4*, 205–212. [CrossRef]
37. Zhulidov, P.A.; Bogdanova, E.A.; Shcheglov, A.S.; Vagner, L.L.; Khaspekov, G.L.; Kozhemyako, V.B.; Matz, M.V.; Meleshkevitch, E.; Moroz, L.L.; Lukyanov, S.A.; et al. Simple cDNA normalization using kamchatka crab duplex-specific nuclease. *Nucleic Acids Res.* **2004**, *32*, e37. [CrossRef]

38. Grabherr, M.G.; Haas, B.J.; Yassour, M.; Levin, J.Z.; Thompson, D.A.; Amit, I.; Adiconis, X.; Fan, L.; Raychowdhury, R.; Zeng, Q.; et al. Trinity: Reconstructing a full-length transcriptome without a genome from RNA-seq data. *Nat. Biotechnol.* **2011**, *29*, 644–652. [CrossRef]
39. Pertea, G.; Huang, X.; Liang, F.; Antonescu, V.; Sultana, R.; Karamycheva, S.; Lee, Y.; White, J.; Cheung, F.; Parvizi, B.; et al. Tigr gene indices clustering tools (tgicl): A software system for fast clustering of largest datasets. *Bioinformatics* **2003**, *19*, 651–652. [CrossRef]
40. Conesa, A.; Götz, S.; García-Gómez, J.M.; Terol, J.; Talón, M.; Robles, M. Blast2go: A universal tool for annotation, visualization and analysis in functional genomics research. *Bioinformatics* **2005**, *21*, 3674–3676. [CrossRef]
41. Götz, S.; García-Gómez, J.M.; Terol, J.; Williams, T.D.; Nagaraj, S.H.; Nueda, M.J.; Robles, M.; Talón, M.; Dopazo, J.; Conesa, A. High-throughput functional annotation and data mining with the blast2go suite. *Nucleic Acids Res.* **2008**, *36*, 3420–3435. [CrossRef]
42. Ashburner, M.; Ball, C.A.; Blake, J.A.; Botstein, D.; Butler, H.; Cherry, J.M.; Davis, A.P.; Dolinski, K.; Dwight, S.S.; Eppig, J.T.; et al. Gene ontology: Tool for the unification of biology. *Nat. Genet.* **2000**, *25*, 25–29. [CrossRef]
43. Trapnell, C.; Williams, B.A.; Pertea, G.; Mortazavi, A.; Kwan, G.; van Baren, M.J.; Salzberg, S.L.; Wold, B.J.; Pachter, L. Transcript assembly and quantification by RNA-seq reveals unannotated transcripts and isoform switching during cell differentiation. *Nat. Biotechnol.* **2010**, *28*, 511–515. [CrossRef]
44. Anders, S.; Huber, W. *Differential Expression of RNA-seq Data at the Gene Level—The Deseq Package*; European Molecular Biology Laboratory (EMBL): Heidelberg, Germany, 2012.
45. Bullard, J.H.; Purdom, E.; Hansen, K.D.; Dudoit, S. Evaluation of statistical methods for normalization and differential expression in mRNA-seq experiments. *BMC Bioinform.* **2010**, *11*, 94. [CrossRef]
46. Noble, W.S. How does multiple testing correction work? *Nat. Biotechnol.* **2009**, *27*, 1135–1137. [CrossRef]
47. McKenna, A.; Hanna, M.; Banks, E.; Sivachenko, A.; Cibulskis, K.; Kernytsky, A.; Garimella, K.; Altshuler, D.; Gabriel, S.; Daly, M.; et al. The genome analysis toolkit: A mapreduce framework for analyzing next-generation DNA sequencing data. *Genome Res.* **2010**, *20*, 1297–1303. [CrossRef]
48. He, Y.Q.; Feng, B.; Guo, Q.S.; Du, Y. Age influences the olfactory profiles of the migratory oriental armyworm mythimna separate at the molecular level. *BMC Genom.* **2017**, *18*, 32. [CrossRef]
49. Qiu, C.Z.; Zhou, Q.Z.; Liu, T.-T.; Fang, S.M.; Wang, Y.W.; Fang, X.; Huang, C.L.; Yu, Q.Y.; Chen, C.H.; Zhang, Z. Evidence of peripheral olfactory impairment in the domestic silkworms: Insight from the comparative transcriptome and population genetics. *BMC Genom.* **2018**, *19*, 788. [CrossRef]
50. Gadenne, C.; Barrozo, R.B.; Anton, S. Plasticity in insect olfaction: To smell or not to smell? *Annu. Rev. Èntomol.* **2016**, *61*, 317–333. [CrossRef]
51. Zhan, W.; Zhang, S.; Geng, H.; Wang, Y.; Guo, K.; Chen, J. Research progress on olfactory recognition proteins of longhorn beetle. *J. Henan Agric. Sci.* **2018**, *47*, 1–6.
52. McKenna, D.D.; Scully, E.D.; Pauchet, Y.; Hoover, K.; Kirsch, R.; Geib, S.M.; Mitchell, R.F.; Waterhouse, R.M.; Ahn, S.-J.; Arsala, D.; et al. Genome of the Asian longhorned beetle (*Anoplophora glabripennis*), a globally significant invasive species, reveals key functional and evolutionary innovations at the beetle–plant interface. *Genome Boil.* **2016**, *17*, 227. [CrossRef]
53. Wang, P.; Lyman, R.F.; Mackay, T.F.C.; Anholt, R.R.H. Natural variation in odorant recognition among odorant-binding proteins in drosophila melanogaster. *Genetics* **2010**, *184*, 759–767. [CrossRef]
54. Wang, P.; Lyman, R.F.; Shabalina, S.A.; Mackay, T.F.C.; Anholt, R.R.H. Association of polymorphisms in odorant-binding protein genes with variation in olfactory response to benzaldehyde in drosophila. *Genetics* **2007**, *177*, 1655–1665. [CrossRef]

forests

MDPI

Article

The Spring Assessing Method of the Threat of *Melolontha* spp. grubs for Scots Pine Plantations

Hanna Szmidla [1,*], Monika Małecka [1], Miłosz Tkaczyk [1], Grzegorz Tarwacki [1] and Zbigniew Sierota [2]

1 Department of Forest Protection, Forest Research Institute, Braci Leśnej 3, Sękocin Stary, 05-090 Raszyn, Poland; m.malecka@ibles.waw.pl (M.M.); m.tkaczyk@ibles.waw.pl (M.T.); g.tarwacki@ibles.waw.pl (G.T.)

2 Department of Forestry and Forest Ecology, Faculty of Environmental Management and Agriculture, Warmia and Mazury University in Olsztyn, Pl. Łódzki 2, 10-727 Olsztyn, Poland; zbigniew.sierota@uwm.edu.pl

* Correspondence: h.szmidla@ibles.waw.pl; Tel.: +48-22-715-0353

Received: 18 March 2019; Accepted: 7 May 2019; Published: 9 May 2019

Abstract: Root-feeding Melolonthinae larvae are a forest pest species in Europe that can exert serious damage. In Poland, they are classified as the most dangerous pest on land dedicated to afforestation and the most serious threat to natural regeneration in the stands. This study was performed in three forest districts in east Poland (Lubartów, Marcule, and Wyszków forest districts) in mixed conifer forests, where the presence of *Melolontha* spp. grubs was evaluated in autumn and spring of 2012 to 2017, respectively. In spring 2012, 2014, and 2016, consecutively 'small sawdust pits' in rows between seedlings were prepared, and in adjacent inter-rows similar control pits without sawdust were marked. In spring and autumn of the following year, sawdust and soil from both types of pits were sieved and Melolonthinae larvae were counted and compared. More grubs were found in sawdust pits in spring than in autumn. In Marcule Forest District (FD) (2014), more grubs were found in inter-rows than in rows with seedlings, when compared to grubs detected using the traditional method of "autumn large pits assessment", recommended by Polish forest rules. The Melolonthinae population size and location of grubs were related to the weather conditions in the evaluated periods, as well. We conclude that to properly assess the cockchafer threat, it is necessary to perform spring assessment and to search in inter-rows. This could be recommended for decisions on control.

Keywords: Cockchafer larvae; forestry; Scots pine damages; small soil pits

1. Introduction

Root-feeding insects are key components of many terrestrial ecosystems. The detrimental effect of below ground insect herbivory in agricultural and forest ecosystems is widely appreciated, because root-damaging pests cause great economic losses [1,2]. Among the many insects that have substantial impacts on forest management in European countries, the genus of *Melolontha* spp. (Coleoptera: Scarabaeidae) are especially important. The cockchafer genus, *Melolontha* spp., is estimated to have been a pest on 200,000 hectares, mainly in Central Europe, over the past 20 years [3–5].

In Poland, the most important species are *Melolontha melolontha* L. (common or May cockchafer), and *M. hippocastani* Fabr. (forest cockchafer) [6–8]. These species occur at varied frequency throughout the country, but are commonly found in the central and southeastern parts of Poland. The area of occurrence of the two species considered as pests has increased in recent years, from 10.4 thousand hectares in 2016 to 18.3 thousand hectares in 2017 [9]. In fact, in certain regions in Poland, the high population density of *Melolontha* spp. grubs makes reforestation or afforestation impossible.

Melolontha spp. have a 3- to 5-year life cycle: In Central Europe, the life cycle is typically four years [10,11]. *Melolontha melolontha* L and *Melolontha hippocastani* Fabr. have very similar biology [12]. During the life cycle, the insect undergoes a complete transformation characterized by varying morphology and behavior at each life stage. The grubs (larvae) are more problematic in forests because they damage the roots of seedlings and trees. [10,13,14]. Dead roots of trees and plants are colonized and decomposed by fungi and bacteria, and released carbon dioxide serves to attract grubs to migrate to the root system and to feed on living roots [15,16]. The most intensive feeding periods for grubs take place after molting (autumn) and after vertical migration to the surface in the spring; it is during these periods that the greatest damage to tree roots occurs. Adult beetles feed in tree crowns, causing defoliation, reduced photosynthetic capacity, and weakness in the trees [17].

Until 2010, pesticides were used to effectively control both adults and larvae [14,18]. Two legal acts adopted in 2009 have had considerable influence on the extent and form of allowable pesticide use: (a) Directive 2009/128/EC of the European Parliament and of the Council on 21 October 2009 established a framework for community action to achieve sustainable use of pesticides; and (b) Regulation (EC) no. 1107/2009 of the European Parliament and of the Council on 21 October 2009 regulated the placement of plant protection products on the market and repealed Council Directives 79/117/EEC and 91/414/EEC. These acts limited chemical control of the cockchafer population, shifting management attention to natural (birds after soil ploughing, use of antifeedants, e.g., buckwheat) or biological (entomopathogenic fungi, parasitic nematodes) methods of control. The permitted insecticides are used as a last resort. In order to practically apply one of these non-chemical methods, it is necessary to know the current threat of the area designated for afforestation or renewal, risk assessment, and decision making.

Globally, field methods used in the monitoring of root pests are highly diverse. To monitor the threat represented by the cockchafer and its larval stage in Australia, 5 to 10 soil cores of a 10 cm depth were taken before pupation [19]. In Germany, a 4×1 m^2 Goettinger frame was used to search for cockchafer larvae in soil [20]. In Austria, Pernfuss et al. [21] dug three square holes per plot (20×20 cm wide and 20 cm deep) and sampled garden chafer (*Phyllopertha horticola*) larvae from these locations before and after control treatment. In Poland, the threat from *Melolontha* spp. larvae to young plantations is assessed in autumn, one year before tree planting. According to the Polish Instruction of Forest Protection [22], this threat assessment is performed between 15 September and 30 October by counting the number of grubs in the soil in at least six large pits ($1 \times 0.5 \times 0.5$ m) placed in each hectare of afforested or restored area. The age of the grubs is determined by the width of the head capsule—one-year up to 2.5 mm, two-year up to 4.0 mm, and three-year and older above 4.0 mm [22]. The number of grubs in one pit exceeding five one-year grubs, four two-year grubs, or three three-year and older grubs specifies the need for protective treatment.

An alternative method has been described where the *Melolontha* spp. threat is assessed in spring in small soil pits ($0.2 \times 0.2 \times 0.3$ m) that are easier to excavate and have been filled with fresh pine sawdust [23]. Sawdust is a nutritional base for colonization by local populations of bacteria and fungi involved in cellulose and lignin degradation [24–27]. This degradation produces: (1) Many antibiotics, inhibitors, and biopesticides, therefore shaping some microbial properties of soils and acting as belowground arthropod attractants [28]; and (2) carbon dioxide, released from the decayed cellulose of sawdust, as well other volatile compounds [16], exert an aggregative impact on the activity of larvae of *Melolontha* spp. [15,16,29,30].

The hypothesis of this study was that the 'small sawdust pits' method allows better determination of the real spring threat of *Melolontha* spp. to seedlings and better indicates the location of grubs in the plantation, compared to the traditional large pit method used in Poland. Additionally, the impact of weather conditions on a local Melolonthinae population was assessed. The results of this work could be used to modify the traditional method of hazard assessment from grubs in afforestation.

2. Materials and Methods

Experimental plots were located in three forest districts (FD): Lubartów (51°23′16.8″ N; 22°37′32.4″ E), Marcule (51°3′58″ N; 21°17′12″ E), and Wyszków (52°43′15″ N; 21°39′03″ E). The field studies were conducted in Lubartów FD in 2012 to 2013, in Marcule FD in 2014 to 2015, and in Wyszków FD in 2016 to 2017. Before planting, the area was an 80 to 100 years old Scots pine in fresh mixed coniferous *Oxalis-Myrtillus* type (OMT) stands, harvested two years before. The area was partially covered with grass and *Myrtillus*, and soil was prepared with a LPZ-75 plough. All evaluated forest districts are situated in the Middle European Plain Province of the Middle Europa Megaregion [31], with similar winter temperatures (IIIrd soil freezing class, according to European Act (PN)-EN-12831) (Figure 1), and annual precipitation classified by Bojarczuk et al. [32] as the continental climatic zone.

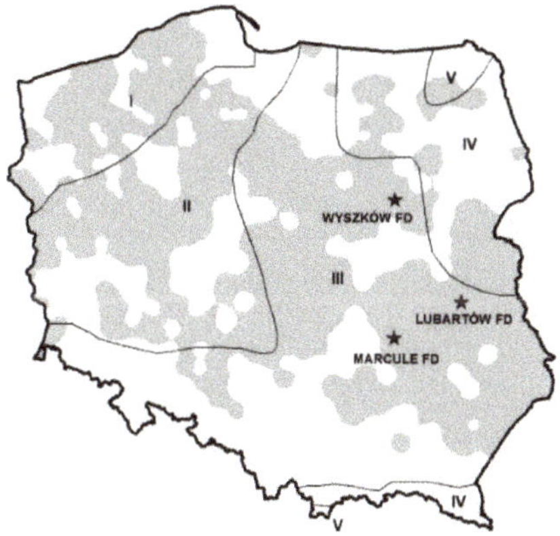

Figure 1. Area of occurrence of Melolonthinae in Poland in 2017 (grey), borders of climatic zones (lines), and locality of experimental plots (stars) in tested forest districts (FD).

One year before the investigations, all experimental areas were considered to have a strong threat from cockchafer larvae to intended Scots pine plantations. The threat was based on the high mean number of cockchafer larvae found in soil checked in the autumn of the year preceding afforestation; this evaluation used one "big soil pit", 1 × 1 × 0.5 m (width/length/depth), for every 5 hectares, according to Polish rules (IOL 2012). The mean number of grubs was approximately 4 three-year grubs/pit, which meant a serious threat and could lead directly to the destruction of the stand [22,33].

In April 2012, 2014, and 2016, respectively, one-year-old seedlings originating from a local open nursery were planted manually, spaced 0.7 m between seedlings in a row and 1.2 m between rows, in each focal area. Within each forest district, three 15-m-long plots were randomly assigned. Each plot contained eight rows of seedlings. In May, 18 holes (0.2 × 0.2 × 0.3 m) were dug between seedlings in each experimental area. The holes, called "small pits", were spaced evenly in every row (Figure 2). Each hole was filled with fresh, moist Scots pine sawdust (0.2 m deep) and covered with 10 cm local soil. Sawdust was sampled from each pit: (i) At the end of September (for an autumn rating), and (ii) at the end of May (for a spring rating, one year after planting). Samples were sieved through sieves with a mesh size of 2 mm. At the same time, 18 similar holes were dug in the interstices, located 50 cm away from the seedlings and two one side of the adjacent interrow, and the soil from these holes was similarly sieved (Figure 2).

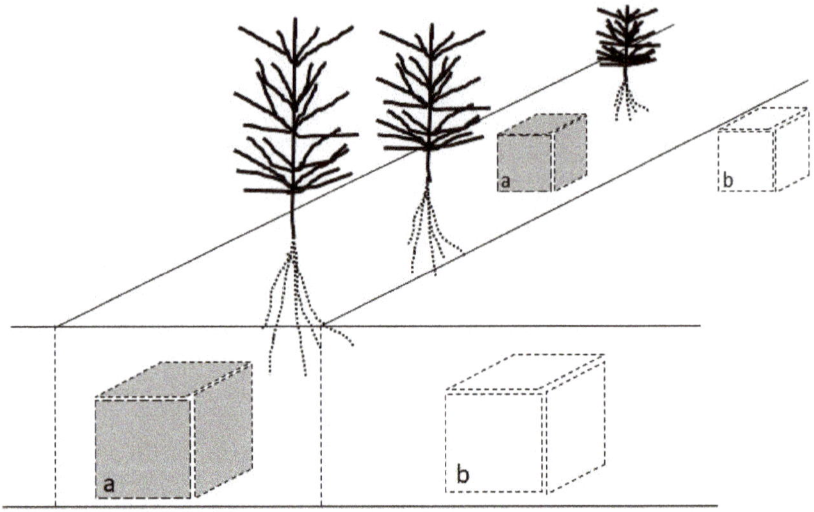

Figure 2. Location of small pits with sawdust (**a**) between seedlings in the row, and control pit without sawdust (**b**) in the interrow.

In the Marcule FD in autumn 2014, we performed an additional experiment in three other plots, in two replications. The additional pits were dug in rows: (i) Between seedlings, and (ii) in the nearest inter-row, at a distance of 50 cm from seedlings, and sieved together in autumn with (iii) a similar pit assessment of the soil directly under neighboring seedlings (across these three treatments there were 108 pits altogether, resulting in sieving of a total of 3×1.3 m^3 of soil).

Larvae remaining on the sieve were counted and compared. The total number of larvae in all small pits, per unit volume (0.012 m^3 or 0.216 m^3 per each area), was converted to an estimate of the number of larvae expected to occupy the volume of the big soil pits (0.5 m^3), according to the Instruction of Forest Protection [22].

The values of monthly air temperature and precipitation in the forest districts for the period of 2011–2017 were obtained courteously from the Institute of Meteorology and Water Management (IMGW Warsaw). For the vegetation growth season (April–October), the values of the Selyaninov hydrothermal coefficient (HTC) were calculated according to the formula:

$$HTC = 10 \times P/\Sigma\ t,$$

where P is the monthly sum of precipitation and Σ t is the sum of the average daily temperatures in the month, respectively, in the vegetation growth period, during which the temperature should be above 10 °C [34]. The Selyaninov HTC coefficient is often used in agriculture and horticulture as a synthetic index of the precipitation and temperature dependence in a given period, defining the weather in the range from hot drought (HTC < 1 means drought) to cold and excessive precipitation [35,36].

Results were analyzed in two ways. First, for every forest district, differences between the number of grubs in control pits and the number of grubs in sawdust pits were determined. In the second part of the analysis, the number of grubs was compared separately between the dates of the performance of small pitches (spring, autumn) for each of the experimental variants. In both cases, the non-parametric Mann-Whitney test was used to evaluate differences, since a Kolmogorov–Smirnov test found that the data deviated significantly from a normal distribution. To evaluate the correlation between the number of grubs and the hydrothermal coefficient, a simple linear regression was used. Analyses were performed in the Statistica v. 13.0 package (Dell Inc., Round Rock, TX, USA).

3. Results

In our research, the *Melolontha* grubs were counted together, regardless of the species and age, assuming that each individual is responsible for damage to the roots. On the areas tested, *M. melolontha* 3-years-old grubs dominated; *M. hippocastanei* were found singly.

The weather conditions in the period 2011–2017 were highly variable in particular months. Total precipitation in autumn 2012 and spring 2013 was rather low, whereas in spring 2014 it was above the median perennial moisture. In August 2015, precipitation was dramatically low, and the air temperature very high. In spring 2016, weather conditions were rather unfavorable for vegetation (due to low rainfall and relatively high temperatures), and spring 2017 was dry and cold, whereas the autumn of 2017 was wet and warm.

The values of the hydrothermal coefficient (HTC) were generally lower than the multi-annual averages for these areas, especially in the particularly dry years of 2012 and 2016 (Figure 3). In 2014 and 2017, the rainfall was higher and the temperature was rather low, which resulted in more favorable conditions for Scots pine cultures.

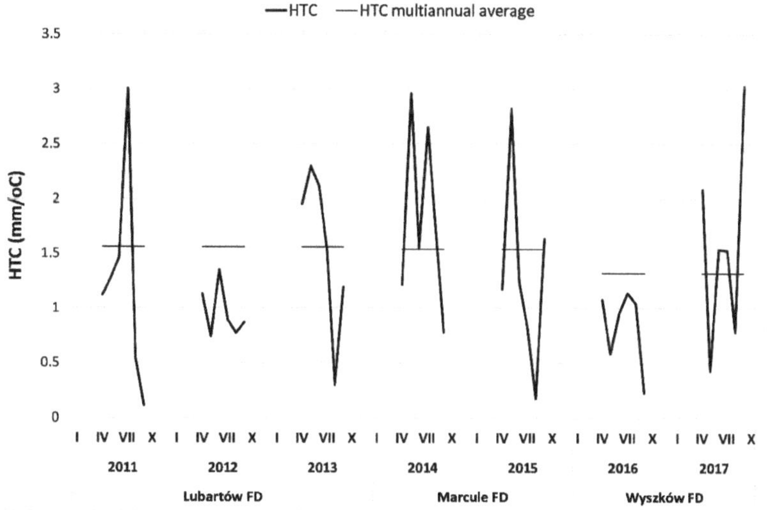

Figure 3. Monthly values of the hydrothermal coefficient (HTC) for research areas (source data: Institute of Meteorology and Water Management Bulletin).

The comparison of the number of grubs of all stages and species found in sawdust pits and sawdust-free pits within each time period, assessed separately for FD districts (Figure 4), showed the following relationships:

- In the Lubartów FD, significant differences between spring pits with sawdust and spring pits without sawdust were observed (p-value = 0.038).
- In the Marcule FD, grubs were significantly more common in control pits made in autumn than in pits made in spring (p-value = 0.002); in pits with sawdust, no differences between seasons were found.
- In the Wyszków FD, cockchafer grubs were significantly more common in spring 2017 in small pits with sawdust than in pits without sawdust (p-value = 0.031).

These results suggest that it is important to perform spring assessment in sawdust pits in all forest districts. In Lubartów FD, the number of cockchafer grubs found in sawdust pits in spring was 4 times greater than in autumn of the preceding year, and in Wyszków FD the number was more than 13 times greater. The results showed that the grubs were gathering only in interrows and were not found both in pits with or without sawdust and under seedlings. The total average number of grubs in interrows between pits with sawdust was 13 for plots (5 when standardized to 0.5 m^2 [22]), and 3 (1) in interrows between pits without sawdust. For comparison, there were 2 grubs found in the large pit according to the traditional IOL method [22] (personal information from FD).

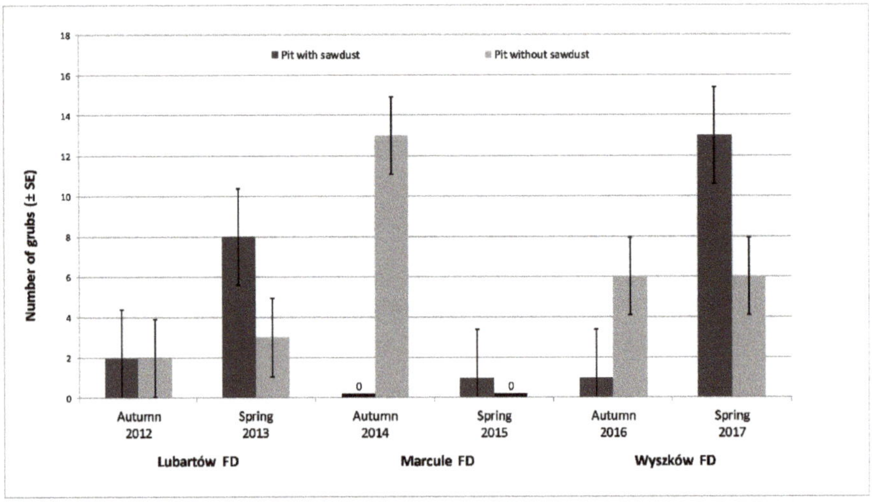

Figure 4. Comparison of the numbers of cockchafers found in pits of different treatments between evaluated periods at particular forest districts (FD).

There was a significant relationship (p-value = 0.049) between the occurrence of grubs in the pits and the hydrothermal coefficient. The value of the correlation coefficient was $R = 0.51$, which indicates the moderate strength of this relationship (Figure 5). The higher value of the hydrothermal coefficient was associated with a higher number of grubs observed in the pits, as result of the better development of plants and their higher attraction.

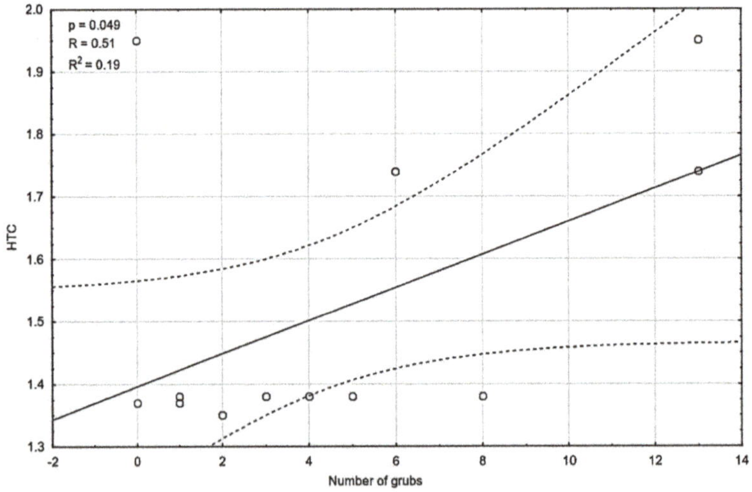

Figure 5. Relationship between the hydrothermal coefficient (HTC) and the total number of grubs in pits.

4. Discussion

Soil is a particularly heterogeneous environment, and so the range of abiotic conditions experienced by root herbivores (e.g., larvae of *Melolonthae*) can be both diverse and complex. *Melolontha* spp. are typical eurytopic organisms; their life cycle is dependent on soil moisture and temperature [17]. Barnett and Johnson [37] described that temperature and soil moisture are responsible for movements and changes in the physiological processes of root herbivores. When the temperature is low, metabolic rates decrease, and, as a consequence, larvae are predicted to have decreased movement compared to optimal temperatures [38].

With low rainfall, the grubs in this study showed a tendency to gather in small sawdust pits, where there was probably higher humidity. Similar results were obtained by Lees [39], who found that larvae of the Coleopteran genus, *Agriotes*, responded in two ways to soil moisture. First, they increased activity and migrated out of dry sand into wet sand, but if soil moisture conditions changed too rapidly, they became trapped and burrowing activity ceased.

In our experiment, we found a significant relationship between the occurrence of grubs in the pits and the hydrothermal coefficient. The higher value of the hydrothermal coefficient was related to a higher number of grubs observed in the pits. This can be explained by the fact that when temperature is low, enzymes react slowly, and, as a consequence, larvae are predicted to display decreased movement compared to times of optimal temperature [38]. An increase of the water content in the soil causes a reduction of oxygen and a significant increase of carbon dioxide levels [37]. This could reduce the movement of most insects not adapted to reduced oxygen contents. In an indirect way, soil moisture can affect the migration of root pests caused by the search for a host plant. Increased soil moisture affects the amount of gaseous molecules, like those released from plant roots and decaying matter, because it can change diffusion rates when compared to standard atmospheric air [40]. Carbon dioxide has been shown to be a major player when it comes to the plant host location by pests [41], and the larvae of *Melolontha* and many other soil insects are attracted to CO_2 [16,29,42]. Carbon dioxide dissolves more quickly in water, which could make it more difficult for pests to move towards the plant roots when soils are saturated.

During this research, we found aggregation of larvae of *Melolontha* in rows close to seedlings when conditions were unfavorable for growth (during low precipitation in spring 2013 and 2017). These conditions affect the amount and quality of food available for pests and also the plant metabolites released into the rhizosphere under stress. For example, Schenk and Jackson [43] found that the absolute rooting depth of plants in moisture-limited environments has a tendency to increase with increasing levels of precipitation. However, at the same time, some plant species have been shown to increase the size of their rooting systems when soils become drier [44]. Jupp and Newman [45] found that some plant systems increased lateral root growth from 300% to 500% when soil is dry. In contrast, dry soil conditions increase the Scots pine's susceptibility to soil acidification, and this could significantly reduce fine root growth and increase root mortality [46]. Released in the process of decomposition of dead roots, carbon dioxide may additionally lure the grubs [16].

Opinion is divided on the impact of drought on plant nutrition. Drought may increase the uptake of nutrients by plant root systems through the aggregation of organic and inorganic materials by capillary forces [44]. On the other hand, drought could increase the concentrations of terpenes and resin acids in gymnosperms [47,48]. Terpenes are the major class of compounds to which 17% of insects respond; compounds from this class act as a chemoattractant to insect species, especially those of the order, *Coleoptera*, and act as potent long-distance signals for foraging insects [49–51].

The results obtained here show that the inspection of pits in autumn of the year of cultivation did not offer satisfactory information about the real threat represented by *Melolontha* spp. to plant cultures in the year following the cultivation. Autumn inspection in the year of cultivation also did not corresponded with results of primary threat assessment in the autumn before spring cultivation. In all cases, inspection performed in the spring of the next year found that cockchafer larvae preferred pits with sawdust over control pits with forest soil and roots. The short period of this research, and the unexpected variability of the weather—exemplified by a dry 2015 followed by a wet 2016—did not allow us to generalize these results beyond the focal experiment. The great number of pits in which there were no grubs suggests that cockchafer larvae occur in clusters. The hypothesis that spring threat assessments and assessment in inter-rows are necessary in order to accurately predict the threat level from grubs was supported. This research showed that a greater number of grubs is correlated with a higher value of the HTC indicator. Sawdust in the pits keeps moisture for a longer time, which is more attractive for grubs in the dry season. The spring assessment period seems to be important due to the migration of grubs toward both developing tree roots with forming mycorrhiza and old, dead roots with residues that give off compounds and indirectly inform grubs of available food [50].

In general, the use of spring risk assessment seems valuable, as indicated by earlier results of mycological [52], bacterial [27], and other sylvicultural research [23,53] in soils subjected to sawdust. Together, these studies confirm that there is increased activity of grubs in these soils after the introduction of sawdust. In addition, small soil pits checked in spring seem to offer more precision for assessment of the severity of the threat of grubs to seedlings, when compared with the more labor-intensive and expensive method prescribed by Polish regulations [22].

5. Conclusions

- A significant relationship between the number of *Melolontha* spp grubs and the hydrothermal coefficient was found.
- It seems that the pits with sawdust checked in spring allowed the number of grubs to be determined more precisely compared to the assessment performed in autumn.

Author Contributions: M.M. and Z.S. generated the data. All authors analyzed and discussed the data. The manuscript was written by H.S., Z.S. and M.M. Statistical analysis using the Statistica v. 13.0 package by: M.T. The general conception of the project was provided by Z.S.

Funding: This work was supported by the State Forest Holding in Poland through Project No. 500 426. (contract number OR.271.3.4.2015).

Acknowledgments: The authors thank the forest staff of the Lubartów, Marcule, and Wyszków forest districts for valuable assistance during fieldwork. The authors would like to thank the anonymous reviewers for their valuable comments and suggestions.

Conflicts of Interest: The authors declare no conflict of interest.

References

1. Hunter, M.D. Root herbivory in forest ecosystems. In *Root Feeders, an Ecosystem Perspective*; Johnson, S.N., Murray, P.J., Eds.; CAB Biosciences: Ascot, UK, 2008; pp. 68–95.
2. Johnson, S.N.; Benefer, C.M.; Frew, A.; Griffiths, B.S.; Hartley, S.E.; Karley, A.J.; Rasmann, S.; Schumann, M.; Sonnemann, I.; Robert, C.A.M. New frontiers in belowground ecology for plant protection from root-feeding insects. *Appl. Soil Ecol.* **2016**, *108*, 96–107. [CrossRef]
3. Keller, S.; Brenner, H. Development of the *Melolontha* populations in the canton Thurgau, eastern Switzerland, over the last 30 years. *IOBC/WPRS Bull.* **2005**, *28*, 31. Available online: http://iobc-wprs.org/pub/bulletins/iobc-wprs_bulletin_2005_28_02.pdf#page=46 (accessed on 18 December 2018).
4. Oltean, I.; Varga, M.; Gliga, S.; Florian, T.; Bunescu, H.; Bodis, I.; Covaci, A. Monitoring *Melolontha melolontha* L. species in 2007, in the nursery from UP IV Bătrâna OS Toplița, Harghita forest district. *Bull. UASVM CN.* **2010**, *67*, 525.
5. Švestka, M. Changes in the abundance of *Melolontha hippocastani* Fabr. and *Melolontha melolontha* (L.) (Coleoptera: Scarabeidae) in the Czech Republic in the period. *J. For. Sci.* **2010**, *56*, 417–428. [CrossRef]
6. Švestka, M. Ecological conditions influencing the localization of egg-laying by females of the cockchafer (*Melolontha hippocastani* F.). *J. For. Sci.* **2007**, *53*, 16–24. [CrossRef]
7. Woreta, D.; Sukovata, L. Effect of food on development of the *Melolonta hippocastani* F. beetles (Coleoptera, Melolonthidae). *For. Res. Pap.* **2010**, *71*, 195–199. [CrossRef]

8. Wagenhoff, E.; Blum, R.; Delb, H. Spring phenology of cockchafers, *Melolontha* spp. (Coleoptera: Scarabaeidae), in forests of south-western Germany: Results of a 3-year survey on adult emergence, swarming flights, and oogenesis from 2009 to 2011. *J. For. Sci.* **2014**, *60*, 154–165. [CrossRef]
9. Sukovata, L. Szkodniki korzeni drzew i krzewów leśnych [Pests of the roots of trees and shrubs]. In *Krótkoterminowa prognoza występowania ważniejszych szkodników i chorób infekcyjnych drzew leśnych w Polsce w 2018 roku [Short-term forecast of the occurrence of major pests and infectious diseases of forest trees in 2018]*; Jabłoński Tomasz, Ed.; Forest Research Institute, Analysis and Reports: Sękocin Stary, Poland, 2018; pp. 24–27. Available online: https://www.bdl.lasy.gov.pl/portal/Media/Default/Publikacje/prognoza_2018.pdf (accessed on 1 December 2018).
10. Švestka, M. Distribution of tribes of cockchafers of the genus *Melolontha* in forests of the Czech Republic and the dependence of their swarming on temperature. *J. For. Sci.* **2006**, *52*, 520–530. [CrossRef]
11. Sierpinska, A. Observations on ecology of common cockchafer (*Melolontha melolontha* L.) and forest cockchafer (*Melolontha hippocastani* Fabr.)—Based on the research conducted in forest district Piotrkow in 2007. *Prog. Plant Prot.* **2008**, *3*, 956–965.
12. Sukovata, L.; Jaworski, T.; Karolewski, P.; Kolk, A. The performance of *Melolontha* grubs on the roots of various plant species. *Turkish J. Agric. For.* **2015**, *39*, 107–116. [CrossRef]
13. Blum, M.S. *Fundamentals of Insects Physiology*; Wiley: New York, NY, USA, 1985; ISBN 0471054682.
14. Malinowski, H. Current problems of forest protection connected with the control of cockchafers (*Melolontha* spp.). *Prog. Plant Prot. w Ochr. Roślin* **2007**, *47*, 314–322.
15. Galbreath, R.A. Orientation of grass grub *Costelytra zealandica* (Coleoptera: Scarabaeidae) to a carbon dioxide source. *N. Z. Entomol.* **1988**, *11*, 6–7. [CrossRef]
16. Weissteiner, S.; Huetteroth, W.; Kollmann, M.; Weißbecker, B.; Romani, R.; Schachtner, J.; Schütz, S. Cockchafer larvae smell host root scents in soil. *PLoS ONE* **2012**, *7*, e45827. [CrossRef] [PubMed]
17. Sierpiński, Z. *Ważniejsze owady-szkodniki korzeni drzew i krzewów leśnych*; Państwowe Wydawnictwo Rolnicze i Leśne: Warszawa, Poland, 1975.
18. Ansari, M.A.; Shah, F.A.; Tirry, L.; Moens, M. Field trials against *Hoplia philanthus* (Coleoptera: Scarabaeidae) with a combination of an entomopathogenic nematode and the fungus *Metarhizium anisopliae* CLO 53. *Biol. Control* **2006**, *39*, 453–459. [CrossRef]
19. Bailey, P.T. *Pests of Field Crops and Pastures: Identification and Control*; CSIRO publishing: Collingwood, Australia, 2007; ISBN 0643099425.
20. Benker, U.; Leuprecht, B. Field experience in the control of common cockchafer in the Bavarian region Spessart. *IOBC/WPRS Bull.* **2005**, *28*, 21–24.
21. Pernfuss, B.; Zelger, R.; Kron-Morelli, R.; Strasser, H. Control of the garden chafer *Phyllopertha horticola* with GranMet-P, a new product made of *Metarhizium anisopliae*. *IOBC/WPRS Bull.* **2005**, *28*, 47–50.
22. *Instrukcja ochrony lasu [Instruction of Forest Protection]*; CILP: Warsaw, Poland, 2012; Available online: https://www.lasy.gov.pl/pl/publikacje/copy_of_gospodarka-lesna/ochrona_lasu/instrukcja-ochrony-lasu/instrukcja-ochrony-lasu-tom-i/view and https://www.lasy.gov.pl/pl/publikacje/copy_of_gospodarka-lesna/ochrona_lasu/instrukcja-ochrony-lasu/instrukcja-ochrony-lasu-tom-ii/view; (accessed on 1 December 2018).
23. Małecka, M.; Sierota, Z.; Tarwacki, G. Effect of sawdust addition into one−year−old Scots pine plantation on number of *Melolontha* grubs. *Sylwan* **2014**, *158*, 604–613.
24. Bååth, E.; Söderström, B. Degradation of macromolecules by microfungi isolated from different podzolic soil horizons. *Can. J. Bot.* **1980**, *58*, 422–425. [CrossRef]
25. Svarstad, H.; Bugge, H.C.; Dhillion, S.S. From Norway to Novartis: Cyclosporin from *Tolypocladium inflatum* in an open access bioprospecting regime. *Biodivers. Conserv.* **2000**, *9*, 1521–1541. [CrossRef]
26. Domsch, K.H.; Gams, W.; Anderson, T.H. *Compendium of Soil Fungi*, 2nd ed.; IHW Verlag: Eching, Germany, 2007.
27. Kubiak, K.; Tkaczyk, M.; Małecka, M.; Sierota, Z. Pine sawdust as stimulator of the microbial community in post-arable afforested soil. *Arch. Agron. Soil Sci.* **2017**, *63*, 427–441. [CrossRef]
28. Wenke, K.; Kai, M.; Piechulla, B. Belowground volatiles facilitate interactions between plant roots and soil organisms. *Planta* **2010**, *231*, 499–506. [CrossRef]
29. Johnson, S.N.; Gregory, P.J. Chemically-mediated host-plant location and selection by root-feeding insects. *Physiol. Entomol.* **2006**, *31*, 1–13. [CrossRef]

30. Weissteiner, S.; Schütz, S. Are different volatile pattern influencing host plant choice of belowground living insects. *Mitt. Dtsch. Ges. Allg. Angew. Ent.* **2006**, *15*, 51–55.
31. Kondracki, J. Geografia Regionalna Polski [Physico-geographical regionalization of Poland]. *Państwowe Wydaw. Nauk. Warszawa* **2002**, 441.
32. Bojarczuk, T.; Bugala, W.; Chylarecki, H. Zrejonizowany dobór drzew i krzewów do uprawy w Polsce. *Arbor. Kórnickie* **1980**.
33. Niemczyk, M.; Karwański, M.; Grzybowska, U. Effect of environmental factors on occurrence of cockchafers (*Melolontha* spp.) in forest stands. *Baltic For.* **2017**, *23*, 334–341.
34. Żmudzka, E. The climatic background of agricultural production in Poland (1951–2000). *Misc. Geogr.* **2004**, *11*, 127–137. [CrossRef]
35. Evarte-Bundere, G.; Evarts-Bunders, P. Using of the hydrothermal coefficient (HTC) for interpretation of distribution of non-native tree species in Latvia on example of cultivated species of genus Tilia. *Acta Biol. Univ. Daugavp* **2012**, *12*, 135–148.
36. Riaznova, A.A.; Voropay, N.N.; Okladnikov, I.G.; Gordov, E.P. Development of computational module of regional aridity for web-GIS "Climate". In Proceedings of the IOP Conference Series: Earth and Environmental Science, Tomsk, Russia, 11–16 July 2016; IOP Publishing: Bristol, UK, 2016; p. 012032.
37. Barnett, K.; Johnson, S.N. Living in the soil matrix: Abiotic factors affecting root herbivores. In *Advances in Insect Physiology*; Johnson, S.N., Hiltpold, I., Turlings, T.C.J., Eds.; Elsevier: Oxford, UK, 2013; Volume 45, pp. 1–52, ISBN 0065-2806.
38. Edwards, A.C.; Cresser, M.S. Freezing and its effect on chemical and biological properties of soil. In *Advances in Soil Science*; Springer: Berlin, Germany, 1992; pp. 59–79.
39. Lees, A.D. On the behaviour of wireworms of the genus *Agriotes* Esch. (Coleoptera, Elateridae): I. Reactions to humidity. *J. Exp. Biol.* **1943**, *20*, 43–53.
40. Payne, D.; Gregory, P.J. *The Soil Atmosphere. Russell's Soil Cond. Plant Growth*; Longman: Harlow, UK, 1988; pp. 298–314.
41. Klingler, J. On the orientation of plant nematodes and of some other soil animals. *Nematologica* **1965**, *11*, 4–18. [CrossRef]
42. Bernklau, E.J.; Bjostad, L.B. Reinvestigation of host location by western corn rootworm larvae (Coleoptera: Chrysomelidae): CO_2 is the only volatile attractant. *J. Econ. Entomol.* **1998**, *91*, 1331–1340. [CrossRef]
43. Schenk, H.J.; Jackson, R.B. Rooting depths, lateral root spreads and below-ground/above-ground allometries of plants in water-limited ecosystems. *J. Ecol.* **2002**, *90*, 480–494. [CrossRef]
44. Gregory, P.J. Roots, rhizosphere and soil: The route to a better understanding of soil science? *Eur. J. Soil Sci.* **2006**, *57*, 2–12. [CrossRef]
45. Jupp, A.P.; Newman, E.I. Morphological and anatomical effects of severe drought on the roots of *Lolium perenne* L. *New Phytol.* **1987**, *105*, 393–402. [CrossRef]
46. Vanguelova, E.I.; Nortcliff, S.; Moffat, A.J.; Kennedy, F. Morphology, biomass and nutrient status of fine roots of Scots pine (*Pinus sylvestris*) as influenced by seasonal fluctuations in soil moisture and soil solution chemistry. *Plant Soil* **2005**, *270*, 233–247. [CrossRef]
47. Tang, C.-S.; Cai, W.-F.; Kohl, K.; Nishimoto, R.K. *Plant Stress and Allelopathy*; American Chemical Society: Washington, DC, USA, 1995.
48. Turtola, S.; Manninen, A.-M.; Rikala, R.; Kainulainen, P. Drought stress alters the concentration of wood terpenoids in Scots pine and Norway spruce seedlings. *J. Chem. Ecol.* **2003**, *29*, 1981–1995. [CrossRef]
49. Dicke, M. Local and systemic production of volatile herbivore-induced terpenoids: Their role in plant-carnivore mutualism. *J. Plant Physiol.* **1994**, *143*, 465–472. [CrossRef]
50. Eilers, E.J.; Talarico, G.; Hansson, B.S.; Hilker, M.; Reinecke, A. Sensing the underground—Ultrastructure and function of sensory organs in root-feeding *Melolontha melolontha* (Coleoptera: Scarabaeinae) larvae. *PLoS ONE* **2012**, *7*, e41357. [CrossRef]
51. Hiltpold, I.; Bernklau, E.; Bjostad, L.B.; Alvarez, N.; Miller-Struttmann, N.E.; Lundgren, J.G.; Hibbard, B.E. Nature, evolution and characterisation of rhizospheric chemical exudates affecting root herbivores. In *Advances in Insect Physiology*; Elsevier: Amsterdam, The Netherlands, 2013; Volume 45, pp. 97–157, ISBN 0065-2806.

52. Kwaśna, H.; Sierota, Z.; Bateman, G.L. Fungal communities in fallow soil before and after amending with pine sawdust. *Appl. Soil Ecol.* **2000**, *14*, 177–182. [CrossRef]
53. Szmidla, H.; Tkaczyk, M.; Monika, M.; Sierota, Z. Assessing the number of Melolonthinae larvae in the sawdust traps in young Scots pine (*Pinus sylvestris* L.) plantations. *Sylwan* **2018**, *162*, 590–597.

forests

Article

Mite Communities (Acari, Mesostigmata) in the Initially Decomposed 'Litter Islands' of 11 Tree Species in Scots Pine (*Pinus sylvestris* L.) Forest

Jacek Kamczyc [1,*], Marcin K. Dyderski [2], Paweł Horodecki [2] and Andrzej M. Jagodziński [2]

1 Department of Game Management and Forest Protection, Faculty of Forestry, Poznań University of Life Sciences, Wojska Polskiego 71c, PL-60625 Poznań, Poland

2 Institute of Dendrology, Polish Academy of Sciences, Parkowa 5, PL-62035 Kórnik, Poland; marcin.dyderski@gmail.com (M.K.D.); pawelhorodecki@gmail.com (P.H.); amj@man.poznan.pl (A.M.J.)

* Correspondence: jacek.kamczyc@mail.up.poznan.pl; Tel.: +48-61-848-78-03

Received: 4 April 2019; Accepted: 7 May 2019; Published: 9 May 2019

Abstract: Replacement of native deciduous forests by coniferous stands was a common result of former European afforestation policies and paradigms of forest management and led to considerable ecological consequences. Therefore, the most popular management strategy nowadays in multi-functional forestry is the re-establishment of mixed or broadleaved forests with native species on suitable habitats. However, our knowledge about the effects of tree species introduced into coniferous monocultures on soil mesofauna communities is scarce. We investigated abundance, species richness and diversity of Mesostigmata mite communities in decomposed litter of seven broadleaved (*Acer platanoides* L., *A. pseudoplatanus* L., *Carpinus betulus* L., *Fagus sylvatica* L., *Tilia cordata* Mill., *Quercus robur* L., *Q. rubra* L.) and four coniferous (*Abies alba* Mill., *Larix decidua* Mill., *Picea abies* [L.] Karst., *Pinus sylvestris* L.) species. We collected 297 litterbags after 6, 12 and 18 months of exposition in Scots pine (*Pinus sylvestris*) monocultures in Siemianice Experimental Forest (SW Poland). Generally, species richness and diversity in litter samples were much lower than in the soil mite pool. The highest abundance was found in *P. sylvestris* and *A. alba* litter, while the lowest was found in *A. platanoides*. The most abundant families were Zerconidae, Parasitidae, Veigaiidae, and Trachytidae. Our study revealed that neither species richness nor diversity were affected, but that mite abundance was affected, by the tree species (litter quality). The mite communities were similarly comprised in both high- and low-quality litter and mite abundance decreased during the decomposition process in nutrient-poor Scots pine forests. Moreover, few mite species benefited from the decomposed litter. Additionally, a litter of various tree species was inhabited mainly by eu- and hemiedaphic mite species. Mite assemblages in *A. alba*, *P. sylvestris*, and *Q. robur* litter had higher abundances. Exposition time seems to be an important driver in shaping the mite community during the early stages of litter decomposition.

Keywords: Acari; mite assemblages; litterbags; coniferous forests; admixture species; litter decay

1. Introduction

In recent centuries, almost all European forest ecosystems have been altered by forest management of varying intensities [1]. Forest management has replaced natural broadleaved forests with coniferous monocultures, mainly due to economic benefits. However, this transformation also has negative ecological consequences, e.g., even-aged coniferous forests provide less diverse habitats for many species [2] and are susceptible to global warming [3,4], pests, diseases or wind-throw, and create more acidic soils [5], which can impact the soil fauna. Therefore, to improve the resistance of the monocultures, additional (admixture) tree species are introduced into the even-aged monocultures. Planting multiple species can gain numerous economic, environmental and social benefits [6], e.g., some species can have nurse effects on other tree species, and mixtures of fast-growing and slower-growing tree species can produce timber and more valuable wood products while reducing risks of soil erosion and providing shelter and protection against frost or pests [7], or mixed forests sustain higher species richness than pure stands [8]. Therefore, the question of how the additional (both coniferous and broadleaved) tree species impact even-aged monocultures are critical in sustainable forestry.

Plant litter decomposition is one of the most important biogeochemical process in the cycling of carbon and nutrients in terrestrial ecosystems [9] and is driven by the interactions among three main factors, i.e., physicochemical environment (climate and soil conditions), litter quality and soil organisms [10]. Previous studies indicated that decomposition decreases with altitude [11] and differs between forests with various stages of anthropogenic impact [12]. This suggests that the environment (climate or/and soil) plays a crucial role in this process. However, there is still a gap in knowledge of how decomposed litter of various tree species affects soil fauna communities in nutrient poor coniferous monocultures.

One of the soil fauna groups related to decomposition processes is mites (Acari). Among them, soil Mesostigmata are very important for soil ecosystems, both in terms of species and function [13]. Most Mesostigmata—free-living predators among the soil mesofauna—have a crucial position in the soil food web, contributing significantly to energy and matter turnover [14]. Moreover, the majority of Gamasina (a subset of Mesostigmata) by affecting population growth of other fauna such as nematodes, Collembola, enchytraeids, insect larva, and other mites, they regulate the population size of soil fauna communities [15]. Thus, they can indirectly induce a strong influence on decomposition dynamics [9]. On the other hand, their abundance might depend on the dynamics of their prey, which have different sensitivity to microhabitat conditions [10]. They are also highly susceptible to anthropogenic and natural disturbances and perturbations [15], which makes them good indicators of ecosystem processes [16].

Studies on decomposition processes associated with soil fauna in forest ecosystems have been conducted in various forest types [10,11,17,18] and with various intensities [19,20]. A majority of the research took place in broadleaved forests, and focused on decomposition of leaves and root litter [9,11,21–23], and only a few studies were carried out in coniferous forests [12,18,22,24].

Previous studies indicated that species richness of various groups of soil invertebrates depends on environmental factors such as light variability, soil temperature and soil and litter chemistry [25]. This may suggest that the litter quality (tree species) is an important driver in shaping soil fauna communities. Additionally, published data focused on the decomposition process of autochthonic (native) [22,26,27], or allochthonic (foreign) litter [28]; however, there is still a gap in knowledge as to how the litter of both coniferous and broadleaved species impact the soil fauna communities. Recent studies have focused on the home-field advantage effect (HFA), which means that the plant detritus in litterbags placed into native ('home') forests decomposes faster than 'foreign' litter, due to well-adapted soil biota communities [29], or on the substrate quality–matrix quality interaction, where the interaction between soil biota and decomposed litter is also affected by variations in litter quality and palatability for decomposers [30]. Although Keiser et al. [31] suggested that microbes are mostly responsible for the HFA effect, recent studies have also focused on animals from higher trophic levels such as decomposer mites (Oribatida) and springtails (Collembola) [32]. These studies indicated that Collembola communities were similar between oak (*Quercus cerris*) and pine (*Pinus sylvestris*) litter, but they differed from black locus (*Robinia pseudoacacia*) litter. On the other hand, they also reported different abundances between the sampling dates. This suggests that the impact of litter on soil fauna communities can be specific and that it changes with time during the decomposition process. Therefore, studies are needed that connect the decomposition process, soil fauna, and 'native' and 'foreign' litter types in nutrient-poor coniferous monocultures. This is very interesting, assuming that the differences in structural compounds between the litters of coniferous and broadleaved species support different energy channels (bacterial and fungal) in the soil food web.

We hypothesized that (1) mite abundance, species richness and diversity of their communities depend on the litter quality (tree species) that is placed in nutrient-poor Scots pine forests, (2) high-quality and easily decomposable litter of various tree species affects mite communities similarly, and (3) mite abundance decreases during the decomposition process, due to low amount of available resources for their prey and changes in available niches.

2. Material and Methods

2.1. Study Site

The study was conducted in the Siemianice Experimental Forest near Biadaszki, SW Poland, which belongs to the Poznań University of Life Sciences. For this study we chose an experimental site with a Scots pine (*Pinus sylvestris*) stand established in 1974. In 1973, prior to establishment of the experiment, the soil was recognized as podsolic, sandy, and nutrient-poor, and vegetation cover was defined as *Leucobryo-Pinetum* [33]. The mature Scots pine (*Pinus sylvestris*) stand was cut, stumps and coarse roots were dug up and removed and deep ploughing was done (60–70 cm). In the spring of 1974 two-year-old Scots pine seedlings were planted in nine different spacings (3 replicates/spacing; area of each plot is 0.11 ha, 27 m × 41 m; 3.07 ha in total with buffer zone), with initial stand densities from 2500 to 20,833 trees per ha. From the onset of the experiment, no cleanings and thinnings were done, and stand density was reduced only as a result of natural mortality.

The climate of the region is transitional (between maritime and continental) with low mean annual precipitation (591 mm) and mean annual temperature of 8.2 °C [34]. During the experiment, mean monthly temperature ranged from −6.6 °C (January 2010; Table A1 in Appendix A) to 19.4 °C (July 2009) and monthly precipitation sum ranged from 9.4 (April 2009) to 224.6 mm (July 2009).

For the decomposition study we chose three research stands (plots) with the same initial density (11,111 trees/ha) and covered with circa (=ca.) 35-year-old stands, to exclude the influence of initial stand density on ecosystem processes [35–37]. In the years 2002–2007, extensive data collection was done on this experimental site. Based on these data, the experimental stands used for the study (three plots) may be described as follows (means ± SE): DBH—9.4 ± 0.28 cm, tree height—12.9 ± 0.15 m, stand basal area—37.4 ± 0.90 m^2/ha, stand density—4908 ± 399 trees/ha, litter biomass of the organic horizon—30.45 ± 2.10 Mg/ha, annual litterfall—2.89 ± 0.16 Mg/ha, pH_{KCl} of Oll horizon—4.04 ± 0.04, pH_{H2O} of Oll horizon—4.39 ± 0.08, pH_{KCl} of Ol horizon—4.03 ± 0.08, pH_{H2O} of Ol horizon—4.71 ± 0.09, pH_{KCl} of Of horizon—3.02 ± 0.03, pH_{H2O} of Of horizon—3.91 ± 0.06. The checklist of mites from the studied forests was presented by Skorupski et al. [38], who recorded 28 species from the soil environment.

2.2. Soil Sampling and Litterbag Experiment Design

The experiment was conducted in three study plots (numbers: 5, 13 and 27) where Scots pine seedlings were planted with an initial spacing of 0.6 m × 1.5 m. Simultaneously with the placement of litterbags we collected 25 soil samples from each plot (75 samples in total) using a steel corer (40 cm^2; soil depth 5 cm), to characterize the species pool, abundance and diversity of Mesostigmata mites. The litter for the litterbag experiment was collected from plots of the common garden experiment located ca. 500 m from the Scots pine forest. The complete characteristics of the litter, which originated from the common garden experiment and was used in the present study, was described by Hobbie et al. [34] The litter included 11 tree species, consisting of seven broadleaved and four coniferous species. The broadleaved species were represented by Norway maple (*Acer platanoides* L.), Sycamore maple (*A. pseudoplatanus* L.), European hornbeam (*Carpinus betulus* L.), European beech (*Fagus sylvatica* L.) and small-leaved lime (*Tilia cordata* Mill.), English oak (*Quercus robur* L.) and invasive Northern red oak (*Q. rubra* L.). The coniferous species included silver fir (*Abies alba* Mill.), European larch (*Larix decidua* Mill.), Norway spruce (*Picea abies* [L.] Karst.) and Scots pine (native, collected from Scots pine forest). The litter for the experiment was oven-dried at 65 °C for at least 3 days to eliminate the living animals including mites and to obtain a constant dry mass. We used this initial mass to calculate the mass loss in relation to the initial litter mass for each litterbag. The homogenous litter of each tree species was placed in nylon bags with the mesh size of 1 mm to allow free access of living animals to migrate into the sample with organic matter. The prepared litterbags (~18 × 18 cm) of 11 tree species were randomly tied to one rope (length ca. 6 m) with equal distances of 0.5 m; therefore, we set up 27 ropes with randomly distributed litterbags. Such prepared litterbags were randomly placed in the experimental *Pinus sylvestris* forest plots.

The experiment started on the 14th of October 2008 and lasted 18 months. Firstly, the ropes with litterbags were laid on the spacing between rows. The distance between the parallel ropes were ca. 70 cm. In total, 75 soil samples and 297 litterbags (11 tree species × 3 plots × 3 replications per plot × 3 sampling periods) were collected in our study. The litterbags overwintered and were sampled in equal numbers (99 litterbags), three times at six-month intervals, on 15 April 2009, 19 October 2009, and after the second overwintering on 16 April 2010.

2.3. Mite Extraction and Identification

Soil samples and litterbags were placed in plastic bags and transported in a portable cooler to the laboratory. Mites were extracted from samples in Tullgren funnels which is recommended for species inventory in highly organic soils such as those in the *Pinus sylvestris* forest floors in this study [39,40]. Extraction started as quickly as possible, within 5 hours after sampling and lasted over a period of 7 days. Among all animals extracted, Mesostigmata mites were selected and identified to species level and developmental stages using taxonomical keys of Karg [41], Ghilarov and Bregetova [42] and Micherdziński [43].

2.4. Data Analysis

Mite abundances coming from the same plots, terms, and litter types were pooled to allow conclusions about diversity within sample plots. This reduced the number of replications per each variant to three but allowed conclusions about diversity of mites at the plot level. In the case of the control sample (soil mite pool), we also pooled data from 25 cores per each plot into single observations. We used them as reference values to describe succession of mites. We evaluated species richness using number of taxa recorded within the study plot, species diversity using Shannon index, and abundance per sample. All mean values are followed by the standard error (SE). To assess differentiation between soil mite communities during the experimental setup (initial species pool) we used the Bray-Curtis dissimilarity index calculated for the binary representation of species composition. For this, we used the vegdist function from the vegan package [44].

To assess the impact of collection date, litter quality (expressed by its identity, which can be linked with measured litter traits) and its decomposition rate (k constant—calculated with use a single-exponential decomposition model proposed by Wieder and Lang [45]) we used generalized linear mixed models (GLMM). We assumed that mite abundance and species richness have Poisson distributions while the Shannon's index has a normal distribution. As we assumed the Poisson distribution of mite abundance, we did not recalculate abundance per sample mass, as Poisson distributions assume integer values.

We evaluated random effects connected with plot identity, to exclude plot-specific factors, which could affect the results. Models were developed using the lme4 package [46], while the statistical significance of variables was calculated using z-values implemented in the lmerTest package [47]. For all GLMMs, we evaluated the parsimony of models using Akaike's Information Criterion (AIC). We also provided AIC_0–AIC of models with intercept and random effects only. To evaluate differences between litter origin and collection dates in the models we used Tukey posteriori tests.

To assess the importance of such factors in shaping mite species communities we used Redundancy Analysis (RDA), implemented in the vegan package [44]. RDA is the method of constrained ordination of the multidimensional dataset (here—abundances of particular mite species). In contrast to unconstrained ordination, RDA also allows evaluation of the importance of environmental variables in ordered sample coordinates within reduced space of the ordination. The importance of factors was tested using permutation analysis of variance (PERMANOVA), also implemented in the vegan package [44]. Prior to analysis, we transformed species abundances using Hellinger's square root transformation. We selected variables for the final models based on AIC. Statistical analyses were conducted using R software [48].

3. Results

3.1. Mite Abundance, Species Richness, and Diversity

In total, 7887 mites were identified and classified into 31 species (Table A2 in Appendix A). The total abundance differed among the sampling dates and also between the soil samples (698 mites) and litterbags (7189 mites in total). The highest total abundance (4326 mites) was recorded in April 2009, six months after the beginning of the experiment; it then decreased from October 2009 (1524) to April 2010 (1339). All collected mites from litterbags represented two suborders: Gamasina (eight families; 6903 individuals (=ind.)) and Uropodina (two families; 286 ind.). Overall, the most abundant families were Zerconidae (3016 mites), Parasitidae (2399), Veigaiidae (1383) and Trachytidae (274). Moreover, among all recorded species, the most abundant were *Zercon peltatus* (41.80% of all mites), *Veigaia nemorensis* (18.68), *Paragamasus jugincola* (Athias-Henriot) (11.52) and *Vulgarogamasus kraepelini* (Berlese) (9.63) as well as *P. runcatellus* (Berlese) (6.01). Our study revealed that proportional abundance of the most abundant species changed in time. In April 2009, *Z. peltatus* reached the proportional abundance of 61.63% of all collected mites, then it decreased in October 2009 to 1.57% and increased in April 2010 to 23.53%. Additionally, the proportional abundance of *V. nemorensis* reached 9.27% of all mites in April 2009; then it increased to 42.19% in October 2009 and decreased to 22.33% in the next spring (April 2010).

Analysis of the soil cores taken at the time the experiment began revealed that mean species richness of mites per plot was 14.7 ± 0.3, species diversity was $H' = 1.950 \pm 0.167$ and mite abundance was 9.3 ± 2.4 ind. sample^{-1}. In general, species richness and diversity in litter samples were much lower than in the soil mite pool (Figure 1). Mean mite abundance depended statistically significantly on collection date and litter origin (Table 1; AIC = 1116.7, AIC_0 = 1984.2). The highest abundance was found in the *P. sylvestris* and *A. alba* litter, while the lowest was in *A. platanoides*. Abundances were statistically significantly higher in April 2009 than at other collection times. Mean species richness of mites depended on collection date and decomposition constant (Table 1; AIC = 428.6, AIC_0 = 433.3). Species richness was the highest in April 2009, while the lowest was in April 2010. Additionally, species richness in October 2009 did not differ statistically significantly from other dates. Although statistically not significant, mite species richness was negatively correlated with decomposition constant *k*. Species diversity of mites was the best explained by the model with decomposition constant only (Table 1; AIC = 76.9, AIC_0 = 77.0); however, influence of decomposition constant was statistically insignificant.

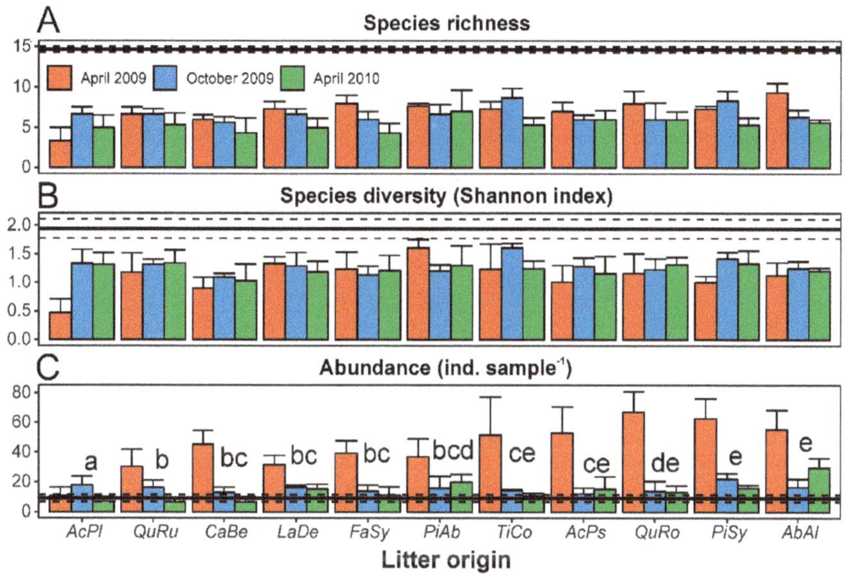

Figure 1. Mean (+SE) species richness (**A**), diversity (**B**) and mite abundance (**C**) in litterbags. The thick horizontal line represents the mean and dashed lines represent the range of SE for the control soil mite pool at the time of experiment setup. Litter origin marked by the same letter did not differ statistically significantly ($p > 0.05$), according to Tukey posteriori test.

Table 1. Generalized linear mixed models explaining mite abundance, species richness, and diversity.

Response	Term	Estimate	SE	*z*	*Pr* (>|*z*|)
Abundance	(Intercept)	4.0945	0.0770	53.1450	<0.0001
	Litter origin—*Acer platanoides*	−1.0376	0.1125	−9.2250	<0.0001
	Litter origin—*Quercus rubra*	−0.6353	0.0978	−6.4980	<0.0001
	Litter origin—*Carpinus betulus*	−0.4374	0.0918	−4.7630	<0.0001
	Litter origin—*Acer pseudoplatanus*	−0.2215	0.0863	−2.5680	0.0102
	Litter origin—*Picea abies*	−0.3214	0.0887	−3.6220	0.0003
	Litter origin—*Abies alba*	0.0099	0.0812	0.1220	0.9031
	Litter origin—*Larix decidua*	−0.4634	0.0926	−5.0060	<0.0001
	Litter origin—*Fagus sylvatica*	−0.4426	0.0920	−4.8110	<0.0001
	Litter origin—*Quercus robur*	−0.0615	0.0826	−0.7440	0.4571
	Litter origin—*Tilia cordata*	−0.2680	0.0874	−3.0670	0.0022
	Date—October 2009	−1.0295	0.0511	−20.1360	<0.0001
	Date—April 2010	−1.1567	0.0536	−21.5720	<0.0001
	Random effect—plot	Variance	0.0070	SD	0.0839
Richness	(Intercept)	2.2331	0.2102	10.6240	<0.0001
	Decomposition constant (*k*)	−1.1424	0.7384	−1.5470	0.1218
	Date—October 2009	−0.0572	0.0937	−0.6100	0.5420
	Date—April 2010	−0.2735	0.0994	−2.7520	0.0059
	Random effect—plot	Variance	0.0197	SD	0.1405
Shannon	(Intercept)	1.3827	0.1961	7.0500	<0.0001
	Decomposition constant (*k*)	−0.6634	0.5996	−1.1060	0.2710
	Random effect—plot	Variance	0.0444	SD	0.2107

RDA of soil mite communities in litterbags (Figure 2) revealed that 89.8% of explained variability was connected with unconstrained factors (i.e., species composition), while constrained factors (i.e., environmental constraints and collection date) explained 10.2%. Constraining by collection date was the only factor in the most parsimonious models—litter type and decomposition constant were not included and not significant. Points representing mite communities sampled in April 2009 and 2010 were separated from plots from October 2009 along the RDA2 axis, while samples from April 2009 were separated from others along RDA1. We did not observe patterns connected with litter origin, except clustering of coniferous species in positive values of RDA1 and RDA2. *Zercon peltatus* C.L. Koch was connected with the April 2009 sampling date while *Veigaia nemorensis* C.L. Koch was connected with October 2009.

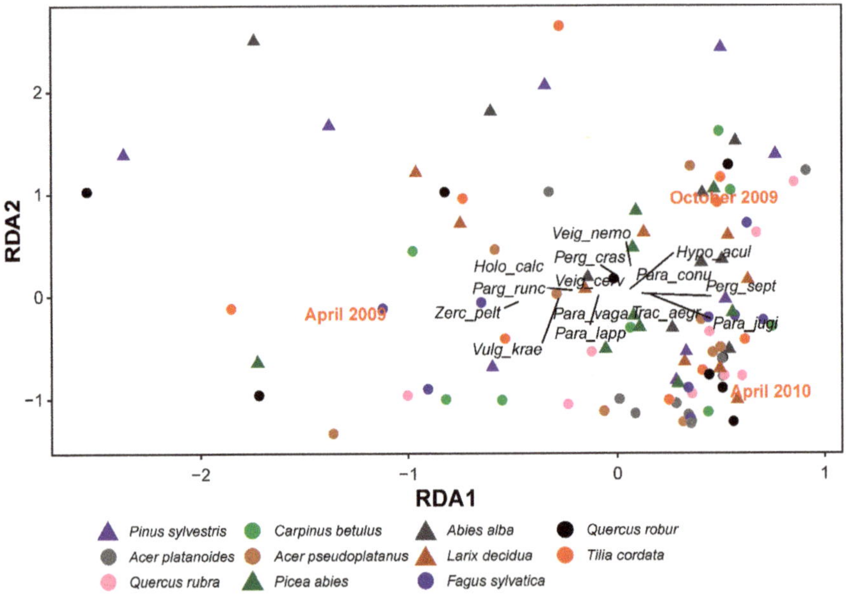

Figure 2. RDA analysis of mite community species composition. Each point represents the community from litterbags per plot per date. Red labels indicate the centroids of factors constraining analysis results (significance test Table 2). Black labels indicate scores of species which occurred at least 10 times among the 16 species; names are first four letters of genera name and first four letters of the species name.

Table 2. PERMANOVA test of the influence of litter identity and collection date on mite species communities in RDA reduced space. AIC_0 refers to the null model (unconstrained analysis).

Term	*df*	Variance	*F*	*Pr* (>*F*)
Collection date	2	0.06917	5.4696	0.001
Residual	96	0.60690	-	-
AIC	−44.43	AIC_0	−37.75	-

3.2. Similarity between Litterbag and Soil Mite Pool Community

At the time of experiment setup, the soil mite community included 22 species (Table A2 in Appendix A). Analysis of community dissimilarity to the initial species pool revealed that mite communities in litter of three tree species was the most distinct from the initial soil mite pool community (Figure 3): *F. sylvatica*, *C. betulus*, and *L. decidua*. A similar level of dissimilarity was found after six months in the case of *A. platanoides*. Most of the species increased their dissimilarity to the control during the experiment, but to lower degrees, reaching dissimilarity indices between 0.29 (*Q. robur*) and 0.46 (*P. sylvestris*). However, most of the species occurring in the litter studied were the same as in soil; species not recorded during the control were rare, and we did not identify any pattern of their richness during litter decomposition (Figure 4).

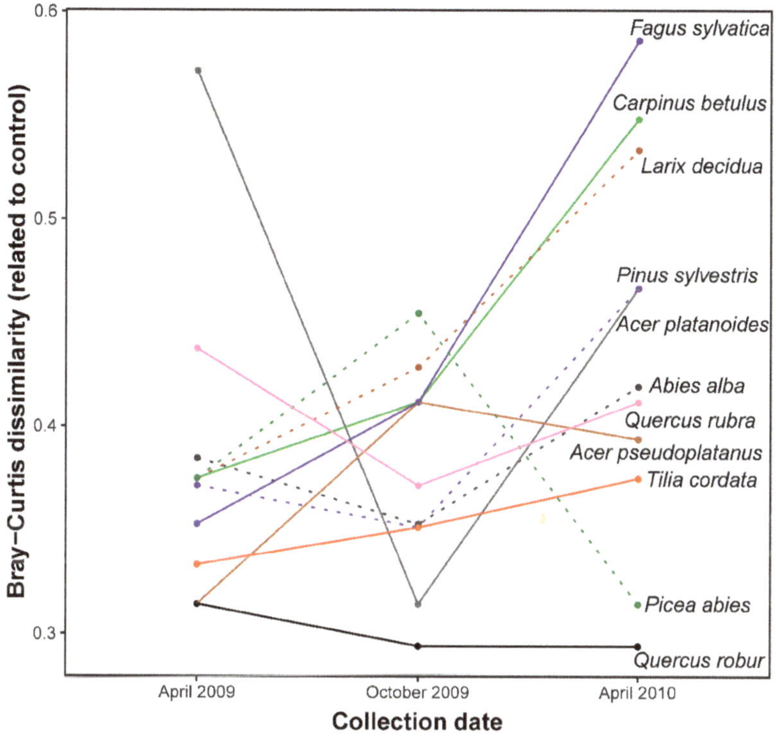

Figure 3. Dissimilarity between mite communities and the control (mite community in the soil during experiment setup), expressed by Bray-Curtis dissimilarity index. Solid line—broadleaved species; dashed line—coniferous species.

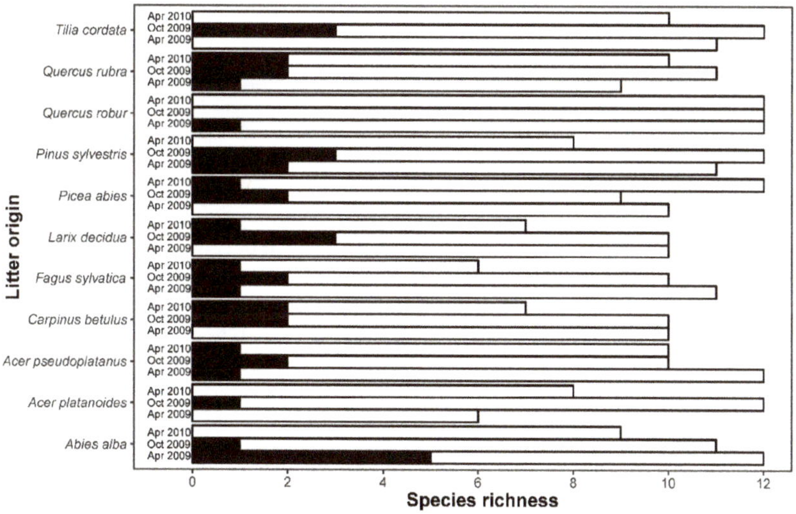

Figure 4. The richness of species noted in the control (during experiment setup—white) and not noted (colonizers—black) in mite communities studied.

3.3. Litter Decomposition

Generally, tree species differed in the litter mass loss ($df = 10$, $F = 10.3829$, $p < 0.0001$) (Figure 5). Among coniferous species the lowest litter mass loss was recorded for *A. alba* (21.21%) and *P. abies* (20.62) after 18 months, with the highest for *P. sylvestris* (27.31). Moreover, litter of *A. alba* decomposed the slowest among conifers studied in April (11.22%) and October 2009 (19.4). Among broadleaved species, the lowest litter mass loss was recorded for *C. betulus* (21.94%) and *F. sylvatica* (22.56), whereas the highest for *A. platanoides* (37.01) in April 2010. Similarly, among all broadleaved species studied the lowest litter mass loss was recorded for *F. sylvatica* in April 2009 (13.97%) and October 2009 (19.3).

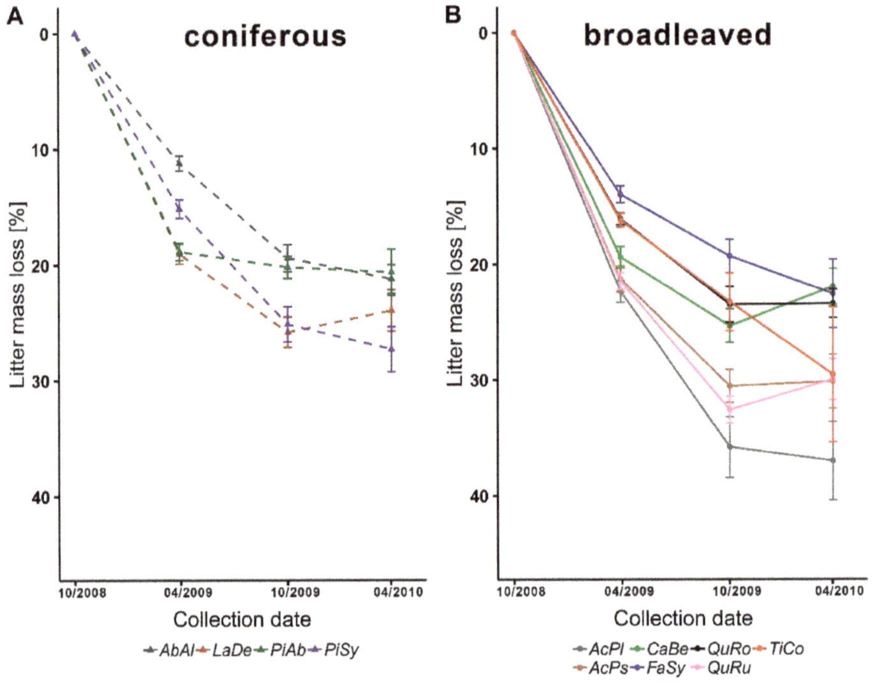

Figure 5. Litter mass loss [%] of coniferous (**A**) and broadleaved (**B**) species in litterbags. Data are presented as mean ± SE.

4. Discussion

Neither species richness nor diversity, but rather mite abundance, was affected by the tree species (litter quality). The mite communities were similarly comprised for both high- and low-quality litter, and the mite abundance decreased during the decomposition process in nutrient-poor Scots pine forests. Moreover, few mite species benefited from the decomposed litter.

The highest decomposition rates were recorded for broadleaved *Acer platanoides* ($k = 0.37$ per year), *Q. rubra* (0.31), *A. pseudoplatanus* (0.30), whereas the lowest were noted for coniferous *A. alba* (0.18), *P. abies* (0.20) and broadleaved *F. sylvatica* (0.19). Similar patterns were recorded by Hobbie et al. [34], who proved slower decay rates for *L. decidua*, *A. alba* and *F. sylvatica* and higher for *C. betulus*, *Q. rubra* and *Acer* spp. (ca. 500 m from our study site). Our results are also partly in line with Horodecki and Jagodziński [49] and Horodecki et al. [28], who recorded lower decomposition rates for litter of *F. sylvatica* and *Q. rubra* in Scots pine stands, but higher rates in mixed stands. Studies on *Quercus prinus* litter demonstrated that the leaves 'disappeared' slowly and that there was minimal faunal influence on decomposition rates [50]. On the other hand, the large increase in colonization of litterbags in the second year did not change the effect of microarthropods on decomposition rates, which remained similar in both years [27]. The studies of Heneghan et al. [23] on the decomposition of *Q. prinus* indicated that microarthropod assemblages did not influence litter mass loss in temperate forest sites.

Our study indicated that the litter quality (litter origin) of various tree species affected only the mite abundances. This is in contrast to Mueller et al. [25], who found that many environmental factors (e.g., light availability, soil properties or nutrient availability), including litter and tree species affected species richness of many invertebrate groups, and that the response of the animal group can be taxon specific. Therefore, our results can suggest that the input of the high-quality litter in the nutrient-poor Scots pine forests do not affect the species richness, and that it only support some of the mite species, that inhabit the soil for a longer time. This result is in line with studies of Gergócs and Hufnagel [32], who indicated that litter type affects the mesofauna and microbiota abundance and the relationships between these two animal groups. Similarly, Gonzáles et al. [22] recorded three times higher mite (Oribatida) density in aspen (*Populus tremuloides* Michx.) than in lodgepole pine (*Pinus contorta* Douglas ex Loudon) litter. On the other hand, faunal abundance was not correlated with mass loss in single-species litter of *Quercus serrata* Murray and *Pinus densiflora* Sieb. et Zucc. [51]. The litter did not impact the density of microinvertebrates, macroinvertebrates, and predators in *Quercus gambelli* Nutt. and *Cecropia scheberiana* Miq. litterbags [52]. Moreover, Gergócs et al. [53] recorded similar Oribatida abundance in oak (*Quercus cerris* L.) and pine (*P. sylvestris*) litter but abundance was reduced by one third in black locust (*Robinia pseudoacacia* L.) litter. Differences in Mesostigmata mite abundances were also recorded between oak (*Q. robur*) and silver birch (*Betula pendula* Roth.), Scots pine (*P. sylvestris*) and alder (*Alnus glutinosa* (L.) Gaertn.) [12]. Urbanowski et al. [12] found ca. 25% higher abundance in oak litter, when compared to all other litter types. In the present study, the difference of mite abundance between oak and Scots pine litter was low (ca. 6% of the abundance in Scots pine litter), which may indicate that high-quality litter similarly affects mite assemblages as 'native' litter in coniferous monocultures.

Our study revealed that the mean species richness and diversity did not depend on litter quality. Similar results were obtained for Oribatida by Gergócs et al. [53], who indicated that oak (*Q. cerris*) and pine (*P. sylvestris*) litter are equally favorable for mites, but both differed from black locust (*R. pseudoacacia*). We have not analyzed *R. pseudoacacia* litter; however, other tree species that were used in the present study had similar characteristic (e.g., *Q. robur*, *C. betulus* or *T. cordata*) and mite species richness did not differ. It is difficult to discuss the species richness of Mesostigmata, because only a few studies have focused on these animals in decomposed litter in litterbags. Previous studies, conducted in reclaimed spoil heap and adjacent forest habitats, indicated that species richness was higher in forest habitats, compared to spoil heap [12]. Moreover, litter type impacts on species richness were found, with significantly higher species richness recorded from oak litter, lower richness from pine and alder and the lowest from silver birch. In contrast, species richness did not differ between oak and pine litter in the present study.

Although the relationships between the litter decomposition and soil microarthropods have been widely studied [54], little is known about the changes in Mesostigmata mite communities in the short term. Repeated measurements of litter mass loss and fauna extraction allow checking whether the changes in species number and abundance in decomposed litter is successional, as has been proposed by Usher [55], seasonal or driven by abiotic factors [24]. Interestingly, our study revealed that the mean species richness in the 'soil mite pool' was higher than in the decomposed litter, regardless of tree species. We found that five mite species, i.e., *Paragamasus jugincola*, *P. runcatellus*, *Z. peltatus*, *Rhodacarus coronatus*, and *V. nemorensis*, reached high proportional abundances, which may indicate that only some species among the 'soil mite pool' benefit from the litter in early stages of decomposition. Some of them, such as Zerconidae or Veigaiidae, represent the K-selected traits with slow development and low dispersal, which are associated with forests [56]. This is in line with Urbanowski et al. [12], who reported a high proportional abundance of *Z. peltatus* within six months in a silver birch forest. Our study indicated that the same mite species dominated the mite community in the litterbags in pooled data; however, their abundance changed slightly. For instance, the nematophagous *Z. peltatus*, which occupied the soil in modest numbers (10% of mites), dominated in the litter (42%). Interestingly, the density of this species peaked at 62% in April 2009. This may suggest the pattern of mite community succession in nutrient-poor Scots pine forests, associated with the availability of nematodes or springtails. These animals are a food source for predatory Mesostigmata [15,57]. This can be explained by the high density of nematophagous *Z. peltatus* [58]. In the present study, the total mite abundance rapidly decreased with sampling dates, especially from April to October 2009, indicating that the availability of the niches changed during the decomposition process. The limiting effect of forest floor litter on soil mites may be determined not by the volume, but rather the quality of food and habitat source [59], and by the fact that plant material of high quality, such as used in our study, is immediately decomposed and utilized by soil fauna. This result is supported by the RDA analysis that revealed the differences between the mite communities in April 2009 and both October 2009 and April 2010 (Figure 2). Our study also indicated the importance of time in the early stages of litter decomposition for mite abundance.

5. Conclusions

Our study indicated that neither species richness nor diversity, but rather the abundance of Mesostigmata mites, was affected by the litter quality added to nutrient-poor Scots pine forests. Moreover, only some species from the 'soil mite pool' colonized the litter, regardless of its quality, and the abundance of few taxa increased rapidly after six months of decomposition. Additionally, litter of various tree species was inhabited mainly by eu- and hemiedaphic mite species. Mite assemblages in *A. alba*, *P. sylvestris*, and *Q. robur* litter had higher abundances. Exposition time seems to be an important driver in shaping the mite community during the early stages of litter decomposition. The impact of the tree species on Mesostigmata species richness and biodiversity depend on many environmental predictors, which are associated with the certain tree species [25]. Our result also highlights the crucial role of tree species and its' environment for the soil fauna, although in early stages of decomposition the changes are reflected in the abundance. This highlights the residence times of broadleaved trees in forest ecosystem, which allow development of their own soil fauna communities. For that reason, broadleaved admixtures in coniferous monocultures are especially important in shaping soil mite biodiversity.

Author Contributions: Conceptualization, A.M.J. and J.K.; Methodology, A.M.J. and J.K.; Software, A.M.J., J.K., M.K.D. and P.H.; Validation, J.K., A.M.J., M.K.D. and P.H.; Formal Analysis, J.K., M.K.D., A.M.J. and P.H.; Investigation, J.K., A.M.J., P.H. and M.K.D.; Resources, J.K., A.M.J., P.H. and M.K.D.; Data Curation, J.K., A.M.J., P.H. and M.K.D.; Writing—Original Draft Preparation, J.K., A.M.J., M.K.D. and P.H.; Writing—Review & Editing, J.K., A.M.J., M.K.D. and P.H.; Visualization, J.K., P.H., M.K.D. and A.M.J.; Supervision, J.K. and A.M.J.; Project Administration: A.M.J. and J.K.; Funding Acquisition: A.M.J. and J.K.

Funding: The study was financially supported by the Institute of Dendrology, Polish Academy of Sciences, Kórnik, Poland, and Faculty of Forestry, Poznań University of Life Sciences, Poznań, Poland. The publication is co-financed within the framework of Ministry of Science and Higher Education programme as "Regional Initiative Excellence" in years 2019–2022, project number 005/RID/2018/19.

Acknowledgments: The authors would like to thank Katarzyna Strzymińska, Bartosz Bartków, Jakub Szeptun, Daniel Szemis for their assistance in laboratory works. We kindly thank Lee E. Frelich (University of Minnesota, Centre for Forest Ecology, USA) for linguistic support. The study was financially supported by the Institute of Dendrology, Polish Academy of Sciences, Kórnik, Poland.

Conflicts of Interest: The authors declare no conflict of interests.

Appendix A

Table A1. Monthly mean temperatures and precipitation sums recorded during the experiment in Biadaszki Forest Experimental Station (ca. 1 km from study site). Setup and collection dates are marked by * and **, respectivelly.

Month and Year	Mean Temperature [°C]	Precipitation Sum [mm]
October 2008 *	8.3	46.9
November 2008	4.9	35.7
December 2008	0.8	25.2
January 2009	−3.5	11.5
February 2009	−1.3	20.1
March 2009	3.5	14.8
April 2009 **	12.0	9.4
May 2009	13.6	96.5
June 2009	15.3	169.3
July 2009	19.4	224.6
August 2009	19.0	84.3
September 2009	15.1	56.9
October 2009 **	6.9	146.6
November 2009	5.6	66.4
December 2009	−0.8	82.3
January 2010	−6.6	61.3
February 2010	−1.6	22.9
March 2010	3.4	50.7
April 2010 **	8.7	70.7

Table A2. Checklist of mite species recorded from soil samples and litterbags with leaf litter of 11 tree species in Scots pine forests. Symbols indicate: "+"—abundance 1–9 ind., "++"—10–99 ind., "+++"—>100 ind. (number of individuals in litterbags were summed from three dates of mite collection).

No.	Mite Species	Soil Pool	Litterbags with Various Litter Types										
			Abies alba	*Acer platanoides*	*Acer pseudoplatanus*	*Carpinus betulus*	*Fagus sylvatica*	*Larix decidua*	*Picea abies*	*Pinus sylvestris*	*Quercus robur*	*Quercus rubra*	*Tilia cordata*
1	*Alliphis halleri* (G. & R. Canestrini, 1881)	-	-	-	-	-	-	-	-	-	-	+	-
2	*Amblyseius* sp.	+	-	-	-	-	-	-	+	-	+	-	-
3	*Arctoseius cetratus* (Sellnick, 1940)	-	-	-	-	-	-	-	-	-	-	-	+
4	*Arctoseius semiscissus* (Berlese, 1892)	-	-	-	-	-	-	-	-	+	-	-	-
5	*Asca aphidioides* (Linnaeus, 1758)	-	+	-	+	+	+	+	-	+	-	+	+
6	*Gamasellodes bicolor* Berlese, 1918	+	+	-	-	-	-	-	+	+	-	+	+
7	*Holoparasitus calcaratus* (C.L. Koch, 1839)	++	+	+	+	+	+	+	+	+	+	-	+
8	*Hypoaspis aculeifer* (Canestrini, 1883)	+	+	+	+	+	+	+	+	+	+	-	+
9	*Hypoaspis praesternalis* Willmann, 1949	-	+	-	-	-	-	-	-	-	-	-	-
10	*Hypoaspis procera* Karg 1965	-	+	-	-	-	-	-	+	+	-	-	-
11	*Hypoaspis vacua* (Michael, 1891)	-	+	-	-	+	-	++	-	-	-	+	-
12	*Laelapsis astronomica* (Koch, 1839)	-	+	-	-	-	-	-	-	-	-	-	-
13	*Oodinychus ovalis* C.L. Koch, 1839	+	-	-	-	+	-	-	-	-	+	+	+
14	*Paragamasus conus* Karg, 1971	++	++	++	++	++	++	++	++	++	+	+	++
15	*Paragamasus jugincola* (Athas-Henriot, 1967)	+++	+++	++	++	++	++	++	+++	++	++	++	++
16	*Paragamasus lapponicus* Tragardh, 1910	+	+	+	+	+	+	-	+	+	++	-	-
17	*Paragamasus vagabundus* (Karg, 1968)	++	+	+	+	+	++	+	++	+	++	+	+
18	*Pargamasus runcatellus* (Berlese, 1903)	+++	++	++	++	++	++	++	++	++	++	++	++
19	*Pergamasus crassipes* Linnaeus, 1758	+	+	+	+	+	+	+	+	+	+	-	+
20	*Pergamasus mediocris* Berlese, 1904	+	-	-	-	-	-	-	-	-	-	-	+
21	*Pergamasus septentrionalis* Oudemans, 1902	-	+	+	+	+	+	+	+	+	+	+	+
22	*Rhodacarellus silesiacus* Willmann, 1936	+	-	-	-	-	-	-	-	+	-	-	-
23	*Rhodacarus coronatus* Berlese, 1921	++	-	+	-	-	-	+	-	-	+	+	+
24	*Trachytes aegrota* (C. L. Koch, 1841)	++	++	++	++	++	++	++	++	++	++	++	+
25	*Trichouropoda obscura* (C.L. Koch, 1836)	+	-	-	-	-	-	-	-	-	-	-	-
26	*Veigaia cervus* (Kramer, 1876)	++	+	+	+	-	++	+	+	+	+	+	-
27	*Veigaia nemorensis* (C. L. Koch, 1839)	++	+++	++	++	+++	++	+++	+++	+++	+++	+++	+++
28	*Vulgarogamasus kraepelini* (Berlese, 1904)	+	+++	+	++	++	++	++	++	+++	++	++	++
29	*Zercon peltatus* C.L. Koch, 1836	++	+++	++	+++	+++	+++	+++	+++	+++	+++	+++	+++
30	*Zercon triangularis* C.L. Koch, 1836	+	-	-	-	+	+	-	-	-	+	+	+
31	*Zercon zelawaiensis* Sellnick, 1944	+	-	-	-	-	-	+	-	-	-	-	-

References

1. Vanbergen, A.J.; Woodcock, B.A.; Watt, A.D.; Niemela, J. Effect of land-use heterogeneity on carabid communities at the landscape scale. *Ecography* **2005**, *28*, 3–16. [CrossRef]
2. Hunter, M.L. *Maintaining Biodiversity in Forest Ecosystems*; Cambridge University Press: Cambridge, UK, 1999.
3. Young, R.; Giese, R. *Introduction to Forest Ecosystem Science and Management*, 3rd ed.; Wiley: Hoboken, NJ, USA, 2003.
4. Dyderski, M.K.; Paź, S.; Frelich, L.E.; Jagodziński, A.M. How much does climate change threaten European forest tree species distributions? *Glob. Chang. Biol.* **2018**, *24*, 1150–1163. [CrossRef]
5. Götmark, F.; Friedman, J.; Kempe, G.; Nord, B. Broadleaved tree species in conifer-dominated forestry: Regeneration and limitation of saplings in southern Sweden. *For. Ecol. Manag.* **2005**, *214*, 142–157. [CrossRef]
6. Alem, S.; Pavlis, J.; Urban, J.; Kucera, J. Pure and mixed plantations of *Eucalyptus camaldulensis* and *Cupressus lusitanica*: Their growth interactions and effect on diversity and density of undergrowth woody plants in relation to light. *Open J. For.* **2015**, *5*, 375.
7. Liu, C.L.C.; Kuchma, O.; Krutovsky, K.V. Mixed-species versus monocultures in plantation forestry: Development, benefits, ecosystem services and perspectives for the future. *Glob. Ecol. Conserv.* **2018**, *15*, 1–13. [CrossRef]
8. Migge, S.; Maraun, M.; Scheu, S.; Schaefer, M. The oribatid mite community (Acarina) of pure and mixed stands of beech (*Fagus sylvatica*) and spruce (*Picea abies*) of different age. *Appl. Soil Ecol.* **1998**, *9*, 115–121. [CrossRef]
9. Wang, S.; Ruan, H.; Wang, B. Effects of soil microarthropods on plant litter decomposition across an elevation gradient in the Wuyi Mountains. *Soil Biol. Biochem.* **2009**, *41*, 891–897. [CrossRef]
10. Fujii, S.; Takeda, H. Succession of soil microarthropod communities during the aboveground and belowground litter decomposition processes. *Soil Biol. Biochem.* **2017**, *110*, 95–102.
11. Franca, M.; Sandmann, D.; Krashevska, V.; Maraun, M.; Scheu, S. Altitude and decomposition stage rather than litter origin structure soil microarthropod communities in tropical montane rainforests. *Soil Biol. Biochem.* **2018**, 263–274.
12. Urbanowski, C.K.; Horodecki, P.; Kamczyc, J.; Skorupski, M.; Jagodziński, A.M. Succession of mite assemblages (Acari, Mesostigmata) during decomposition of tree leaves in forest stands growing on reclaimed post-mining spoil heap and adjacent forest habitats. *Forests* **2018**, *9*, 718. [CrossRef]
13. Petersen, H.; Luxton, M. A comparative analysis of soil fauna populations and their role in decomposition processes. *Oikos* **1982**, *39*, 288–388. [CrossRef]
14. Ruf, A.; Beck, L. The use of predatory soil mites in ecological soil classification and assessment concepts, with perspectives for oribatid mites. *Ecotoxicol. Environ. Saf.* **2005**, *62*, 290–299. [CrossRef]
15. Koehler, H.H. Predatory mites (Gamasina, Mesostigmata). *Agric. Ecosyst. Environ.* **1999**, *74*, 395–410. [CrossRef]
16. Gulvik, M. Mites (Acari) as indicators of soil biodiversity and land use monitoring: A review. *Pol. J. Ecol.* **2007**, 415–440.
17. Gan, H.; Zak, D.R.; Hunter, M.D. Chronic nitrogen deposition alters the structure and function of detrital food webs in a northern hardwood ecosystem. *Ecol. Appl.* **2013**, *23*, 1311–1321. [CrossRef]
18. Berg, M.P.; Kniese, J.P.; Bedaux, J.J.M.; Verhoef, H.A. Dynamics and stratification of functional groups of micro- and mesoarthropods in the organic layer of a Scots pine forest. *Biol. Fertil. Soils* **1998**, *26*, 268–284. [CrossRef]
19. Magcale-Macandog, D.B.; Manlubatan, M.B.T.; Javier, J.M.; Edrial, J.D.; Mago, K.S.; De Luna, J.E.I.; Nayoos, J.; Porcioncula, R.P. Leaf litter decomposition and diversity of arthropod decomposers in tropical Muyong forest in Banaue, Philippines. *Paddy Water Environ.* **2018**, *16*, 265–277. [CrossRef]
20. Vasconcelos, H.L.; Laurance, W.F. Influence of habitat, litter type, and soil invertebrates on leaf-litter decomposition in a fragmented Amazonian landscape. *Oecologia* **2005**, *144*, 456–462. [CrossRef]
21. Elkins, N.Z.; Whitford, W.G. The role of microarthropods and nematodes in decomposition in a semi-arid ecosystem. *Oecologia* **1982**, *55*, 303–310. [CrossRef]
22. González, G.; Seastedt, T.R.; Donato, Z. Earthworms, arthropods and plant litter decomposition in aspen (*Populus tremuloides*) and lodgepole pine (*Pinus contorta*) forests in Colorado, USA. *Pedobiologia* **2003**, *47*, 863–869. [CrossRef]

23. Heneghan, L.; Coleman, D.C.; Zou, X.; Crossley, D.A.; Haines, B.L. Soil microarthropod contributions to decomposition dynamics: Tropical–temperate comparisons of a single substrate. *Ecology* **1999**, *80*, 1873–1882.
24. Lindo, Z.; Winchester, N.N. Oribatid mite communities and foliar litter decomposition in canopy suspended soils and forest floor habitats of western redcedar forests, Vancouver Island, Canada. *Soil Biol. Biochem.* **2007**, *39*, 2957–2966. [CrossRef]
25. Mueller, K.E.; Eisenhauer, N.; Reich, P.B.; Hobbie, S.E.; Chadwick, O.A.; Chorover, J.; Dobies, T.; Hale, C.M.; Jagodziński, A.M.; Kałucka, I.; et al. Light, earthworms, and soil resources as predictors of diversity of 10 soil invertebrate groups across monocultures of 14 tree species. *Soil Biol. Biochem.* **2016**, *92*, 184–198. [CrossRef]
26. Blair, J.M.; Crossley, D.A.; Callaham, L.C. Effects of litter quality and microarthropods on N dynamics and retention of exogenous 15N in decomposing litter. *Biol. Fertil. Soils* **1992**, *12*, 241–252. [CrossRef]
27. Seastedt, T.R.; Crossley, D.A. Nutrients in forest litter treated with naphthalene and simulated throughfall: A field microcosm study. *Soil Biol. Biochem.* **1983**, *15*, 159–165. [CrossRef]
28. Horodecki, P.; Nowiński, M.; Jagodziński, A.M. Advantages of mixed tree stands in restoration of upper soil layers on postmining sites: A five-year leaf litter decomposition experiment. *Land Degrad. Dev.* **2019**, *30*, 3–13. [CrossRef]
29. Gholz, H.L. Long-term dynamics of pine and hardwood litter in contrasting environments: Toward a global model of decomposition. *Glob. Change Biol.* **2000**, *6*, 751–765. [CrossRef]
30. Freschet, G.T.; Aerts, R.; Cornelissen, J.H.C. Multiple mechanisms for trait effects on litter decomposition: Moving beyond home-field advantage with a new hypothesis. *J. Ecol.* **2012**, *100*, 619–630. [CrossRef]
31. Keiser, A.D.; Strickland, M.S.; Fierer, N.; Bradford, M.A. The effect of resource history on the functioning of soil microbial communities is maintained across time. *Biogeosciences* **2011**, *8*, 1477–1486. [CrossRef]
32. Gergócs, V.; Hufnagel, L. The effect of microarthropods on litter decomposition depends on litter quality. *Eur. J. Soil Biol.* **2016**, *75*, 24–30. [CrossRef]
33. Ceitel, J. Zmiany mikroklimatu przygruntowej warstwy powietrza oraz morfologii drzew ze wzrostem upraw sosnowych założonych w różnych więźbach początkowych; Katedra Hodowli Lasu, Akademia Rolnicza w Poznaniu. Ph.D. Thesis, Department of Silviculture, Agricultural University, Poznań, Poland, 1982.
34. Hobbie, S.E.; Reich, P.B.; Oleksyn, J.; Ogdahl, M.; Zytkowiak, R.; Hale, C.; Karolewski, P. Tree species effects on decomposition and forest floor dynamics in a common garden. *Ecology* **2006**, *87*, 2288–2297. [CrossRef]
35. Jagodziński, A.M.; Oleksyn, J. Ecological consequences of silviculture at variable stand densities. I. Stand growth and development. *Sylwan* **2009**, *153*, 75–85.
36. Jagodziński, A.M.; Oleksyn, J. Ecological consequences of silviculture at variable stand densities. II. Biomass production and allocation, nutrient retention. *Sylwan* **2009**, *153*, 147–157.
37. Jagodziński, A.M.; Oleksyn, J. Ecological consequences of silviculture at variable stand densities. III. Stand stability, phytoclimate and biodiversity. *Sylwan* **2009**, *153*, 219–230.
38. Skorupski, M.; Dalik, J.; Ceitel, J. Species composition of mites from the order Mesostigmata in experimental pine and spruce tree stands planted in various initial spacing. *Sci. Pap. Agric. Univ. Poznan For.* **2003**, 49–56.
39. Crossley, D.A.; Blair, J.M. Proceedings of the International Workshop on Modern Techniques in Soil Ecology Relevant to Organic Matter Breakdown, Nutrient Cycling and Soil Biological Processes A high-efficiency, "low-technology" Tullgren-type extractor for soil microarthropods. *Agric. Ecosyst. Environ.* **1991**, *34*, 187–192. [CrossRef]
40. Edwards, C.A. The assessment of populations of soil-inhabiting invertebrates. *Agric. Ecosyst. Environ.* **1991**, *34*, 145–176. [CrossRef]
41. Karg, W. *Acari (Acarina), Milben Parasitiformes (Anactinochaeta), Cohors Gamasina Leach. Raubmilben. Die Tierwelt Deutschlands*; VEB Gustav Fischer Verlag: Jena, Germany, 1993.
42. Ghilarov, M.C.; Bregetova, N.G. *Opredelitel obitajuscich v Pocve Klescej-Mesostigmata (Key to the Soil Mites—Mesostigmata)*; Nauka: Leningrad, Russia, 1977.
43. Micherdziński, W. *Die Familie Parasitidae Oudemans, 1901 "(Acarina, Mesostigmata)"*; Państwowe Wydawnictwo Naukowe: Kraków, Poland, 1969.
44. Oksanen, J.; Blanchet, F.; Legendre, P.; Michin, P.; O'Hara, R.; Simpson, G.; Solymos, P.; Henry, M.; Stevens, H.; Wagner, H. "vegan" 2.3.3.–Community Ecology Package. Available online: https://cran.r-project.org/web/packages/vegan/vegan.pdf (accessed on 8 May 2019).
45. Wieder, R.K.; Lang, G.E. A critique of the analytical methods used in examining decomposition data obtained from litter bags. *Ecology* **1982**, *63*, 1636–1642. [CrossRef]

46. Bates, D.; Mächler, M.; Bolker, B.; Walker, S. Fitting linear mixed-effects models using lme4. *J. Stat. Softw.* **2015**, *67*, 1–48. [CrossRef]
47. Kuznetsova, A.; Brockhoff, P.B.; Christensen, R.H.B. lmerTest package: Tests in linear mixed effects models. *J. Stat. Softw.* **2017**, *82*, 1–26. [CrossRef]
48. R Core Team R: A Language and Environment for Statistical Computing. Available online: https://www.R-project.org/ (accessed on 14 March 2019).
49. Horodecki, P.; Jagodziński, A.M. Tree species effects on litter decomposition in pure stands on afforested post-mining sites. *For. Ecol. Manag.* **2017**, *406*, 1–11. [CrossRef]
50. Seastedt, T.R. The role of microarthropods in decomposition and mineralization processes. *Annu. Rev. Entomol.* **1984**, *29*, 25–46. [CrossRef]
51. Kaneko, N.; Salamanca, E. Mixed leaf litter effects on decomposition rates and soil microarthropod communities in an oak–pine stand in Japan. *Ecol. Res.* **1999**, *14*, 131–138. [CrossRef]
52. González, G.; Seastedt, T.R. Comparison of the abundance and composition of litter fauna in tropical and subalpine forests. *Pedobiologia* **2000**, *44*, 545–555. [CrossRef]
53. Gergócs, V.; Rétháti, G.; Hufnagel, L. Litter quality indirectly influences community composition, reproductive mode and trophic structure of oribatid mite communities: A microcosm. *Exp. Appl. Acarol.* **2015**, *67*, 335–356. [CrossRef]
54. Kampichler, C.; Bruckner, A. The role of microarthropods in terrestrial decomposition: A meta-analysis of 40 years of litterbag studies. *Biol. Rev.* **2009**, *84*, 375–389. [CrossRef] [PubMed]
55. Usher, M.B. Studies on a wood-feeding termite community in Ghana, West Africa. *Biotropica* **1975**, *7*, 217–233. [CrossRef]
56. Minor, M.A.; Cianciolo, J.M. Diversity of soil mites (Acari: Oribatida, Mesostigmata) along a gradient of land use types in New York. *Appl. Soil Ecol.* **2007**, *35*, 140–153. [CrossRef]
57. Walter, D.E.; Proctor, H. *Mites: Ecology, Evolution & Behaviour–Life at a Microscale|David Evans Walter, Springer*; CABI Publishing: New York, NY, USA, 1999.
58. Martikainen, E.; Huhta, V. Interactions between nematodes and predatory mites in raw humus soil: A microcosm experiment. *Rev. DÉcologie Biol. Sol* **1990**, *27*, 13–20.
59. Hasegawa, M.; Okabe, K.; Fukuyama, K.; Makino, S.; Okochi, I.; Tanaka, H.; Goto, H.; Mizoguchi, T.; Sakata, T. Community structures of Mesostigmata, Prostigmata and Oribatida in broad-leaved regeneration forests and conifer plantations of various ages. *Exp. Appl. Acarol.* **2013**, *59*, 391–408. [CrossRef] [PubMed]

forests

MDPI

Article

Simple Is Best: Pine Twigs Are Better Than Artificial Lures for Trapping of Pine Weevils in Pitfall Traps

Michal Lalík [1,2,*], Jaroslav Holuša [2], Juraj Galko [1], Karolína Resnerová [2], Andrej Kunca [1], Christo Nikolov [1], Silvia Mudrončeková [3] and Peter Surový [2]

1 National Forest Centre, Forest Research Institute Zvolen, T. G. Masaryka 22, 96001 Zvolen, Slovakia
2 Department of Forest Protection and Entomology, Faculty of Forestry and Wood Sciences, Czech University of Life Sciences Prague, Kamýcka 1176, 16500 Prague, Czech Republic
3 Research Station of State Forests of TANAP, 05960 Tatranská Lomnica, Slovakia
* Correspondence: michal.lalik@nlcsk.org; Tel.: +42-19-0374-6426

Received: 26 June 2019; Accepted: 22 July 2019; Published: 29 July 2019

Abstract: The large pine weevil *Hylobius abietis* (Linnaeus 1758) is the main pest of coniferous seedlings in Europe and causes substantial damage in areas that have been clear-cut or otherwise disturbed. We compared the efficacy of different attractants for the capture of *H. abietis* adults in white pitfall traps. The field experiment was performed from mid-April to the end of August 2018 at six plots in Central Europe located in spruce stands that had been clear-cut. At each plot, we compared five attractants: one pine twig with ethanol, Hylodor, alpha-pinene + ethanol, turpentine oil and ethanol (separated), and turpentine oil + ethanol (not separated). Traps without attractant served as a control. Six traps for each attractant or control were distributed at each plot. Of the total number of *H. abietis* adults trapped, 43.3%, 20.5%, 17.9%, 9.8%, 8.5% and 0.5% were captured in traps with pine twigs with ethanol, alpha-pinene, Hylodor, turpentine, oil + ethanol (separated), turpentine oil + ethanol (combined), and no attractant, respectively. The bottom of each trap contained propylene glycol to kill and preserve beetles. The small number of beetles captured in the control traps confirms that the propylene glycol:water mixture did not influence the trapping of *H. abietis*. The use of pitfall traps with a suitable attractant (especially pine twigs and ethanol) should be useful for monitoring of *H. abietis*, because it is simple and cost-effective. The use of such pitfall traps to control *H. abietis* by mass trapping would require 50 to 100 traps per ha.

Keywords: Hylodor; alpha-pinene; turpentine oil; ethanol; propylene-glycol; Norway spruce

1. Introduction

The large pine weevil *Hylobius abietis* (Linnaeus 1758) (Coleoptera: Curculionidae) is the main pest of coniferous seedlings in Europe and is especially damaging in areas where seedlings have been planted following clearcutting or other disturbances [1–3]. In such areas, the presence of fresh stumps maintains high numbers of *H. abietis*. Over the last 100 years, foresters have used various methods to protect seedlings from the damage caused by *H. abietis* feeding [1,3,4].

The basic protection method is to prevent *H. abietis* from feeding on seedlings. Feeding barriers have been commonly used in Germany since 1920 [1]. Plastic collars were developed at the end of the 1970s [5] and have been in use since that time. During the late 1980s and the beginning of the 1990s, researchers developed and tested several other shields, including stockings [6], plastic fibre wrappings ("BEMA") [7], and coated barriers [8]. Feeding can also be prevented by applying various coatings to the seedling stem. The first coating used was latex, which prevented feeding on the bark [9]. Protection can also be provided by a sand coating, which involves an initial application of acrylate glue and a subsequent application of fine sand (grain size <0.2 mm). The treated seedlings can then be planted in stands [10]. Another option is to coat seedlings up to a height of 15 cm with a special flexible wax, which can protect the seedlings for 2 years [11].

All of the methods mentioned in the previous paragraph protect seedlings but do not reduce the numbers of *H. abietis* in the environment. From the 1950s to 1970s, the development of barriers for the protection of seedlings slowed, because insecticides with dichlorodiphenyltrichloroethane (DDT) were used. Because DDT can harm humans [12] and other non-targets [13], pesticides containing DDT were gradually restricted and finally forbidden. After insecticides with DDT were banned, research concerning the development of other insecticides increased. Pyrethrins and synthetic pyrethroids are sold as commercial pesticides used to control pest insects in agriculture, homes, communities, restaurants, hospitals, schools, and as a topical head lice treatment [14]. Because of physiological effects of synthetic pyrethroids on *H. abietis* [15–18], synthetic pyrethroids are used in the field [19–21].

The mass trapping of *H. abietis* using attracting materials was practised before seedling protection. Mass trapping appears in the literature by 1839 [22]. The materials used to attract *H. abietis* included "trap barks", "trap logs", or fresh branches. In each case, the bark, logs, or branches (which were obtained from host species) were deployed in the field and then removed along with attracted beetles after some period of time. A disadvantage was that the attracted adults remained alive and had to be collected and killed. Trap barks were made more attractive by free placing pine twigs inside two pieces of spruce barks [23].

The labour needed to collect *H. abietis* from trap barks can be reduced by treating the bark with a synthetic pyrethroid or other insecticide such that the beetles are killed. In Europe, however, this method is currently only being used in a few countries [3,24,25]. The main reasons for the decreased use of this seedling protection method is the difficulty in preparing trap barks and the need for their frequent inspection and replacement.

Another method of catching *H. abietis* adults without the need to collect them is to use pitfall traps. Low numbers of traps are used (20–30 traps per ha), for monitoring [26]. The installation of pitfall traps is simple, and their efficacy is increased by addition of an attractant. Many kinds of attractants have been used, including wooden cylinders and discs [27–30], but the original lure from trap barks, i.e., pine twigs [23], has been forgotten. Only a single modern work uses twigs from Norway spruce in addition to bark [31].

Alpha-pinene combined with ethanol is highly attractive to *H. abietis*, which is understandable, because alpha-pinene is one of the compounds in the resin of coniferous [32]. Traps containing alpha-pinene with ethanol are more effective than traps that contain only alpha-pinene or only ethanol [33]. These attractants have usually involved the use of two "evaporators" (the containers holding the attractant and releasing its odours) per pitfall trap, i.e., alpha-pinene in one evaporator and 70% ethanol in a second evaporator [32,34–36]. These chemicals have also been applied to increase the attractiveness of bark and other natural materials [29].

Pitfall traps with attractants can be used both to monitor and control pest populations [27,30]. Such traps, along with chemical protection of seedlings, have been widely used in Poland [37]. Traps IBL-4 containing Hylodor, the aggregation pheromone of *H. abietis* [38], caught significantly more of *H. abietis* than natural (branches, discs) baits [30].

The pitfall traps used to trap *H. abietis* are usually simple containers or tubes [31–35,39]. The containers or tubes are buried in the soil, but their tops, which extend above the soil surface, have holes at the soil surface that enable beetles to enter. A recent trend is to create more complex constructions, such as a funnel with holes and a bottle [36], or other types of traps used in Poland [38].

The goal of this work was to evaluate the ability of simple pitfall traps (small buckets) to capture *H. abietis* adults. The traps contained a natural attractant (a pine twig), artificial attractants based on natural compounds (alpha-pinene, turpentine oil, and ethanol), or a commercially available aggregation pheromone (Hylodor).

2. Materials and Methods

2.1. Localities

The experiment was conducted in Norway spruce (*Picea abies* (L.) Karst) forests at four localities in the Czech Republic and at two localities in Slovakia (Figure 1). Norway spruce represented at least 90% of the trees (Table 1).

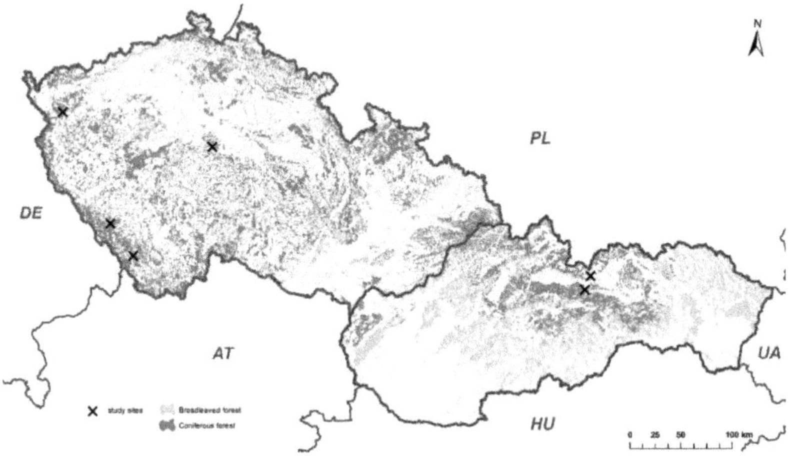

Figure 1. The six localities in the Czech Republic and Slovakia used in the current study.

Table 1. Description of study localities.

Country/Locality	Geographical Coordinates	Area (ha)	Altitude	Aspect	Establishment of Cleared Area	Spruce Share (%)
CZ/Arnoštov	48.8890528N 13.9529108E	0.74	944–946	Southeast	Felling September 2017	95
SK/Vyšné Hágy	49.120814N 20.102610E	21.5	1216–1220	South	Wind-throw May 2014	95
CZ/Kašperské Hory	49.1343394N 13.5876222E	0.81	780–785	Southwest	Felling October 2017	100
CZ/Kostelec n. Č. L	49.9198067N 14.840943E	0.62	402–404	South	Felling December 2017	90
SK/Liptovská Teplička	48.995592N 20.038704E	10.4	1115–1127	North	Bark-beetle disturbance February 2018	100
CZ/Mariánske Lázně	50.0246936N 12.7173419E	0.45	819–820	Plane	Bark-beetle disturbance May 2017	100

The trees at the locality had been removed for felling (harvest) or because of wind damage or bark beetle outbreak (Table 1). After the trees were removed, there were about 400 stumps per ha.

2.2. Attractants and Evaporators

Pine twigs and four chemical attractants were tested. All twigs (50 mm long and 12 ± 2 mm in diameter) were cut from a single 50-year-old Pinus sylvestris (L.) tree on 10 April 2018. The age of the twigs was 4–6 year. The twigs were stored in hermetically sealed polyethylene bags at −20 °C. One thawed twig was placed per trap, and a 20-mL bottle containing ethyl alcohol was placed next to the twig. The lid of the bottle had six holes with a diameter of 2 mm and one hole with a diameter of 4 mm; a polypropylene string that was attached to the larger hole and to the trap rim was used to suspend the bottle in the trap.

Four kinds of evaporators were used to disperse the chemical attractants. One of these was the Hylodor (A) (Z.D. Chemipan), which is a commercially available evaporator plus attractant (a mixture of organic and inorganic substances) produced in Poland (Table 2). The other type of evaporator consisted of a polypropylene tube fabricated by the Fytofarm Ltd. Co. (Bratislava, Slovakia)

(Table 2). The polypropylene tubes were used for the other three kinds of attractants and the control. After the tubes were filled, they were head sealed, and the content was released only through the polypropylene walls.

Table 2. Descriptions of the attractants.

Attractant	Daily Vapour (g) **	Attractant Composition	Ratio of Components
A—Hylodor	0.012	mixture of organic and inorganic substances	-
B—alpha-pinene + ethanol *	0.070	mixture of alpha-pinene (1R)-(+)-alpha-pinene (98%) CAS7785-70-8, EINECS: 232087-8 + ethanol CAS: 64-17-5, EINECS: 200-578-6	1:2 ***
C—Turpentine oil + ethanol separated *	1st tube: 0.023 2nd tube: 0.241	1st tube: ethanol CAS: 64-17-5, EINECS: 200-578-6 2nd tube: turpentine, oil CAS: 7785-70-8, EINECS: 232087-8	100% 100%
D—Turpentine oil + ethanol *	0.093	mixture: turpentine, oil CAS: 7785-70-8, EINECS: 232087-8 + ethanol CAS: 64-17-5, EINECS: 200-578-6	2:1 ****
PC—pine branch + ethanol in a bottle	0.402	ethanol CAS: 64-17-5, EINECS: 200-578-6	-

* The evaporators for these attractants were polypropylene tubes (diameter = 23 mm, height = 49 mm, volume = 12.5 ml, weight without attractant = 3.5 to 4.0 g). Each tube contained 6 mL of attractant. ** Vapour in the laboratory condition at 19.54 ± 1.73 °C, 36.53 ± 3.12%; *** Used according to [33]; **** Used according to [29].

2.3. Pitfall Traps and Experimental Design

Pitfall traps consisted of 1.2-litre buckets with 10 holes (10 mm in diameter) in the upper part of the trap (Figure 2). The traps had lids and were buried in the ground so that the holes, through which beetles could pass, were at the soil surface. The bottom of each trap contained 200 mL of propylene glycol (CAS: 57-55-6, EINECS: 200-338-0) mixed with water (1:1 ratio) and Tween 80 wetting agent (Carl Roth, CAS 9005-65-6), which killed and preserved the beetles that fell into the trap. One evaporator, in which the attractant was placed, was also hung from the upper rim of each trap so that it rested along the inner wall of the trap and above the propylene glycol-water mixture.

Figure 2. Photographs of representative pitfall traps and different type lure (**a**–**d**).

A total of 36 pitfall traps were placed in one plot (22.0 m × 28.5 m) at each locality. The distance between the traps was 5–6 m depending on the obstacles in the field. Among the 36 traps in each plot, 24 contained chemical attractants, 6 contained one pine twig + ethanol in a bottle, and 6 were without any attractant and served as negative control. The traps were distributed in the plot so that a specific type of attractant was not located near the same attractant type (Figure 3).

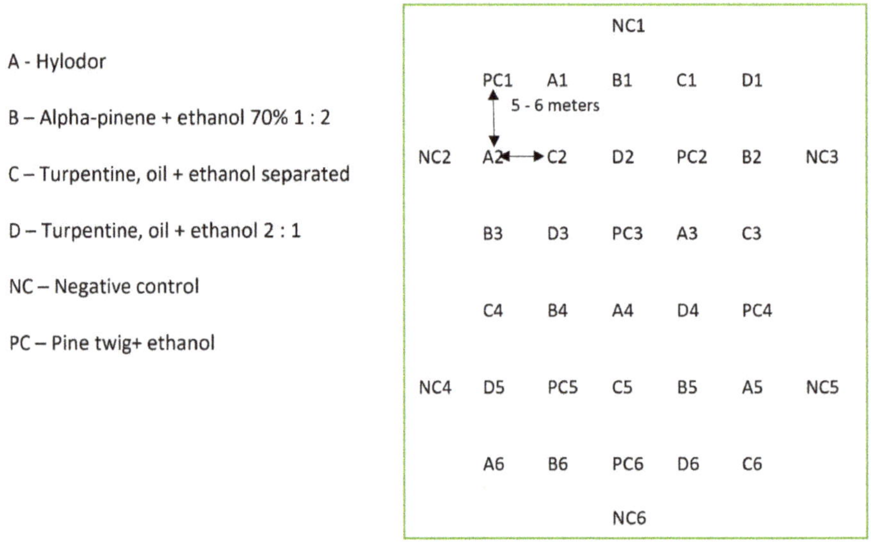

Figure 3. Locations of the pitfall traps in a plot (the attractant is indicated by the letters, and the trap replicate is indicated by the number).

The experiment began on 13 April 2018 and ended on 31 August 2018. Traps were checked every week at each plot on the same day of the week; all trapped beetles were removed and counted according to species and sex. For traps with twigs, twigs were replaced, and fresh ethanol was added to the bottles every week. For traps with synthetic attractants, the attractants were replaced during the 8th and 16th week of the experiment.

2.4. Statistical Analyses

The cumulative number of *H. abietis* adult males and adult females caught per trap across all trap types did not fit a normal distribution (Figure 4), and the data failed the test for homogeneity of variance concerning the effects of attractant and locality (Bartlett's K-squared = 206.72, df = 5, p-value $<0.001 = 2.2 \times 10^{-16}$). As a consequence, the general linear model (GLM) rather than ANOVA was used. The complete model is presented in Appendix A.

The results of the GLM model were tested using ANOVA for detection of differences and their sizes among different treatments and localities. Statistical analyses were performed in R software (RStudio Version 1.1.423, © 2009–2018 RStudio, Inc., package lme4, version 1.1-19, Boston, MA, USA).

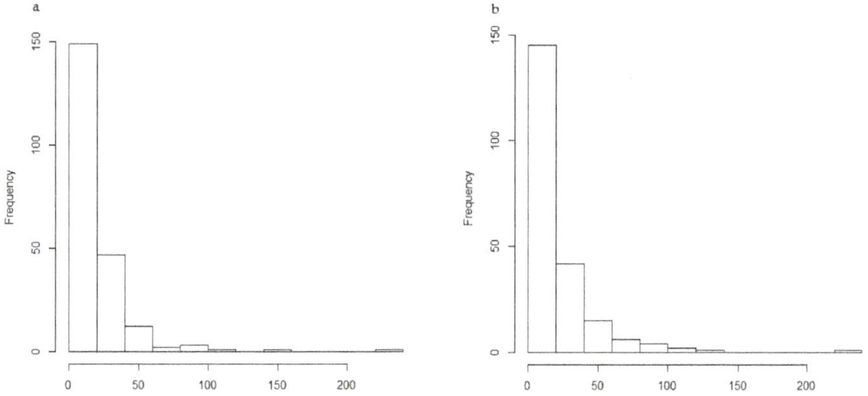

Figure 4. Frequency distribution of numbers of *H. abietis* adult males (**a**) and females (**b**) caught per trap across all attractant types.

3. Results

The main species captured in our pitfall traps were *H. abietis* and *H. pinastri*. In total, 8266 adults of *H. abietis* and 2040 adults of *H. pinastri* were caught in all pitfall traps at all localities. The number of *H. abietis* caught per locality and per attractant type was correlated with the number of *H. pinastri* caught per locality and per attractant type ($y = 10.08 + 0.20x$; $r = 0.90$; $p = 0.0000$; $r^2 = 0.81$). The number of trapped adults (total of both species) was highest with a pine twig + ethanol (Table 3). Among the chemical attractants, the number of trapped adults (total of both species) was highest with alpha-pinene + ethanol followed by turpentine oil and ethanol (separated); the number was lowest with turpentine oil + ethanol (combined) (Table 3).

Table 3. Numbers of *H. abietis/H. pinastri* trapped as affected by locality and attractant.

-	Attractant					
Locality	**Hylodor**	**Alpha-Pinene + Ethanol**	**Turpentine, Oil + Ethanol Separated**	**Turpentine, Oil + Ethanol Combined**	**Negative Control**	**Pine Twig + Ethanol**
Arnoštov	269/38	405/74	137/9	102/2	8/0	681/144
Vyšné Hágy	150/72	265/96	130/48	59/9	1/0	449/185
Kašperské Hory	34,379	273/54	205/63	205/76	23/0	656/202
Kostelec n. Č.L	228/72	286/65	200/51	139/37	0/0	367/95
Liptovská Teplička	456/88	416/80	103/18	137/35	9/0	1310/226
Mariánské Lázně	34/7	47/17	32/10	22/18	2/0	117/69
Total	1480/356	1692/387	807/199	664/177	43/0	3580/921

The *H. abietis* sex ratio differed among localities. Two localities had a 50:50 sex ratio, two had more males than females, and two had more females than males (Table 4). Across all localities, the average number of adults trapped per ha was >14,000 for *H. abietis* and >3000 for *H. pinastri* (Table 4). The number of *H. abietis* adults trapped per ha significantly differed among localities and was highest at Liptovská Teplička and lowest at Mariánské Lázně (Table 4).

Table 4. Numbers of *H. abietis* males and females and *H. pinastri* (both sexes) trapped at individual localities (* average).

Locality	Number of *H. abietis* Males Trapped	Number of *H. abietis* Females Trapped	Sex Ratio of Trapped Males:Females	Total Number of *H. abietis* Trapped	Number of *H. abietis* Trapped/Ha	Number of *H. pinastri* Trapped	Number of *H. pinastri* Trapped/Ha
Arnoštov	480	1122	30:70	1602	16,584	267	2764
Vyšné Hágy	456	598	43:57	1054	10,910	411	4255
Kašperské Hory	979	726	57:43	1705	17,650	474	4907
Kostelec n. Č.L.	610	610	50:50	1220	12,630	320	3313
Liptovská Teplička	1212	1219	50:50	2431	25,165	447	4627
Mariánské Lázně	140	114	55:45	254	2630	121	1253
Total or average	3877	4389	47:53 *	1378	14,261 *	340	3520 *

The numbers of trapped individuals significantly differed among localities and attractants (Figures 5 and 6 and Appendix A). Trapping was highest with a pine twig ± ethanol. Trapping was generally similar with Hylodor and alpha-pinene + ethanol but was higher with Hylodor at some localities (Kašperské Hory, Liptovská Teplička) and was higher with alpha-pinene + ethanol at other localities (Arnoštov, Vyšné Hágy, Kostelec n. Č.L., Mariánské Lázně). Trapping was significantly lower with turpentine oil + ethanol (separated) and with turpentine oil + ethanol (combined) than with a pine twig ± ethanol, Hylodor, or alpha-pinene + ethanol.

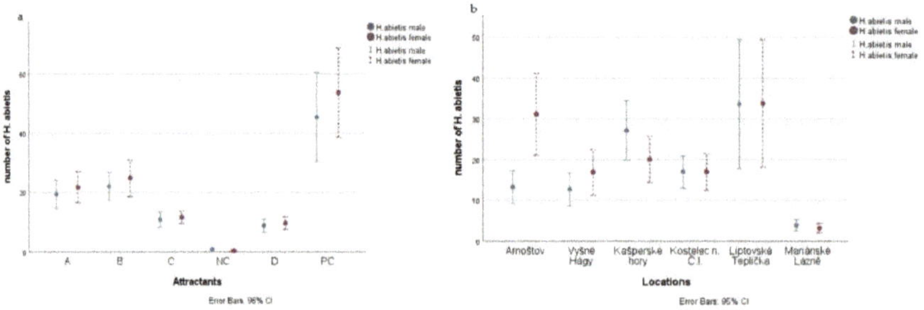

Figure 5. Number (mean ± SE) of *H. abietis* trapped with individual attractants across all localities (**a**) and at individual localities across all attractant treatments including the control (**b**) (for attractant abbreviations, see Figure 1).

At all localities, trapping was highest at the beginning of May and then gradually decreased (Figure 6). At the end of the experiment, only a few individuals were trapped at each locality (Figure 6).

Pairwise comparisons were made of the efficacy of attractants in trapping *H. abietis* (based on the number of individuals by sex trapped across all localities) (Tables 5 and 6). Almost all pairs were statistically different. For both males and females, trapping was always highest with a pine twig with ethanol and was always lowest with the no attractant control. Trapping did not significantly differ between turpentine oil + ethanol (separated) vs. turpentine oil + ethanol (combined) for both males and females or between turpentine oil + ethanol (separated) vs. Hylodor for males.

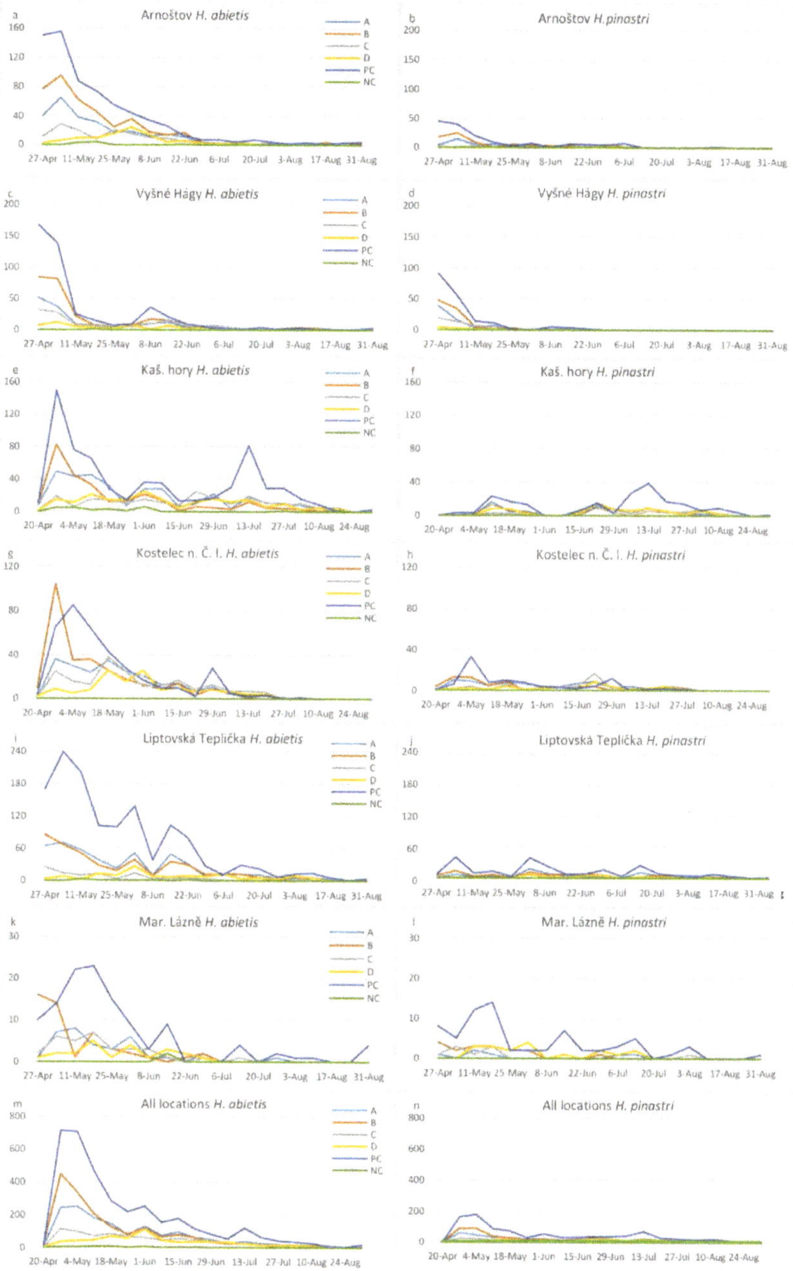

Figure 6. Mean numbers of *H. abietis* (**a**,**c**,**e**,**g**,**i**,**k**,**m**) and *H. pinastri* (**b**,**d**,**h**,**j**,**l**,**n**) adults trapped over time as affected by attractant and locality.

Table 5. Pairwise comparison of attractant efficacy in trapping male *H. abietis* (based on number of individuals trapped across all localities).

Attractant Pair	Absolute Difference	Standard Error	z-Value	p-Value
B-A	0.4400	0.1471	2.991	<0.02791 *
C-A	−0.4595	0.1844	−2.492	0.10974
NC-A	−2.5390	0.4241	−5.987	<0.001 ***
D-A	−0.7198	0.2005	−3.591	<0.00368 **
PC-A	0.9268	0.1355	6.838	<0.001 ***
C-B	−0.8995	0.1712	−5.254	<0.001 ***
NC-B	−2.9789	0.4185	−7.118	<0.001 ***
D-B	−1.1598	0.1884	−6.155	<0.001 ***
PC-B	0.4868	0.1170	4.162	<0.001 ***
NC-C	−2.0794	0.4330	−4.802	<0.001 ***
D-C	−0.2603	0.2188	−1.190	0.82316
PC-C	1.3863	0.1614	8.591	<0.001 ***
D-NC	1.8192	0.1614	4.133	<0.001 ***
PC-NC	3.4657	0.4146	8.360	<0.001 ***
PC-D	1.6466	0.1795	9.171	<0.001 ***

(Signif. codes: 0 '***'; 0.001 '**'; 0.01 '*').

Table 6. Pairwise comparison of attractant efficacy in trapping female *H. abietis* (based on number of individuals trapped across all localities).

Attractant Pair	Absolute Difference	Standard Error	z-Value	p-Value
B-A	0.39330	0.09315	4.222	<0.001 ***
C-A	−0.77405	0.12813	−6.041	<0.001 ***
NC-A	−4.56954	0.71076	−6.429	<0.001 ***
D-A	−1.08830	0.14341	−7.589	<0.001 ***
PC-A	0.92557	0.08506	10.882	<0.001 ***
C-B	−1.16736	0.12138	−9.618	<0.001 ***
NC-B	−4.96284	0.70957	−6.994	<0.001 ***
D-B	−1.48160	0.13741	−10.782	<0.001 ***
PC-B	0.53227	0.07450	7.145	<0.001 ***
NC-C	−3.79549	0.71501	−5.308	<0.001 ***
D-C	−0.31425	0.16316	−1.926	0.337
PC-C	1.69963	0.11528	14.744	<0.001 ***
D-NC	3.48124	0.71790	4.849	<0.001 ***
PC-NC	5.49512	0.70856	7.755	<0.001 ***
PC-D	2.01388	0.13205	15.251	<0.001 ***

(Signif. codes: 0 '***'; 0.001 '**'; 0.01 '*').

4. Discussion

During the growing season of 2018, we recorded the capture of *H. abietis* in pitfall traps deployed in recently cleared spruce stands in six Central European localities where altitudes ranged from 400 to 1220 m a.s.l. The large-scale experiment (covering an area of 550 km × 200 km) was performed in two countries (Figure 1). Buckets of 1.2-litre were used as traps with 10-mm-diameter entrance holes (Figure 2). These pitfall traps were easy to construct and deploy. However, pitfall traps are not selective, even with use of entrance holes. They can also catch other invertebrates, but impact is weak [31]. They have to be expose only for necessary period and have not be forgotten in forests [40].

All of the attractants in our experiment showed some ability to attract *H. abietis* and *H. pinastri*. *H. abietis* is very abundant in Central Europe [3]. *H. pinastri* is also common [4,41–44] but is less abundant than *H. abietis* [45–47]. We assume that fewer *H. pinastri* than *H. abietis* were attracted to pitfall traps with pine twigs because *H. pinastri* prefers spruce [48] while *H. abietis* prefers pine [49].

For *H. abietis*, trapping was always highest with a pine twig with ethanol. It should be recognized that daily vapours of studied attractants are low in comparison with published data [33,35,36]. Experimental evaporators have comparable vaporization as manufactured Hylodor. Ref. [33] had higher vapour, but used attractants that in the bottle and filter paper. This method is not used in practice. We wanted to test the methods usable for forest practice and easy to use. For foresters, the industrially produced evaporators are the best. Therefore, we used Hylodor and other types of evaporator consisting of a polypropylene tube manufactured by Fytofarm Ltd. Co. The thermoplastic material used ensures regular evaporation for as long as possible [50].

Natural materials (trap barks) for attracting *H. abietis* began to be used in the first half of 19th century [22] and were frequently used by the beginning of the 20th century [1]. A disadvantage of trap barks is the need for their frequent replacement (once every 2 weeks), and the need to remove the attracted *H. abietis* adults every 2–5 days. Another disadvantage of trap barks is their high price, which in Slovakia is about 3.5 € for one piece of bark [51]. For monitoring, about 25 pieces of trap bark/ha are needed. The price of such monitoring (excluding the maintenance costs) would be 175 €/ha per month. In contrast to trap barks, pitfall traps do not require regular visits. During our experiment, we emptied them once per week, but only one inspection per month would be needed if attractants in tubes were used. The alpha-pinene + ethanol attractant costs 2.5 € per trap and can last for 6–8 weeks, and the cost of one bucket used for the pitfall trap is 0.33 €. In this case, the costs excluding maintenance costs are 70.75 € with 25 pitfall traps/ha.

There are additional problems with the use of trap barks and other natural materials as pest attractants [33]. The types of natural materials used to attract pests can differ in their monoterpene content [52,53]. Another problem is that wooden attractants dry, which gradually reduces their attractiveness [45]. When pests attack wooden attractants, they feed on them and disturb the material. This increases the release of compounds that attract *H. abietis* [32] but decreases the longevity of the attractant. As a consequence, we replaced the pine twigs in our experiment every week. [31], who used source spruce twig replaced it also each week. Although we do not know anything about population densities in their study areas, they caught 10x less beetles in comparison with our results [31].

The second most effective attractant in the current study was alpha-pinene, which is present in conifers and has a resin-like odour. This compound is usual in that its content within a tree species is constant regardless [54]. The ratio of fragrant compounds in coniferous species is largely genetically driven and is not affected by other factors [55]. In the evaporator, alpha-pinene was mixed with ethanol, because such a combination is six times more attractive to *H. abietis* than alpha-pinene alone, and 10 times more attractive than ethanol alone [33]. In addition, the effect of the combination in the field is synergistic [32].

The third most successful attractant was the commercially available Hylodor. This attractant, which is produced in Poland, is described as an aggregation pheromone, but its composition is unknown. Kuźmiński and Bilon [30] showed that the number of beetles trapped was only slightly higher with Hylodor than with logs or wooden discs.

The fourth highest number of catches was obtained when turpentine oil and 70% ethanol were placed in separate tubes. Turpentine oil is extracted from resin by distillation and contains mainly alpha-pinene and beta pinene [29]. Although this lure has a low price [31,56], it trapped only low numbers of *Hylobius* spp. adults in the current study. The last attractant that was used in this study was turpentine oil mixed with ethanol in a 2:1 ratio. This attractant also trapped only a low number of *Hylobius* spp. adults. The concentration of turpentine oil and ethanol used in our experiment was equal to that used by [29]. We suspect that such a strong concentration could repel rather than attract, and that a smaller proportion of turpentine might be more effective [56].

The negative control (without attractant) pitfall traps in the current study trapped only a few individuals, which probably were not attracted to the traps but simply entered them accidently. The small number of beetles captured in these traps confirms that the propylene glycol:water mixture used to kill and preserve the trapped beetles did not function as an attractant or as a repellent, and did not therefore influence the trapping of *H. abietis* [56].

In total, we captured 8266 adults of *H. abietis* and 2040 adults of *H. pinastri* at six localities. Ref. [57] stated that a trap can attract adults from a distance of 2.5 m in all directions. In this study, we found that regardless of the lure used, a pitfall trap can attract adults at distances greater than 2.5 m. With a trap situated at the edge of the plot, we caught on average 290 ± 203 adults of *H. abietis* and 74 ± 52 adults of *H. pinastri*. In centrally located traps, we caught on average 250 ± 156 and 59 ± 41 adults of *H. abietis* and *H. pinastri* per trap, respectively. If we assume that each trap in a plot can capture beetles within a 3-m radius with baited traps (approximately 600 m^2), we will get an "*H. abietis* trapping area" of approximately 28.0 × 34.5 m (966 m^2). When we calculated the number of catches per ha, we found several thousand to several tens of thousands of *H. abietis* adults per ha and thousands of *H. pinastri* adults per ha depending on the locality (Table 4). In most localities, we found more than 10,000 beetles per ha; i.e., the density of *H. abietis* coincided with that stated by [58] (10,000–18,000 adults/ha). Other authors have also indicated that *H. abietis* numbered about 10,000/ha [36,59], and [4] found that *H. abietis* numbers exceeded 14,000/ha at three localities. We therefore assume that we caught the majority of the local population of *H. abietis* as well as of *H. pinastri*. This is also supported by the fact that in the second half of the summer, we did not catch any beetles, although the adults of new generation, which have a flight at the autumn, should be caught [60].

Clearcutting is almost always performed in autumn or winter, and the remaining stumps are surrounded by older stands. Larvae develop on the roots of fresh pine or spruce stumps and exceptionally also on the roots of 1-year-old stumps [54]. At mid-elevations in Central Europe, stumps older than 1 year are not attractive (personal observation Holuša, Galko, Lalík) [24]. This is also the reason why the number of beetles caught was the least at the locality Mariánské Lázně (the original forest was cut in May of previous year). In Scandinavia, stumps can remain attractive for 3 years [27], which was confirmed at the Vyšné Hágy locality, which is 1216–1220 m a.s.l., where the stumps were 4 years old, and the beetle was still abundant in the cleared area (personal observation). It is clear that beetles trapped in cleared areas could have originated from the surrounding areas. To select suitable localities, *H. abietis* adults use their olfactory sensors at the base of their antennae. At longer distances, adults orientate according to attractants produced by host tree species [61,62]. On clear cuttings, they probably distinguish shape of seedlings and older trees while flying [63]. At short distances, they react to pheromones [64]; male pheromones cause adults to aggregate, and female pheromones attract males [65]. The females can crawl or fly considerable distances to locate oviposition sites [66]. *H. abietis* can fly up to 2000 m in a single event, and can fly more than 80,000 m over their entire life [67].

Over the entire experiment, we caught 4389 females and 3877 males of *H. abietis*. Based on these values, the overall female:male sex ratio was 53:47, although males were more abundant than females at some plots. [32], who used pitfall traps baited with three types of attractants (alpha-pinene, ethanol, and a combination of these two) reported an *H. abietis* sex ratio similar to that in our study; the number of females was greater than the number of males irrespective of the combination of lures.

5. Conclusions

Forestry managers use monitoring data of *H. abietis* to plan methods for reducing the damage on plants. In the case of critical numbers exceeded the feeding barriers or insecticides are applied. The results show that pine weevil populations can be monitored using minimally modified buckets as pitfall traps. The cost of such traps is low, and the installation is simple. A natural material, namely a pine twig (+ ethanol), was the most successful attractant, which was consistent with previous reports. The pine twig + ethanol attractant is easy to prepare and is substantially cheaper than artificial attractants. Replacement of the twig after 1 week eliminates eventual shifts in the composition of volatile compounds, and the replacement of the twig every 2 weeks may be sufficient. During the replacement of twigs and the refilling of ethanol, trapped adults are removed from the trap to avoid attracting carrion beetles. When performed by a two-member team, the replacement required about 20 min per locality. On a per ha basis, the inspection of all traps installed (i.e., 300 traps/ha) would require about 3 h. From a practical point of view, the number of traps can be reduced, because traps at the edge caught more beetles, and the distance between the traps could probably be increased to 10–20 m, i.e., trap density could probably be reduced to 50–100 traps/ha. The results also suggest that pitfall traps with pine twigs + ethanol might be used for mass trapping. Such mass trapping would be very simple and based on our results all beetles could be trapped during period of April to June. This is especially attractive because the use of chemical insecticides is increasingly restricted in the Europe. As noted earlier, the monitoring and mass trapping of *H. abietis* is substantially cheaper with pitfall traps than with trap barks.

Author Contributions: Data curation, M.L., K.R. and J.H.; Formal analysis, P.S.; Methodology, M.L., J.G., A.K., C.N., S.M.; Writing—original draft, M.L., J.G. and J.H.; Writing—review and editing, M.L. and J.H.

Funding: This work was supported by the Slovak Research and Development Agency under the contract No. APVV-16-0031, APVV-18-0086 and by the Ministry of Agriculture and Rural Development of the Slovak Republic based on the item No. 08V0301. This work was also supported by the Czech University of Life Sciences Prague project No. IGA C_01_18 and by the grant "Advanced research supporting the forestry and wood-processing sector's adaptation to global change and the 4th industrial revolution", No. CZ.02.1.01/0.0/0.0/16_019/0000803 financed by OP RDE.

Acknowledgments: The authors thank Bruce Jaffee (USA) for editorial and linguistic improvement of manuscript, and Tomáš Holík, Jan Pešl, Tomáš Fiala, Nikola Bohatá, and Jana Kubov for support with field work.

Conflicts of Interest: The authors declare no conflict of interest.

Appendix A

glm (formula = hAbietisMale ~ atraktant * lokalita, family = "poisson")

Deviance Residuals:

Min	1Q	Median	3Q	Max
−10.2299	−1.1360	−0.1412	0.8858	10.0012

Coefficients:

Table A1. Full GLM model of tested factors, locality and attractant, and their interactions.

	Estimate	Std.	Error	z value Pr(>|z|)
(Intercept)	2.539×10^{0}	1.147×10^{-1}	2.2134×10^{1}	$< 2 \times 10^{-16}$
atraktantB	4.400×10^{-1}	1.471×10^{-1}	2.991×10^{0}	2.778×10^{-3}
atraktantC	-4.595×10^{-1}	1.844×10^{-1}	-2.492×10^{0}	1.2685×10^{-2}
atraktantCon.	-2.539×10^{0}	4.241×10^{-1}	-5.987×10^{0}	2.13×10^{-9}
atraktantD	-7.198×10^{-1}	2.005×10^{-1}	-3.591×10^{0}	3.30×10^{-4}
atraktantPC	9.268×10^{-1}	1.355×10^{-1}	6.838×10^{0}	8.00×10^{-12}
lokalitaVyšné Hágy	-2.532×10^{-1}	1.735×10^{-1}	-1.459×10^{0}	1.44503×10^{-1}
lokalitaKašperské hory	9.775×10^{-1}	1.346×10^{-1}	7.264×10^{0}	3.75×10^{-13}
lokalitaKostelec n.C.l.	5.213×10^{-1}	1.448×10^{-1}	3.600×10^{0}	3.18×10^{-4}
lokalitaLiptovská Teplička	1.035×10^{0}	1.335×10^{-1}	7.753×10^{0}	8.99×10^{-15}
lokalitaMarianske lázne	-1.386×10^{0}	2.565×10^{-1}	-5.405×10^{0}	6.49×10^{-8}
atraktantB:lokalitaVyšné Hágy	2.616×10^{-1}	2.168×10^{-1}	1.207×10^{0}	2.27423×10^{-1}
atraktantC:lokalitaVyšné Hágy	5.409×10^{-1}	2.580×10^{-1}	2.096×10^{0}	3.6046×10^{-2}
atraktantCon.:lokalitaVyšné Hágy	-1.539×10^{0}	1.094×10^{0}	-1.406×10^{0}	1.59605×10^{-1}
atraktantD:lokalitaVyšné Hágy	-6.189×10^{-2}	3.069×10^{-1}	-2.02×10^{-1}	8.40180×10^{-1}
atraktantPC:lokalitaVyšné Hágy	2.161×10^{-1}	2.018×10^{-1}	1.071×10^{0}	2.84314×10^{-1}
atraktantB:lokalitaKašperské hory	-5.949×10^{-1}	1.799×10^{-1}	-3.307×10^{0}	9.43×10^{-4}
atraktantC:lokalitaKašperské hory	-9.515×10^{-2}	2.181×10^{-1}	-4.36×10^{-1}	6.62641×10^{-1}
atraktantCon.:lokalitaKašperské hory	-1.302×10^{-1}	5.062×10^{-1}	-2.57×10^{-1}	7.96947×10^{-1}
atraktantD:lokalitaKašperské hory	1.300×10^{-1}	2.325×10^{-1}	5.59×10^{-1}	5.75957×10^{-1}
atraktantPC:lokalitaKašperské hory	-3.462×10^{-1}	1.615×10^{-1}	-2.143×10^{0}	3.2100×10^{-2}
atraktantB:lokalitaKostelec n.C.l.	-3.867×10^{-1}	1.920×10^{-1}	-2.014×10^{0}	4.3966×10^{-2}
atraktantC:lokalitaKostelec n.C.l.	2.709×10^{-1}	2.264×10^{-1}	1.197×10^{0}	2.31322×10^{-1}
atraktantCon.:lokalitaKostelec n.C.l.	-1.582×10^{1}	5.208×10^{2}	-3.0×10^{-2}	9.75762×10^{-1}
atraktantD:lokalitaKostelec n.C.l.	2.667×10^{-2}	2.522×10^{-1}	1.06×10^{-1}	9.15798×10^{-1}
atraktantPC:lokalitaKostelec n.C.l.	-6.026×10^{-1}	1.784×10^{-1}	-3.378×10^{0}	7.30×10^{-4}
atraktantB:lokalitaLiptovská Teplička	-3.988×10^{-1}	1.755×10^{-1}	-2.273×10^{0}	2.3056×10^{-2}
atraktantC:lokalitaLiptovská Teplička	-1.078×10^{0}	2.458×10^{-1}	-4.385×10^{0}	1.16×10^{-5}
atraktantCon.:lokalitaLiptovská Teplička	-7.476×10^{-1}	5.563×10^{-1}	-1.344×10^{0}	1.79030×10^{-1}
atraktantD:lokalitaLiptovská Teplička	-5.353×10^{-1}	2.475×10^{-1}	-2.163×10^{0}	3.0552×10^{-2}
atraktantPC:lokalitaLiptovská Teplička	1.965×10^{-1}	1.567×10^{-1}	1.254×10^{0}	2.09960×10^{-1}
atraktantB:lokalitaMarianske lázne	8.855×10^{-2}	3.336×10^{-1}	-2.65×10^{-1}	7.90676×10^{-1}
atraktantC:lokalitaMarianske lázne	-9.352×10^{-15}	4.123×10^{-1}	0×10^{0}	1.000000×10^{0}
atraktantCon.:lokalitaMarianske lázne	2.877×10^{-1}	8.558×10^{-1}	3.36×10^{-1}	7.36764×10^{-1}
atraktantD:lokalitaMarianske lázne	5.480×10^{-1}	3.941×10^{-1}	1.390×10^{0}	1.64403×10^{-1}
atraktantPC:lokalitaMarianske lázne	2.877×10^{-1}	2.943×10^{-1}	9.77×10^{-1}	3.28344×10^{-1}

Table A2. Significant factors and their interaction.

(Intercept)	***
atraktantB	**
atraktantC	*
atraktantCon.	***
atraktantD	***
atraktantPC	***
lokalitaVyšné Hágy	-
lokalitaKašperské hory	***
lokalitaKostelec n. C.l.	***
lokalitaLiptovská Teplička	***
lokalitaMarianske lázne	***
atraktantB:lokalitaVyšné Hágy	-
atraktantC:lokalitaVyšné Hágy	*
atraktantCon.:lokalitaVyšné Hágy	-
atraktantD:lokalitaVyšné Hágy	-
atraktantPC:lokalitaVyšné Hágy	-
atraktantB:lokalitaKašperské hory	***
atraktantC:lokalitaKašperské hory	-
atraktantCon.:lokalitaKašperské hory	-
atraktantD:lokalitaKašperské hory	-
atraktantPC:lokalitaKašperské hory	*
atraktantB:lokalitaKostelec n. C.l.	*
atraktantC:lokalitaKostelec n. C.l.	-
atraktantCon.:lokalitaKostelec n. C.l.	-
atraktantD:lokalitaKostelec n. C.l.	-
atraktantPC:lokalitaKostelec n. C.l.	***
atraktantB:lokalitaLiptovská Teplička	*
atraktantC:lokalitaLiptovská Teplička	***
atraktantCon.:lokalitaLiptovská Teplička	-
atraktantD:lokalitaLiptovská Teplička	*
atraktantPC:lokalitaLiptovská Teplička	-
atraktantB:lokalitaMarianske lázne	-
atraktantC:lokalitaMarianske lázne	-
atraktantCon.:lokalitaMarianske lázne	-
atraktantD:lokalitaMarianske lázne	-
atraktantPC:lokalitaMarianske lázne	-

Signif. codes: 0 '***' 0.001 '**' 0.01 '*'.

(Dispersion parameter for poisson family taken to be 1)
Null deviance: 4695.37 on 215 degrees of freedom
Residual deviance: 734.05 on 180 degrees of freedom
AIC: 1638.40
Number of Fisher Scoring iterations: 13

References

1. Escherich, K. *Die Forstinsekten Mitteleuropas*, 2nd ed.; Paul Parey: Berlin, Germany, 1923; p. 663.
2. Day, K.R.; Leather, S.R. Threats to forestry by insect pests in Europe. In *Forests and Insects*; Watt, A.D., Stork, N.E., Hunter, M.D., Eds.; Chapman & Hall: London, UK, 1997; pp. 177–205.
3. Långstöm, B.; Day, K.R. Damage, control and management of weevil pests, especially *Hylobius abietis*. In *Bark and Wood Boring Insects in Living Trees in Europe: A Synthesis*; Lieutier, F., Day, K.R., Battisti, A., Grégoire, J.-C., Evans, H.F., Eds.; Springer: Dordrecht, The Netherlands, 2004; pp. 415–444.
4. Eidmann, H.H. Hylobius Schönh. In *Die Forstschädlinge Europas*, 2nd ed.; Schwenke, W.K., Ed.; Paul Parey: Hamburg/Berlin, Germany, 1974; pp. 275–293.
5. Lindström, A.; Hellqvist, C.; Gyldberg, B.; Långström, B.; Mattsson, A. Field performance of a protective collar against damage by *Hylobius abietis*. *Scand. J. For. Res.* **1986**, *1*, 3–15. [CrossRef]

6. Eidmann, H.H.; von Sydow, F. Stockings for protection of containerised seedlings against pine weevil (*Hylobius abietis* L.) damage. *Scand. J. For. Res.* **1989**, *4*, 537–547. [CrossRef]
7. Hagner, M.; Jonsson, C. Survival after planting without soil preparation for pine and spruce seedlings protected from *Hylobius abietis* by physical and chemical shelters. *Scand. J. For. Res.* **1995**, *10*, 225–234. [CrossRef]
8. Eidmann, H.H.; Nordenhem, H.; Weslien, J. Physical protection of conifer seedlings against pine weevil feeding. *Scand. J. For. Res.* **1996**, *11*, 68–75. [CrossRef]
9. Zumr, V.; Stary, P. LATEX paint as an antifeedent against *Hylobius abietis* (L.) (Col., Curculionidae) on conifer seedlings. *Anz. Schädl. Pflanzenschutz Umweltschutz* **1995**, *21*, 42–43. [CrossRef]
10. Nordlander, G.; Nordenhem, H.; Hellqvist, C. A flexible sand coating (Conniflex) for the protection of conifer seedlings against damage by the pine weevil *Hylobius abietis*. *Agric. For. Entomol.* **2009**, *11*, 91–100. [CrossRef]
11. Kvaae.no. Available online: http://kvaae.no/what/ (accessed on 15 May 2019).
12. Harte, J.; Holdren, C.; Schneider, R.; Shirley, C. *Toxics A to Z, a Guide to Everyday Pollution Hazards*; University of California: Oxford, UK, 1991; p. 478. ISBN 9780520072244.
13. Rattner, B.A. History of wildlife toxicology. *Ecotoxicology* **2007**, *18*, 773–783. [CrossRef]
14. Bradberry, S.M.; Cage, S.A.; Proudfoot, A.T.; Vale, J.A. Poisoning due to pyrethroids. *Toxicol. Rev.* **2005**, *24*, 93–106. [CrossRef]
15. Pszczolkowski, M.A.; Dobrowolski, M. Circadian dynamics of locomotor activity and deltamethrin susceptibility in the pine Weevil *Hylobius abietis*. *Phytoparasitica* **1999**, *27*, 19–25. [CrossRef]
16. Dobrowolski, M. The susceptibility of the large pine weevil (*Hylobius abietis* L.) to insecticides and the role of the oxidative metabolism in the developing of the pest resistance to DDT and pyrethroids. *Folia For. Polon.* **2000**, *42*, 83–94.
17. Lempérière, G.; Julien, J. Protection against the pine weevil-Efficiency of a carbosulfan-based systemic insecticide. *Rev. For. Fr.* **2003**, *55*, 129–140. [CrossRef]
18. Rose, D.; Matthews, G.A.; Leather, S.R. Sub-lethal responses of the large pine weevil, *Hylobius abietis*, to the pyrethroid insecticide lambda-cyhalothrin. *Physiol. Entomol.* **2006**, *31*, 316–327. [CrossRef]
19. Glovacka, B.; Lech, A.; Wilczynski, W. Application of deltamethrin for spraying or dipping to protect Scots pine seedlings againts *Hylobius abietis* L. and logs against *Tomicus piniperda* L. *Ann. Sci. For.* **1991**, *48*, 113–117. [CrossRef]
20. Torstensson, L.; Börjesson, E.; Arvidsson, B. Treatment of bare root spruce seedlings with permethrin against pine weevil before lifting. *Scand. J. For. Res.* **1999**, *14*, 408–415. [CrossRef]
21. Rose, D.; Matthews, G.A.; Leather, S.R. Recognition and avoidance of insecticide-treated Scots Pine (*Pinus sylvestris*) by *Hylobius abietis* (Coleoptera: Curculionidae): Implications for pest management strategies. *Agric. For. Entomol.* **2005**, *7*, 187–191. [CrossRef]
22. Ratzeburg, J.T.C. *Die Forst- Insekten. Erster Teil–Die Käfer*; 2 Auflage; Nicolai'sche Buchhandlung: Berlin, Germany, 1839; p. 247.
23. Pfeffer, A.; Horák, E.; Kudela, M.; Muller, J.; Novakova, E.; Stolina, M. *Ochrana Lesů*; Státní Zemědělské Nakladatelství: Praha, Czech Republic, 1961; p. 838.
24. Modlinger, R.; Knížek, M. Klikoroh borový *Hylobius abietis* (L.). *Lesnická Pr.* **2009**, *88*, 1–4.
25. Galko, J.; Gubka, A.; Vakula, J. Praktické Skúsenosti s Využitím Lapacích Kôr na Zníženie Škôd Spôsobených Tvrdoňom Smrekovým na Mladých Výsadbách Ihličnatých Drevín. In *Aktuálne Problémy v Ochrane Lesa*; Kunca, A., Nový Smokovec, S., Kunca, A., Eds.; Národné Lesnícke Centrum: Zvolen, Slovakia, 2012; pp. 60–64.
26. Galko, J.; Vakula, J.; Kunca, A.; Rell, S.; Gubka, A. *Ochrana Lesa;* Ochrana lesa proti tvrdoňom a lykokazom na sadeniciach; Úrad pre Normalizáciu, Metorologiu a Skúšobníctvo: Bratislava, Slovakia, STN 48 2714; 2016; p. 8.
27. Nordenhem, H. Age, sexual development, and seasonal occurrence of the pine weevil *Hylobius abietis*. *J. Appl. Entomol.* **1989**, *108*, 260–270. [CrossRef]
28. Skłodowski, J.J.W.; Gadziński, J. Effectiveness of beetle catches in two types of traps for *Hylobius abietis* L. *Sylwan* **2001**, *6*, 55–63.
29. Moreira, X.; Costas, R.; Sampedro, L.; Zas, R. A simple method for trapping *Hylobius abietis* (L.) alive in northern Spain. *For. Syst.* **2008**, *17*, 188–192. [CrossRef]

30. Kuźmiński, R.; Bilon, A. Evaluation of effectiveness of selected types of traps used in capturing of large pine weevil–*Hylobius abietis* (L.). *Acta Sci. Pol.—Silv. Colendarum R. Ind. Lig.* **2009**, *8*, 19–26.
31. Zumr, V.; Stary, P.P. Field experiments with different attractants in baited pitfall traps for *Hylobius abietis* L. (Col., Curculionidae). *J. Appl. Entomol.* **1992**, *113*, 451–455. [CrossRef]
32. Tilles, D.A.; Sjödin, K.; Nordlander, G.; Eidmen, H.H. Synergism between ethanol and conifer host volatiles as attractants for the pine weevil, *Hylobius abietis* L. (Coleoptera: Curculionidae). *J. Econ. Entomol.* **1986**, *79*, 970–973. [CrossRef]
33. Nordlander, G. A method for trapping hylobius abietis (L.) with a standardized bait and its potential for forecasting seedling damage. *Scand. J. For. Res.* **1987**, *2*, 199–213. [CrossRef]
34. Erbilgin, N.; Szele, A.; Klepzig, K.D.; Raffa, K.F. Trap Type, Chirality of α-Pinene, and Geographic Region Affect Sampling Efficiency of Root and Lower Stem Insects in Pine. *J. Econ. Entomol.* **2001**, *94*, 1113–1121. [CrossRef] [PubMed]
35. Nordlander, G. Limonene inhibist attraction to Alpha-pinene in the pine weevils *Hylobius abietis* and *H. pinastri*. *J. Chem. Ecol.* **1990**, *16*, 1307–1320. [CrossRef] [PubMed]
36. Olenici, N.; Duduman, M.L.; Teodosiu, M.; Olenici, V. Efficacy of artificial traps to prevent the damage of conifer seedlings by large pine weevil (*Hylobius abietis* L.)—A preliminary study. *Bull. Transilv. Univ. Bras.* **2016**, *9*, 9–20.
37. Stocki, J.S. *The Use of Pheromones and Pheromone Traps in Forest Protection in Poland in the Years 1980–1997*; Practice oriented results on the use and production of Neem ingredients and pheromones VIII; Kleeberg, H., Zebitz, C.P.W., Eds.; Druck & Graphics: Giessen, Germany, 2000; pp. 128–133.
38. Skrzecz, I. Non-target insects in the pine weevil (*Hylobius abietis* L.) traps with Hylodor dispenser. *Folia For. Polon. Ser. A For.* **2003**, *35*, 27–35.
39. Voolma, K.; Siida, I.; Sibul, I. Forest insects attracted to ground traps baited with turpentineand ethanol on clear-cuttings. *Nor. J. Entomol.* **2001**, *48*, 103–110.
40. Wheater, C.P.; Bell, J.R.; Cook, P.A. *Practical Field Ecology: A Project Guide*; Wiley-Blackwell: Chichester, UK, 2011; p. 388.
41. Örlander, G.; Nilsson, U.; Nordlander, G. Pine weevil abundance on clear-cuttings of different ages: A 6-year study using pitfall traps. *Scand. J. For. Res.* **1997**, *12*, 225–240. [CrossRef]
42. Saalas, U. *Suomen Metsähyönteiset*; Finnish Forest Insects; Wsoy: Porvoo, Finland, 1949; p. 719.
43. Ozols, G. *Prides un Egles Dedrofagien Kukaini Latvijas Mezos*; (Dendrophagus insects of pine and spruce in Latvian forests); Zinātne: Riga, Latvia, 1985; p. 208.
44. Ehnström, B.; Axelsson, R. *Insekts Gnag I Bark Och Ved*; SLU: Uppsala, Sweden, 2002.
45. Längström, B. Abundance and seasonal activity of adult *Hylobius*-weevils in reforestation areas during the first years following final felling. *Commun. Lnst. For. Fenn.* **1982**, *106*, 1–23.
46. Maavara, V.; Merihein, A.; Parmas, H.; Parmasto, E. *Metsakaitse*; Valgus: Tallinn, Estonia, 1961; p. 733.
47. Luik, A.; Voolma, K. Some aspects of the occurrence, biology and cold-hardiness of *Hylobius* weevils. In *Insects Affecting Reforestation: Biology and Damage*; Alfaro, R.I., Glover, S.G., Eds.; Pacific and Yukon Region, Forestry Canada: Victoria, BC, Canada, 1989; pp. 28–33.
48. Viiri, H.; Miettinen, O. Feeding preferences of *Hylobius pinastri* Gyll. *Balt. For.* **2013**, *19*, 161–164.
49. Leather, S.R.; Ahmed, S.J.; Hogan, L. Adult feeding preferences of the pine weevil, *Hylobius abietis* (Coleoptera: Curculionidae). *Eur. J. Entomol.* **1994**, *91*, 385–389.
50. Varkonda, Š.; Florian, Š. *Odparník na regulované uvoľnovanie prchavých látok*; (Evaporator for controlled release of volatile substances) Utility model, SK 4570 U; Úrad Priemyselného Vlastníctva: Bratislava, Slovakia, 2006; p. 3.
51. Galko, J.; Kunca, A.; Ondruš, M.; Špilda, I.; Rell, S. Zhodnotenie a porovnanie nákladov na rôznu formu ošetrenia ihličnatých sadeníc proti tvrdoňovi smrekovému. In *Aktuálne Problémy v Ochrane Lesa*; Kunca, A., Nový Smokovec, S., Kunca, A., Eds.; Národné Lesnícke Centrum: Zvolen, Slovakia, 2015; pp. 101–105.
52. Thorin, J.; Nömmik, H. Monoterpene composition of cortical oleoresin from different clones of *Pinus sylvestris*. *Phytochemistry* **1974**, *13*, 1879–1881. [CrossRef]
53. Yazdani, R.; Nilsson, J.E. Cortical monoterpene variation in natural populations of *Pinus sylvestris* in Sweden. *Scand. J. For. Res.* **1986**, *1*, 85–93. [CrossRef]
54. Nordlander, G. Host finding in the pine weevil *Hylobius abietis*: Effects of conifer volatiles and added limonene. *Entomol. Exp. Appl.* **1991**, *59*, 229–237. [CrossRef]

55. Squillace, A.E. Analyses of monoterpenes of conifers by gas-liquid chromatography. In *Modem Methods in Forest Genetics*; Miksche, J.P., Ed.; Springer: New York, NY, USA, 1976; pp. 139–157.
56. Voolma, K.; Sibul, I. The effect of collecting fluid on the catch results of *Hylobius abietis* in pitfall traps. *Agron. Res.* **2006**, *4*, 457–460.
57. Saintonge, F.X.; Malphettes, C.B. Un piege pour surveiller les populations d'hylobes (*Hylobius abietis* L.) (Coleop.: Curc.)? Etudes de Cemagref. *Sér. For.* **1991**, *6*, 138–155.
58. Charitonova, N.Z. *Bolšoj Sosnovyj Dolgonosik I Borba s Nim*; 1. Vyd; Lesnaja Promyšlennost Progress: Moskva, Russia, 1965; p. 88. ISBN 934-0-41-595-768-24.
59. Nordlander, G.; Örlander, G.; Langvall, O. Feeding by the pine weevil *Hylobius abietis* in relation to sun. exposure and distance to forest edges. *Agric. For. Entomol.* **2003**, *5*, 191–198. [CrossRef]
60. Bejer-Petersen, B.; Juutinen, P.; Kangas, E.; Bakke, A.; Butovitsch, V.; Eidmann, H.; Heqvist, K.J.; Lekander, B. Studies on *Hylobius abietis* L. I. Development and life cycles in the Nordic countries. *Acta Entomol. Fenn.* **1962**, *17*, 1–107.
61. Selander, J.; Kalo, P.; Kangas, E.; Pertunnen, V. Olfactory behavior of *Hylobius abietis* L. (Col., Curculionidae). I. Response to several terpenoid fractions isolated from Scots pine phloem. *Ann. Entomol. Fenn.* **1974**, *39*, 40–45.
62. Nordenham, H.; Eidmann, H.H. Response of the pine weevil *Hylobius abietis* L. (Col., Curculionidae) to host volatiles in different phases of its adult life cycle. *J. Appl. Entomol.* **1991**, *112*, 353–358. [CrossRef]
63. Björklund, N.; Nordlander, G.; Bylund, H. Olfactory and visual stimuli used in orientation to conifer seedlings by the pine weevil, *Hylobius abietis*. *Physiol. Entomol.* **2005**, *30*, 225–231. [CrossRef]
64. Tilles, D.A.; Eidmann, H.H.; Solbreck, B. Mating stimulant of the pine weevil *Hylobius abietis* (L.). *J. Chem. Ecol.* **1988**, *14*, 1495–1503. [CrossRef] [PubMed]
65. Selander, J. Evidence of pheromone-mediated behaviour in the large pine weevil, *Hylobius abietis* (Coleoptera, Curculionidae). *Ann. Ent. Fenn.* **1978**, *44*, 105–112.
66. Mráček, Z.; Šrůtka, P. Stav znalostí a možnostech obrany proti klikorhu borovému (*Hylobius abietis* L.) v zemích Evropy. *Zprávy Lesnického Výzk.* **1984**, *29*, 21–25.
67. Solbreck, C.; Gyldberg, B. Temporal flight pattern of the large pine weevil, *Hylobius abietis* L. (Coleoptera, Curculionidae), with special reference to the influence of weather. *Z. Angew. Entomol.* **1979**, *88*, 532–536. [CrossRef]

forests

MDPI

Article

Current and Future Distribution of *Ricania shantungensis* (Hemiptera: Ricaniidae) in Korea: Application of Spatial Analysis to Select Relevant Environmental Variables for MaxEnt and CLIMEX Modeling

Sunghoon Baek [1], Min-Jung Kim [1] and Joon-Ho Lee [1,2,*]

1 Entomology program, Department of Agricultural Biotechnology, Seoul National University, Seoul 08826, Korea; shbaek007@hotmail.com (S.B.); 2017-24294@snu.ac.kr (M.-J.K.)

2 Research Institute of Agriculture and Life Sciences, Seoul National University, Seoul 08826, Korea

* Correspondence: jh7lee@snu.ac.kr; Tel.: +82-880-4705

Received: 10 May 2019; Accepted: 6 June 2019; Published: 7 June 2019

Abstract: Since the first report on its occurrence in 2010, *Ricania shantungensis* Chou & Lu in Korea has quickly spread. This pest population in agricultural areas has increased by over 100% each year and has caused serious economic damage in the last few years. This study was conducted to predict the potential habitat and the current and future distribution of *R. shantungensis* in Korea using CLIMEX and the Maximum Entropy Model (MaxEnt), and to suggest a new parameter selection method for both modeling programs. Weights of variables used in CLIMEX and those used in MaxEnt were determined using spatial association indices of spatial analysis by distance indices (SADIE). Weather data of Zhejiang province in China and those of all Korean territories were compared with Climate Matching in CLIMEX. MaxEnt was applied and evaluated with 295 data points on the presence and absence of *R. shantungensis* and eight environmental variables that were preselected by spatial and correlation tests. In MaxEnt, maximum temperature of the warmest month, annual mean temperature, mean temperature of the coldest month, and precipitation of the driest month were determined to be the most important variables affecting the distribution of *R. shantungensis* in Korea. The results of this study indicated that *R. shantungensis* had a higher probability of occurrence in western areas than in eastern areas of Korea, and showed great potential to spread eastward. These results are expected to be helpful for managing *R. shantungensis* in Korea and selecting relevant environmental variables for species distribution modeling.

Keywords: *Ricania shantungensis*; CLIMAX; MaxEnt; SADIE; species distribution model

---

1. Introduction

Since *Ricania shantungensis* Chou & Lu (Hemiptera: Ricaniidae) was first reported in Gongju and Yesan of Chungcheong-do in 2010 in Korea, this insect pest has quickly spread southward to Jeonla-do and northward to Gyunggi-do [1]. It is now slowly spreading toward the middle parts of Gyungsang-do [2]. Overall, the population of *R. shantungensis* in agricultural areas has increased by over 100% each year from 2015 to 2017, causing serious economic damage [3]. This pest directly causes damage by sucking plant saps and laying eggs. It also indirectly induces sooty mold disease on leaves through its excretions [1]. Accordingly, there has been an increasing demand for knowledge about the current and future distribution of *R. shantungensis* in order to determine how environmental conditions influence its occurrence and improve the efficiency of this species' management. There has been a report [4] in which the habitat suitability of *R. shantungensis* in Korea was estimated with the Maximum

Entropy Model (MaxEnt). In that report, the predicted habitats were able to explain the distribution of *R. shantungensis* at that time. However, the current distribution of *R. shantungensis* appears to be different from their prediction in areas of Korea where this pest has newly spread. In addition, they did not explain how the variables selected in the model affected its occurrence. Moreover, to date, the future distribution of *R. shantungensis* has not been predicted.

MaxEnt estimates the distribution of a species by finding the distribution that is closest to the geographically uniform distribution, which is determined by the environmental characteristics of locations where it is present with given environmental variables [5]. The characteristics of MaxEnt are the use of data on presence and pseudo-absence, the determination of environmental and/or biological factors related to target species' distribution, and a correlative parameter selection process [5–7]. Its correlative parameter selection process has received major criticisms: ecologically unrelated variable selection with target species distribution [8], a relatively low prediction ability in spreading species [9], and highly biased results in unsampled areas [10].

Due to the inherent issues with correlative methods for species distribution models (SDMs), deterministic methods have also been used for SDMs. Among the deterministic methods, CLIMEX is one of the most popular methods due to its good prediction of target species' distribution and relative ease of use [11–13]. CLIMEX estimates the potential distribution and risk of a species in relation to climate by using a deterministic method that requires pre-determined parameters regardless of target species distribution in the areas of interest, which is the opposite of the parameter selection process in MaxEnt [13]. CLIMEX is composed of the CLIMEX Model and Climate Matching [11,13]. Both the CLIMEX Model and Climate Matching in CLIMEX require biological information (e.g., growth-related indices such as temperature, moisture, radiation, substrate, light, and diapause, and stress-related indices such as cold, heat, dry, wet, and stress interactions) to determine the model parameters and the weights of parameters, respectively [13]. Biological information on many invasive species is limited because they have not been studied well because they are often less economically important in their areas of origin. This is likely to be the case for *R. shantungensis.* There is no available information on the distribution of *R. shantungensis*, except for that in Korea, to estimate CLIMEX Model parameters. Only a limited amount of information is available for determining the parameters themselves. However, Climate Matching requires less information, such as information on the climate of the areas of origin and interest and the weights of the model parameters, as compared to CLIMEX Modeling [13]. The origin of *R. shantungensis* was assumed to be the Zhejiang or Shandong provinces in the eastern coastal areas of China [14]. However, *R. shantungensis* has been recorded and managed as an important economic pest in Zhejiang province only [15]. Thus, the occurrence of *R. shantungensis* in Korea could be estimated by comparing information on the climate of the Zhejiang province of China with information on the climate in areas of Korea.

Climate matching in CLIMEX could provide information on the potential habitats of *R. shantungensis* in Korea. However, it may not exactly match the actual distribution of *R. shantungensis* because prediction in the model is only determined by climate factors. In addition, the model predicts potential habitats using a deterministic process. In contrast, MaxEnt can evaluate both climate and non-climate factors and select correlative variables within its distribution. Thus, the distribution predicted by this model should match the actual distribution well, although the reliability of the parameters selected by MaxEnt remains to be proven. In other words, prediction in CLIMEX and MaxEnt is restricted to the fundamental and realized niches of a target species, respectively [8]. Therefore, results from CLIMEX could help us to predict the directions of spread of *R. shantungensis* in Korea in the near future from the current distribution that has been predicted by MaxEnt. Moreover, results from both CLIMEX and MaxEnt could minimize the drawbacks of each modeling process and support each other by considering the ecological aspects in CLIMEX and the current distribution in MaxEnt for *R. shantungensis* if both CLIMEX and MaxEnt are applied together. However, there is no standardized process for determining the relevant variables in MaxEnt and the weights of the parameters in CLIMEX.

Therefore, this study was conducted to evaluate the potential habitat of *R. shantungensis* with CLIMEX, to predict the current distribution of the possible presence of *R. shantungensis* with MaxEnt, to suggest a method for selecting parameters in MaxEnt and the parameters' weights in CLIMEX, and finally to predict the future distribution of the possible presence of *R. shantungensis* in Korea.

2. Materials and Methods

2.1. Data Collection for the Presence and Absence of R. shantungensis

In order to collect data on the presence and absence of *R. shantungensis*, a map of Korea was divided into 30 km by 30 km grids, and at least one location point per grid was surveyed for the occurrence of *R. shantungensis* to minimize unsampled areas (Figure 1). First, data on presence were obtained from previously published data [4] and unpublished reports by the Gyunggi and Chungnam Agricultural Research and Extension Services in 2017. Grid locations where the presence of *R. shantungensis* was not checked were then surveyed in 2017. The survey was mainly conducted along rest areas of highways within a grid because *R. shantungensis* had been first found at rest areas of highways in newly invaded areas [4]. If rest areas were not present within a grid, then major host plants (i.e., *Robinia pseudoacacia*, *Diospyros kaki*, *Castanea* spp., *Prunus serrulata*, and so on) at temporary parking places abutting forested areas were randomly observed within the grid. The absence of *R. shantungensis* was determined after checking at least 20 trees. Whole branches within observers' reach were observed in each tree. A total of 295 data points (149 previously reported points and 146 newly surveyed points) were collected, and *R. shantungensis* was found at 175 points among them (Figure 1).

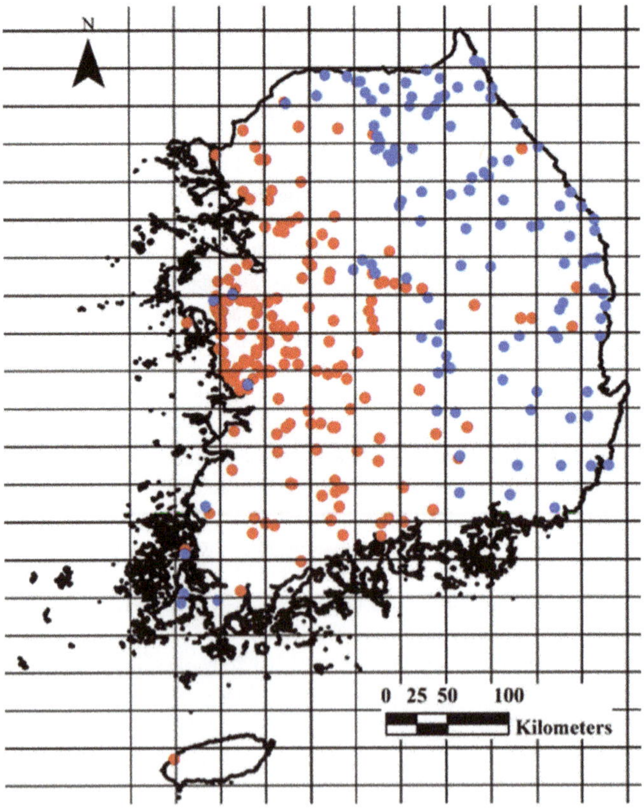

Figure 1. Imaginary grids (30 km × 30 km) and observed points for the presence of *Ricania shantungensis* in Korea (●, Presence; ●, Absence).

2.2. Environmental, Traffic, Footprint, and Landcover Data Collection

Numerical climate data, composed of monthly temperatures (i.e., maximum, minimum, and mean) and precipitation for the period 1981–2010 from 73 meteorological stations, were obtained from the Korea Meteorological Administration (KMA) (http://www.kma.go.kr). Climatic data were interpolated throughout all areas of Korea using inverse distance weighting with an option of estimating a point value with the nearest five meteorological stations' data. The estimated monthly temperature and precipitation data were transformed into 19 bioclimatic variables in the format of WorldClim (http://www.worldclim.org) in DIVA-GIS 7.5 [16]. All variables were created in the same resolution (i.e., 1 km by 1 km) and file format (i.e., ASCII) by considering the resolution of the available projected map and environmental data. Future climate data were also downloaded from the KMA. These data were developed by the representative concentration pathway (RCP) 8.5 scenario for temperature and precipitation based on historical data (1986–2005) in Korea. In this study, climate data (i.e., temperature and precipitation) of RCP 8.5 for the 2030s (2031–2040), the 2050s (2051–2060), the 2070s (2071–2080), and the 2090s (2091–2100) were used to estimate the future distribution of *R. shantungensis* in Korea.

A digital elevation model (DEM) was downloaded from the website of the National Spatial Data Infrastructure Portal in Korea (http://www.nsdi.go.kr) to create elevation data. Because these elevation data were created with a different resolution (i.e., 0.8 km by 0.8 km) from that of the 19 bioclimatic variables, data were resampled by the numeric value of the nearest neighbor cell using the resample tool in ArcGIS 10.1 (ESRI; Redlands, CA, USA).

Traffic data were downloaded from the website of the Traffic Monitoring System in Korea (http://www.road.re.kr). For analysis, the downloaded traffic amount at 3477 points was interpolated with the ordinary Kriging method and mapped in ASCII file format with a resolution of 1 km by 1 km in ArcGIS 10.1. Footprint data were also downloaded from the website of National Aeronautics and Space Administration (NASA) Earth Observing System Data and Information System (http://earthdata.nasa.gov). Worldwide data were clipped with the mask of the Korean territory in ArcGIS 10.1.

Landcover data furnished by the Environmental Geographic Information Service in Korea (http://egis.me.go.kr) were downloaded and resampled by the attribute value with the largest area within a 1 km by 1 km cell in ArcGIS 10.1 because the resolution of the downloaded data (i.e., 5 m by 5 m) was much higher than that of the other variables.

A total of 23 variables (Table 1), including 19 bioclimatic variables, were prepared to describe the current distribution of *R. shantungensis* in Korea.

2.3. Spatial Association between Environmental Variables and Distribution of R. shantungensis

In order to accurately predict the distribution of *R. shantungensis* in Korea, variables selected for modeling should be biologically related to *R. shantungensis*. If distributions of selected variables are spatially related to the distribution of *R. shantungensis*, possibilities of biological relationships between *R. shantungensis* and its surrounding environmental variables are expected to be increased [17]. Thus, spatial analysis by distance indices (SADIE) [18] was used to measure spatial associations between data on the occurrence of *R. shantungensis* and 22 environmental variables. The Landcover variable could not be applied for spatial analysis because it was in the form of categorical data. SADIE quantifies the contribution of counts at each location to a patch (i.e., a region of relatively large counts close to one another) and a gap (i.e., a region of relatively small counts close to one another) [19]. For the spatial association test in SADIE, the variables that are compared should have coincident coordinates, and all data should have clustering indices (i.e., a degree of contribution at each location) [19]. To match the coordinates of each point of data on the occurrence of *R. shantungensis* and the 22 environmental variables, point data were extracted from middle points of each grid (Figure 1) in ArcGIS 10.1. For this process, a krigged map for *R. shantungensis* occurrence was used. Spatial indices of each point of 23 sets were estimated in SADIE. The overall spatial association (X) is the mean of the local correlation coefficient between the clustering indices of the two sets; $X > 0$ for a positive spatial association, $X = 0$

for no spatial association, and $X < 0$ for a negative spatial association [19]. Positive X values indicate the coincidence of a patch cluster for one set with a patch cluster for the other or the coincidence of two gaps, and negative X values are indicated by a patch coinciding with a gap [19]. The associated probability (P) was also calculated based on randomization tests [19]. The null hypothesis is that the spatial arrangement of the count data between two data sets is random [20]. In this study, $X > 0$ indicates a positive spatial association, $X = 0$ indicates no spatial association, and $X < 0$ indicates a negative spatial association between the occurrence of *R. shantungensis* and its surrounding environments. All SADIE statistics were calculated using SADIEShell version 1.22 (Rothamsted Experimental Station, Harpenden, Herts, UK).

Table 1. A list of the 23 variables, including 19 bioclimatic variables, used in the prediction of the current distribution of *R. shantungensis*.

Variables	Unit	Range
Bio1 = Annual mean temperature	°C	2.6–16.9
Bio2 = Mean diurnal range (Mean of monthly (max temp.–min temp.))	°C	5.0–12.9
Bio3 = Isothermality (Bio2/Bio7 × 100)	-	19.6–32.9
Bio4 = Temperature seasonality (standard deviation × 100)	-	707.6–1072.1
Bio5 = Max temperature of warmest month	°C	17.7–30.7
Bio6 = Min temperature of coldest month	°C	−18.8–4.1
Bio7 = Temperature annual range (Bio5–Bio6)	°C	25.2–40.4
Bio8 = Mean temperature of wettest quarter	°C	13.0–25.4
Bio9 = Mean temperature of driest quarter	°C	−11.2–10.5
Bio10 = Mean temperature of warmest quarter	°C	13.0–25.9
Bio11 = Mean temperature of coldest quarter	°C	−11.2–8.3
Bio12 = Annual precipitation	°C	995–1918
Bio13 = Precipitation of wettest month	mm	208–403
Bio14 = Precipitation of driest month	mm	15–47
Bio15 = Precipitation seasonality (coefficient of variation)	-	59.5–108.0
Bio16 = Precipitation of wettest quarter	mm	557–1023
Bio17 = Precipitation of driest quarter	mm	54–178
Bio18 = Precipitation of warmest quarter	mm	530–910
Bio19 = Precipitation of coldest quarter	mm	54–182
Elevation	m	−3–1819
Traffic volume	ea	834–176,191
Footprint	-	12–100
Landcover	-	-

2.4. Prediction of Habitat Suitability of *R. shantungensis* with CLIMEX

The Climate Matching function in CLIMEX predicts the habitat suitability of a target species in a country of interest (i.e., Korea in this study and Away in CLIMEX) by comparing the climate of the origin area (i.e., Zhejiang in this study and Home in CLIMEX) with that of Away [8]. Metadata provided by the CLIMEX program and 30 years (1981–2010) of climate data from the KMA were used as the climate data of Zhejiang in China and all Korean territories, respectively. The similarity level of climates between Home and Away was determined by the Composite Match Index (CMI), which is calculated using seven components (i.e., Maximum temperature (I_{tmax}), Minimum temperature (I_{tmin}), Average temperature (I_{tav}), Total rainfall (I_{rain}), Rainfall pattern (I_{rpat}), Relative humidity (I_{hum}), and Soil moisture (I_{smst})) and the individual weight of each component [13]. If the CMI value is closer to 1 than 0, climates of Away areas are more similar to those of the Home area [11].

Among the seven components, only five related to temperature and precipitation were used for Climate Matching in this study. Between I_{tav} and I_{tmin}, I_{tav} was selected and used for the analysis because I_{tav} was highly correlated (correlation coefficient (r) > 0.9) with I_{tmin} and the X value of I_{tav} in spatial association tests was higher than that of I_{tmin} (Table 2). Weights of the four remaining variables were determined by using the *X* values of related environmental factors in spatial association tests between environmental factors and *R. shantungensis* occurrence as follows:

$$\text{Weight of } I_{tav} = [X_{Bio1}/(X_{Bio1} + X_{Bio5})] * [\{(X_{Bio1} + X_{Bio5})/2\}/\{(X_{Bio12} + X_{Bio13} + X_{Bio14} + X_{Bio16} + X_{Bio17} + X_{Bio18} + X_{Bio19})/7\}] \quad (1)$$

$$\text{Weight of } I_{tmax} = [X_{Bio5}/(X_{Bio1} + X_{Bio5})] * [\{(X_{Bio1} + X_{Bio5})/2\}/\{(X_{Bio12} + X_{Bio13} + X_{Bio14} + X_{Bio16} + X_{Bio17} + X_{Bio18} + X_{Bio19})/7\}] \quad (2)$$

$$\text{Weight of } I_{rain} = [X_{Bio12}/\{X_{Bio12} + (X_{Bio13} + X_{Bio14} + X_{Bio16} + X_{Bio17} + X_{Bio18} + X_{Bio19})/6\}] * [\{(X_{Bio12} + X_{Bio13} + X_{Bio14} + X_{Bio16} + X_{Bio17} + X_{Bio18} + X_{Bio19})/7\}/\{(X_{Bio1} + X_{Bio5})/2\}] \quad (3)$$

$$\text{Weight of } I_{rpat} = [\{X_{Bio12} + (X_{Bio13} + X_{Bio14} + X_{Bio16} + X_{Bio17} + X_{Bio18} + X_{Bio19})/6\}/X_{Bio12}] * [\{(X_{Bio12} + X_{Bio13} + X_{Bio14} + X_{Bio16} + X_{Bio17} + X_{Bio18} + X_{Bio19})/7\}/\{(X_{Bio1} + X_{Bio5})/2\}] \quad (4)$$

where X_{Bioi} indicates the *X* value of bioclimatic factor Bio*i* in the spatial association test with *R. shantungensis* occurrence. Bio1, Bio5, Bio12, and Bio13, 14, 16, 17, 18, and 19 were related to I_{tav}, I_{tmax}, I_{rain}, and I_{rpat}, respectively. These bioclimatic factors were not only related to components of Climate Matching in CLIMEX but also spatially associated with *R. shantungensis* occurrence. From these processes, the actual weights that were used of each component were 0.61, 0, 0.39, 0.48, 0.49, 0, and 0 for I_{tmax}, I_{tmin}, I_{tav}, I_{rain}, I_{rpat}, I_{hum}, and I_{smst}, respectively.

Table 2. Spatial association between the 22 environmental variables and the possible presence of *R. shantungensis* in Korea.

Environmental Variables	*X* [1]	*P*	Environmental Variables	*X*	*P*
Bio 1	0.2466	0.0006	Bio 12	0.3011	0.0011
Bio 2	−0.0984	0.8373	Bio 13	0.2142	0.0207
Bio 3	−0.1692	0.9442	Bio 14	0.4088	<0.0001
Bio 4	0.0413	0.3443	Bio 15	−0.016	0.5625
Bio 5	0.3881	<0.0001	Bio 16	0.2688	0.0035
Bio 6	0.2491	0.0046	Bio 17	0.2937	0.0027
Bio 7	0.0498	0.3169	Bio 18	0.3835	<0.0001
Bio 8	0.3461	<0.0001	Bio 19	0.2715	0.0052
Bio 9	0.1235	0.1078	Elevation	−0.2611	0.9956
Bio 10	0.3812	<0.0001	Traffic	0.0304	0.3830
Bio 11	0.2293	0.0151	Footprint	0.0833	0.2153

[1] Index of association (*X*) with its associated probability (*P*). For a two-tailed test at the 95% confidence level, $P < 0.025$ indicates a significant positive association and $P > 0.975$ indicates a significant negative association.

2.5. Prediction of Current Distribution for R. shantungensis with MaxEnt

Because MaxEnt requires data on presence [21], of the 295 data points, 175 data points on presence were used for prediction. The 132 data points from the new field survey were used as training data and the other (from the published paper) 43 data points were used as test data because the selected parameters should explain both the past (i.e., the test in this study) and the current data (i.e., training in this study) to avoid one of the criticisms of MaxEnt (i.e., that it is suitable for species of equilibrium status [9]). Thus, the test data in this study were not used to evaluate the model's performance. Instead, they were used to select and exclude variables unrelated to the presence of *R. shantungensis* in Korea. In the average nearest neighbor test in ArcGIS 10.1, the training data showed a significantly ($p < 0.01$) clustered pattern, which could cause overfitting problems [22] against environmental variables. Thus, by reducing the spatial cluster of the data on the presence of *R. shantungensis*, spatial filtering [23] was executed by the expected mean distance (0.138124 decimal degrees) in the average nearest neighbor test (a hypothetical random distribution in the region of interest) using the 'spatially rarefy occurrence data' tool in the SDM toolbox of ArcGIS 10.1. Ultimately, 72 points among the 132 occurrence points were used as training data for MaxEnt modeling, and 43 points were used for test data.

To eliminate multi-collinear variables, which can lead to overfitting problems [24] or an ambiguous interpretation, a correlation test was executed in ArcGIS 10.1. If the correlation coefficient (r) was greater than 0.9 ($r > 0.9$, $r < -0.9$), only one variable was used for further analyses among pairs of correlated variables by considering the X values of spatial association tests among correlated variables (Table 1). Through these procedures, 13 variables (i.e., Bio1, Bio2, Bio3, Bio5, Bio6, Bio7, Bio12, Bio14, Bio17, Bio18, Footprint, Traffic volume, and Landcover) among the original 23 were preferentially selected. To avoid another criticism of MaxEnt (i.e., variable selection being irrelevant to the target species' distribution due to MaxEnt's processor) [8], spatially associated variables (i.e., Bio1, Bio5, Bio6, Bio12, Bio14, Bio17, Bio18, and Landcover) were used for MaxEnt modeling.

MaxEnt software version 3.4.1 [25] was initially executed with the default parameter settings (e.g., feature type, regularization multiplier, and 10,000 background points) using training data, test data, and eight pre-selected variables. In all executions of MaxEnt modeling, a jackknife test was conducted to measure variable importance and exclude unrelated variables. To reduce model complexity and eliminate unrelated variables, variables were excluded from the model under two conditions: First, the value of test gain was less than zero; and second, the gain value without one variable was higher than that in the case when all variables were used. These processes were repeated until there was no variable to be discarded. Through these stepwise procedures, seven variables (i.e., Bio1, Bio5, Bio6, Bio12, Bio14, Bio 17, and Bio18) were finally selected, and the estimated occurrence model for *R. shantungensis* was built.

2.6. Comparison between the Expected Occurrence by MaxEnt and the Actual Occurrence of R. shantungensis

In order to compare the absence of *R. shantungensis* between the observed and predicted values by the MaxEnt model, the predicted values of the MaxEnt model were extracted with 120 observed points (i.e., observed absence points in this study) in ArcGIS 10.1. The area under the curve (AUC) score of MaxEnt provides the predictive ability of the developed model. However, the actual absence points of *R. shantungensis* in Korea were aggregated in the western parts. They were not used in the MaxEnt modeling process. Thus, data on absence need to be evaluated to verify the prediction reliability of the MaxEnt model that is suggested in this study. If the extracted value was more than 0.5 at one point, *R. shantungensis* was considered to be present at that point. The prediction abilities of the MaxEnt model for observed occurrence were statistically compared to random probability (i.e., 50%) using a Chi-square test [26].

2.7. Prediction of Future Distributions of R. shantungensis in Korea

MaxEnt was also used to predict the future probability of *R. shantungensis* presence in Korea with the seven selected variables (i.e., Bio1, Bio5, Bio6, Bio12, Bio14, Bio 17, and Bio18) of the RCP 8.5 climate change scenario for the 2030s (2031–2040), the 2050s (2051–2060), the 2070s (2071–2080), and the 2090s (2091–2100).

3. Results

3.1. Habitat Suitability of R. shantungensis with CLIMEX

In all Korean territories, the CMI values were higher than 0.7, indicating that climatic conditions in all locations of Korea were similar to those of Zhejiang in China. However, the similarity of climatic conditions was lower in the eastern part of Korea than in the western part of Korea, where *R. shantungensis* mainly occurred (Figure 2).

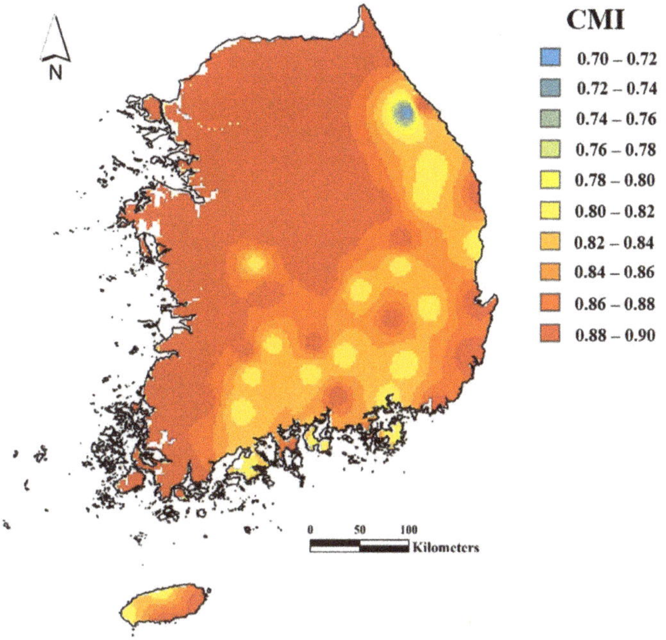

Figure 2. The climate similarity (i.e., the climate matching index (CMI) in CLIMEX) map in Korea compared to one of the most favorable climatic conditions for *R. shantungensis* populations: Zhejiang province in China. CMI > 0.5 indicates a suitable habitat for *R. shantungensis*.

3.2. Prediction of Current and Future Distribution of R. shantungensis with MaxEnt

The AUC score of the training data in the MaxEnt model in this study was 0.78, which was larger than 0.75, indicating reliable predictive ability (Figure 3). The AUC score of the test data was 0.79, slightly higher than that of the training data. The contributions of Bio5, Bio1, Bio6, Bio14, Bio18, Bio 17, and Bio2 in this model were 41.2, 15.9, 15.7, 12.6, 7.1, 5.5, and 2.0%, respectively. According to the response curves for the variables related to the presence of *R. shantungensis* in Korea, each variable affected the possible presence of *R. shantungensis* in a different way (Figure 4): The maximum temperature of the warmest month (i.e., Bio5) showed a generally sigmoidal effect. The minimum temperature of the coldest month (i.e., Bio6) showed a sigmoidal effect until −6 °C, a negative effect between −6 °C and 2 °C, then an exponential effect after 2 °C. The precipitation of the warmest quarter (i.e., Bio18) showed two peaks around minimum and 825 mm. The precipitation of the driest month (i.e., Bio 14) showed endurance at a certain level of precipitation, and then an exponential effect. However, the precipitation of the driest quarter (i.e., Bio 17) showed a negative effect. Both the annual mean temperature (i.e., Bio1) and annual precipitation (i.e., Bio12) were favorable for the occurrence of *R. shantungensis* in Korea by maintaining an approximately 61% occurrence possibility.

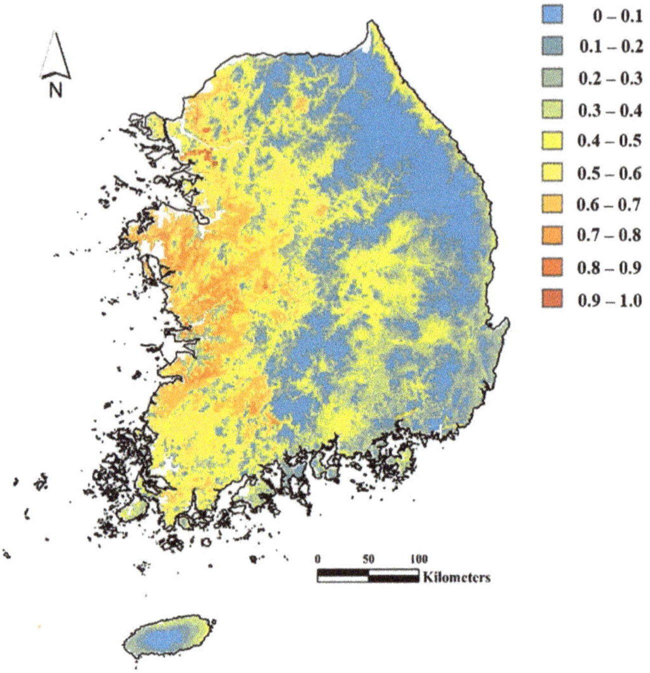

Figure 3. Presence probability maps of *R. shantungensis* in Korea using the Maximum Entropy Model (MaxEnt).

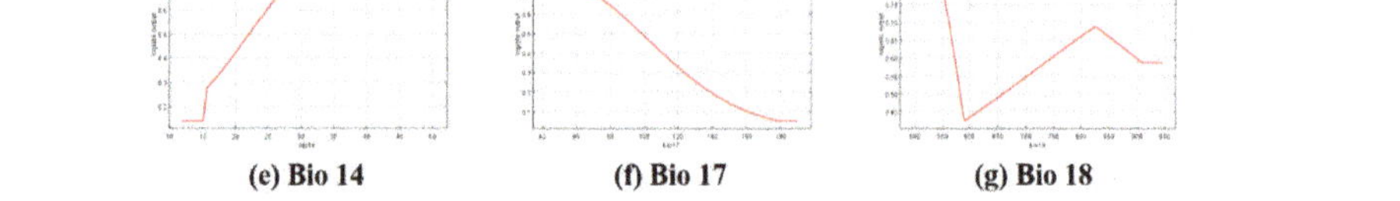

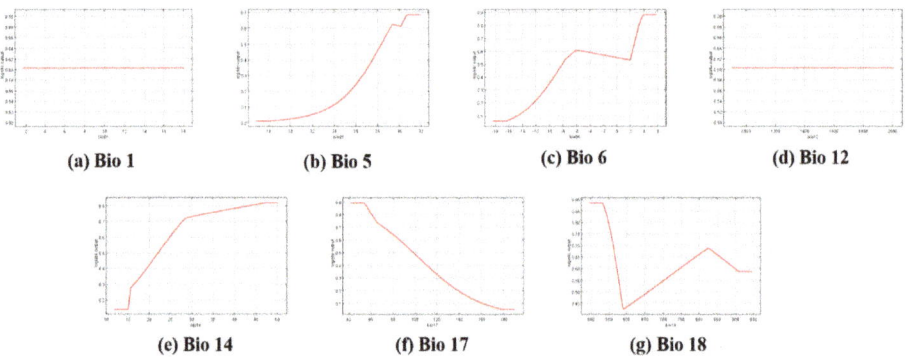

Figure 4. The response curves for each variable related to the presence of *R. shantungensis* in Korea: (**a**) Bio 1; (**b**) Bio 5; (**c**) Bio 6; (**d**) Bio 12; (**e**) Bio 14; (**f**) Bio 17; (**g**) Bio 18.

The correctly predicted rate of MaxEnt in actually absent points was 86.7% by matching 102 points among 120, which was statistically (x^2 = 64.5, p < 0.001) meaningful. Future presence probability maps of *R. shantungensis* in Korea were also developed with MaxEnt by using climatic data under the RCP 8.5 scenario for the 2030s, 2050s, 2070s, and 2090s (Figure 5). In the future distribution of *R. shantungensis*, Gyungsang-Do and Gangwon-Do were expected to have a high occurrence probability. Currently, *R. shantungensis* has been found at only a few points in Gyungsang-Do and Gangwon-Do in a survey of 2017. In the 2070s and 2090s, the occurrence probability of *R. shantungensis* is expected to be low in southwest Korea, where, to date, the occurrence of this pest has been severe.

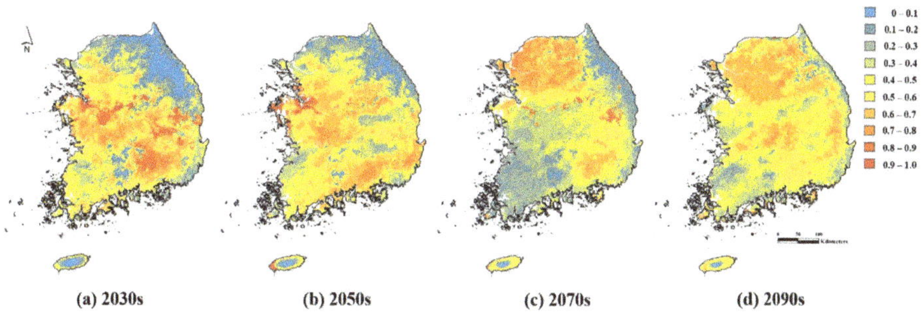

Figure 5. Future presence probability maps of *R. shantungensis* in Korea using MaxEnt: (**a**) 2030s; (**b**) 2050s; (**c**) 2070s; (**d**) 2090s.

4. Discussion

This study predicted the potential habitat suitability and current distribution probability of *R. shantungensis* in Korea using CLIMEX and MaxEnt. Both models, in general, well-described the observed distribution pattern of *R. shantungensis* in Korea. Accordingly, spatial analysis could be a useful tool for finding reliable variables to explain the distribution of a target species. Moreover, this study forecasted the future occurrence of *R. shantungensis* and found relevant environmental variables with the occurrence of *R. shantungensis* in Korea. These results could be helpful for preparing management strategies for *R. shantungensis* and national long-term agricultural plans in Korea.

The predicted current distribution of *R. shantungensis* in this study was generally similar to that of the previous study [4], but showed opposite results in a few areas (e.g., Kangwon-do parts overlapping with Kyunggi-do, eastern parts in Gyeongsang-do, and northern parts of Gyunggi-do) despite the fact that both studies used the same program (MaxEnt) and similar environmental variables. These areas were newly invaded areas by *R. shantungensis* within Korea after the study of Kim et al. [4]. These differences could potentially be caused by the correlative parameter selection process of MaxEnt [9]. The MaxEnt model for this spreading insect could have a high probability of selecting non-relevant variables due to the correlative modeling process in MaxEnt [8]. Kim et al. [4] did not speculate whether this pest distribution in Korea had reached equilibrium status. Kim et al. [4] also reported that precipitation and mean temperature of the warmest quarter, forest type, and landcover were the most important variables affecting *R. shantungensis*. Although these variables were able to predict *R. shantungensis* well at that time, they failed to accurately describe its distribution in 2017. Moreover, Kim et al. [4] did not explain why these variables were selected and how these variables affected the distribution of *R. shantungensis*. However, important variables in this study, i.e., maximum temperature of the warmest month, annual mean temperature, mean temperature of the coldest month, and precipitation in the driest month, could explain the data reported in Kim et al. [4] and the current data, indicating the strong relationship of these variables with the occurrence of *R. shantungensis*. Thus, results of this study would be more reliable for predicting the possible presence of *R. shantungensis* in Korea.

Correlative models, including MaxEnt, have the greatest advantage in that they can result in a higher matching rate between observed and predicted data than models using deterministic processors. However, correlative models have been subject to criticism [8–10]. This study attempted to minimize the disadvantages of MaxEnt. By using spatial analysis in variable selection processes, only variables that were relevant to *R. shantungensis* occurrence were used in the modeling process. Moreover, this study showed that the selected variables and their parameters could explain the occurrence of *R. shantungensis* in both the past and present by using two separate sets of data, although *R. shantungensis* is still spreading in Korea [1–3]. Multiple studies [27–29] have reported on the general characteristics of suitable variables related to the occurrence of a target species. Bradie and Leung [27] reviewed published results from MaxEnt and found that temperature and precipitation among environmental variables were generally related to target species' occurrence. Petitpierre et al. [28] concluded that the variables having a direct impact on species physiology and fitness were related to species' occurrence by reviewing MaxEnt studies. Braunisch et al. [29] suggested that variables related to target species occurrence should be selected in MaxEnt modeling, although they discussed methods to select variables in the case of an uncertain relationship between a species' occurrence and its environments. The variables selected in this study also met the criteria of previous studies [27–29]. This study also showed a high accuracy of 86.7% at points where the data were not used in MaxEnt modeling. It is possible that the accuracy of the model's prediction could be increased by data on absence [30] even though there is the potential for false absences [31]. However, the results of this study were able to predict the presence and absence of *R. shantungensis* well. Finally, the results from CLIMEX, one of the deterministic models, also showed that western parts of Korea were relatively more suitable for *R. shantungensis* than eastern parts, similar to the results from MaxEnt.

The results of this study indicated that seven environmental variables (i.e., Bio1, Bio5, Bio6, Bio12, Bio 14, Bio 17, and Bio18) were important for predicting the occurrence of *R. shantungensis*. It is reasonable to conclude that temperature-related variables (i.e., Bio1, Bio5, and Bio6) positively affect *R. shantungensis* occurrence because the metabolic rate with increasing temperature is normally increased up to its upper lethal temperature, and this is reflected in the developmental rate in insects due to their ectodermal characteristics [32]. It was proven that elevation, which is negatively correlated with Bio5, also negatively affected the occurrence of *R. shantungensis* [33]. Kim et al. [34] reported that high humidity was helpful for increasing the survival of *R. shantungensis* eggs during winter. Thus, Bio14 and 17 would be expected to have positive effects on *R. shantungensis* occurrence in Korea, while long-term rain during winter was predicted to negatively affect *R. shantungensis* occurrence in the MaxEnt modeling. Long-term and frequent rain during winter might cause overwintering eggs to die due to the ecological characteristics of *R. shantungensis*; eggs are laid between cracks of branches and covered with wools made by their mothers. Precipitation during summer (i.e., Bio18) showed two peaks at low precipitation (i.e., 500–530 mm) as well as at around 825 mm of precipitation. These two peaks could be caused more by the frequency of heavy rain in Korea during summer rather than absolute amounts of precipitation because heavy rain could affect the flight activities and survival of adults of *R. shantungensis*.

The results of this study indicated that the maximum temperature of the warmest month was the most important variable for predicting the occurrence of *R. shantungensis*, as it had a 41.2% contribution rate in the MaxEnt model. This month (i.e., August in Korea) matches the pre-ovipositional period of *R. shantungensis* in Korea. Female adults lay eggs inside cracks and cover their eggs with wool that they make [1]. Moreover, they should find a suitable habitat for the development of their progenies. These oviposition behaviors of *R. shantungensis* are expected to consume a large amount of energy. Thus, *R. shantungensis* adults may require significant feeding activity before oviposition and a high temperature would be helpful for their activities and ovary development. Their high energy requirement is also proven by their relatively long pre-oviposition period (i.e., over one month) considering their body sizes and temperature [1]. The annual mean temperature (15.9% in contribution rate) and precipitation (2% in contribution rate) were shown to constantly and positively contribute to *R. shantungensis* occurrence. This result is consistent with CLIMEX results showing that all areas in the Korean territory were a suitable habitat for *R. shantungensis*. Both mean temperature of the coldest month (15.7% in contribution rate) and precipitation of the driest month (12.6% in contribution rate) were related to overwintering of *R. shantungensis*. These variables were proven to be highly related to the occurrence of *R. shantungensis* in MaxEnt and SADIE as well as its ecology. Thus, prediction of *R. shantungensis* occurrence could be enabled in other non-invaded countries using these major variables that are related to its occurrence [21,28,35].

The occurrence of *R. shantungensis* as an invasive species was expected to be highly related to human-related variables (i.e., Traffic and Footprint in this study) because insect spread is generally correlated with human-related factors [36]. Moreover, there was a report that showed high contamination rates of *R. shantungensis* at rest areas of highways and roadsides [4]. However, this study showed that these human-related variables were not significantly ($p < 0.05$) related to the occurrence of *R. shantungensis* in Korea. This could be related to the high mortality of *R. shantungensis* eggs in cut trees [34]. Kang et al. [2] also reported that *R. shantungensis* was slowly spreading toward the middle parts of Gyungsang-Do, along the coastal areas of southern Korea, even though the direction of spread was toward Busan-Si with very high Traffic and Footprint values. In this study, one of the habitat-related variables, Landcover, was not selected by MaxEnt because the gain value was higher without than with this variable. This result could be related to the ecology of *R. shantungensis*, which has very broad host ranges [4]. In fact, this pest has broken out in various types of landcover, such as cities, orchards, agricultural fields, and forests. Thus, *R. shantungensis* would be expected to spread continuously, except for under conditions of transplanted trees contaminated with *R. shantungensis*.

Although a rapid spread is not to be expected, the spreading of *R. shantungensis* to Gyungsang-Do appeared to be inevitable because the habitat suitability of this region for *R. shantungensis* in CLIMEX was high (i.e., a CMI value > 0.7). The distribution probability of these areas in MaxEnt was the highest in Korea in the 2030s. Gyungsang-Do is economically and agriculturally important in Korea. It has large apple, persimmon, and chestnut export agricultural complexes. These plants are known to be good host plants for *R. shantungensis*. Thus, management of *R. shantungensis* is needed to minimize its economic impacts and slow its spread in Korea based on the current distribution, and the predicted distribution in the 2030s, of *R. shantungensis*.

5. Conclusions

An increasing number of invasive species are expected to continue to occur internationally due to increased trades, transportation, and human activities, as well as global warming [37]. Thus, the demand for species distribution models (SDMs) would be increased to determine current distribution, predict future distribution and habitat suitability, and identify associations between occurrence and the surrounding environment related to the occurrence of a target species. For more reliable SDMs, the selection of variables should be related to the characteristics of a target species. As an example, spatial association analysis was applied in this study. Parameters selected by spatial association analysis resulted in increased prediction accuracy and reliability for the occurrence of *R. shantungensis* in Korea. The results from CLIMEX and MaxEnt predicted that *R. shantungensis* currently has a higher occurrence probability in western areas than in eastern areas in Korea. However, it has great potential to spread eastward in the future. These results will be helpful for developing relevant management strategies and national long-term agricultural plans in Korea.

Author Contributions: Conceptualization, S.B. and J.-H.L.; methodology, S.B.; validation, S.B. and J.-H.L.; formal analysis, S.B. and M.-J.K.; investigation, S.B. and M.-J.K.; writing—original draft preparation, S.B.; writing—review and editing, S.B. and J.-H.L. All of the authors approved the final version.

Funding: This work was carried out with the support of "Research Program for Forest Science & Technology Development (Project No. FE0100-1988-01)" funded by National Institute of Forest Science, Republic of Korea. This work was partially supported by BK 21 plus.

Acknowledgments: We thank the officers from the Gyunggi and Chungnam Agricultural Research and Extension Service for surveying *R. shantungensis* occurrence.

Conflicts of Interest: The authors declare no conflicts of interest. The funders had no role in the design of the study; in the collection, analyses, or interpretation of data; in the writing of the manuscript, or in the decision to publish the results.

References

1. Jo, S.J. Study on the control and ecology of *Ricania shantungensis*. *J. Tree Health* **2014**, *19*, 35–44.
2. Kang, D.W.; Ha, J.B.; Lee, S.B.; Han, I.Y.; Kwon, J.H.; Choi, Y.J. Occurrence characteristics and distribution of *Ricania shantungensis* in Gyungsangnam-Do. In Proceedings of the 2016 Fall Meeting of Korean Society of Applied Entomology, Buyeo Lotte Resorts, Buyeo, Korea, 20–21 October 2016.

3. Hong, S.J.; Lee, K.J.; Kim, S.T.; No, H.I.; Jung, J.Y. Current distribution and status of *Ricania shantungensis* in Korea. In Proceedings of the 2017 Fall Meeting of Korean Society of Applied Entomology, Wellihillipark Resort, Hoengseoung, Korea, 26–27 October 2017.
4. Kim, D.E.; Lee, H.; Kim, M.J.; Lee, D.-H. Predicting the potential habitat, host plants, and geographical distribution of *Pochazia shantungensis* (Hemiptera: Ricaniidae) in Korea. *Kor. J. Appl. Entomol.* **2015**, *54*, 179–189. [CrossRef]
5. Phillips, S.J.; Anderson, R.P.; Schapire, R.E. Maximum entropy modeling of species geographic distributions. *Ecol. Modell.* **2006**, *190*, 231–259. [CrossRef]
6. Pearson, R.G. Species' Distribution Modeling for Conservation Educators and Practitioners. Synthesis. American Museum of Natural History. Available online: http://ncep.amnh.org (accessed on 17 September 2018).
7. Elith, J.; Phillips, S.J.; Hastie, T.; Dudik, M.; Chee, Y.E.; Yates, C.J. A statistical explanation of MaxEnt for ecologists. *Divers. Distrib.* **2011**, *17*, 43–57. [CrossRef]
8. Zeng, Y.; Low, B.W.; Yeo, D.C.J. Novel methods to select environmental variables in MaxEnt: A case study using invasive crayfish. *Ecol. Model.* **2016**, *341*, 5–13. [CrossRef]
9. Elith, J.; Leathwick, J.R. Species distribution models: Ecological explanation and prediction across space and time. *Ann. Rev. Ecol. Evol. Syst.* **2009**, *40*, 677–697. [CrossRef]
10. Pearson, R.G.; Raxworthy, C.J.; Nakamura, M.; Peterson, A.T. Predicting species distributions from small numbers of occurrence records: a test case using cryptic geckos in Madagascar. *J. Biogeo.* **2007**, *34*, 102–117. [CrossRef]
11. Jung, J.-M.; Lee, W.-H.; Jung, S. Insect distribution in response to climate change based on a model: Review of function and use of CLIMEX. *Entomol. Res.* **2016**, *46*, 223–235. [CrossRef]
12. Seo, C.-W.; Park, Y.-R.; Choi, Y.-S. Comparison of species distribution models according to location data. *J. Korean Soc. Geospat. Inform. Syst.* **2008**, *16*, 59–64.
13. Sutherst, R.W.; Maywald, G.F.; Kriticos, D.J. *CLIMEX Version 3 User's Guide*; Hearne Scientific Software Pty Ltd.: Melbourne, Australia, 2007.
14. Kwon, D.H.; Kim, S.-J.; Kang, T.-J.; Lee, J.H.; Kim, D.H. Analysis of the molecular phylogenetics and genetic structure of an invasive alien species, *Ricania shantungensis*, in Korea. *J. Asia Pac. Entomol.* **2017**, *20*, 901–906. [CrossRef]
15. Chou, I.; Lu, J.S.; Huang, J.; Wang, S. *Economic Insect Fauna of China, Fasc. 36*; Homoptera Fulgoroidae. Science Press: Beijing, China, 1985.
16. Hijmans, R.J.; Guarion, L.; Mathur, P. DIVA-GIS. Free GIS for Biodiversity Research, Version 7.5. Available online: http://www.diva-gis.org/ (accessed on 13 September 2018).
17. Baek, S. Spatial and Temporal Distribution of Hemlock Woolly Adelgid (Hemiptera: Adelgidae) ovisacs And Its Associations with the Environment. Master Thesis, West Virginia University, Morgantown, WV, USA, 2015.
18. Perry, J.N. Spatial analysis by distance indices. *J. Anim. Ecol.* **1995**, *64*, 303–314. [CrossRef]
19. Perry, J.N.; Dixon, P. A new method for measuring spatial association in ecological count data. *Ecoscience* **2002**, *9*, 133–141. [CrossRef]
20. Park, Y.-L.; Perring, T.M.; Farrar, C.A.; Gispert, C. Spatial and temporal distributions of two sympatric *Homalodisca* spp. (Hemiptera: Cicadellidae): Implications for areawide pest management. *Agric. Ecosyst. Environ.* **2006**, *113*, 168–174. [CrossRef]
21. Phillips, S.J.; Dudík, M. Modeling of species distributions with Maxent: New extensions and a comprehensive evaluation. *Ecography* **2008**, *31*, 161–175. [CrossRef]
22. Kramer-Schadt, S.; Niedballa, J.; Pilgrim, J.D.; Schröder, B.; Lindenborn, J.; Reinfelder, V.; Stillfried, M.; Heckmann, I.; Scharf, A.K.; Augeri, D.M.; et al. The importance of correcting for sampling bias in MaxEnt species distribution models. *Divers. Distrib.* **2013**, *36*, 1366–1379. [CrossRef]
23. Boria, R.A.; Olson, L.E.; Goodman, S.M.; Anderson, R.P. Spatial filtering to reduce sampling bias can improve the performance of ecological niche models. *Ecol. Model.* **2014**, *275*, 73–77. [CrossRef]
24. Dormann, C.F.; Elith, J.; Bacher, S.; Buchman, C.; Carl, G.; Carré, G.; García Marquéz, J.R.; Gruber, B.; Lafourcade, B.; Leitão, P.J.; et al. Collinearity: A review of methods to deal with it and a simulation study evaluating their performance. *Ecography* **2013**, *36*, 27–46. [CrossRef]

25. Phillips, S.J.; Dudík, M.; Schapire, R.E. Maxent Software for Modeling Species Niches and Distributions (Version 3.4.1). Available online: http://biodiversityinformatics.amnh.org/open_source/maxent/ (accessed on 13 September 2018).
26. SAS Institute. *SAS/STAT User's Guide, Version 9.3*; SAS Institute: Cary, NC, USA, 2011.
27. Bradie, J.; Leung, B. A quantitative synthesis of the importance of variables used in MaxEnt species distribution models. *K. Biogeogr.* **2017**, *44*, 1344–1361. [CrossRef]
28. Petitpierre, B.; Broennimann, O.; Kueffer, C.; Daehler, C.; Guisan, A. Selecting predictors to maximize transferability of species distribution models: Lessons from cross-continental plant invasions. *Global Ecol. Biogeogr.* **2017**, *26*, 275–287. [CrossRef]
29. Braunisch, V.; Coppers, J.; Arlettaz, R.; Suchant, R.; Schmid, H.; Bollmann, K. Selecting from correlated climate variables: A major source of uncertainty for predicting species distributions under climate change. *Ecography* **2013**, *36*, 971–983. [CrossRef]
30. Brotons, L.; Thuiller, W.; Araujo, M.B.; Hirzel, A.H. Presence-absence versus presence-only modeling methods for predicting bird habitat suitability. *Ecography* **2004**, *27*, 437–448. [CrossRef]
31. Hirzel, A.; Guisan, A. Which is the optimal sampling strategy for habitat suitability modeling. *Ecol. Model.* **2002**, *157*, 331–341. [CrossRef]
32. Chapman, R.F. *The Insects Structure and Function*, 3rd ed.; Harvard University Press: Cambridge, MA, USA, 1982.
33. Choi, Y.S.; Seo, H.Y.; Jo, S.H.; Hwang, I.S.; Yee, Y.S.; Park, D.G. Host preference of *Ricania* spp. (Hemiptera: Ricaniidae) at different developmental stages. *Kor. J. Appl. Entomol.* **2017**, *56*, 319–329.
34. Kim, D.H.; Yang, C.R.; Kim, H.H.; Seo, M.H.; Yun, J.B. Effect of moisture content of pruned blueberry and peach twigs on hatchability of *Ricania shantungensis* (Hemiptera: Ricaniidae) eggs. *Kor. J. Appl. Entomol.* **2017**, *56*, 357–363.
35. Duque-Lazo, J.; van Gils, H.; Groen, T.A.; Navarro-Cerrilo, R.M. Transferability of species distribution models: The case of *phytophthora cinnamomi* in Southwest Spain and Southwest Australia. *Ecol. Model.* **2016**, *320*, 62–70. [CrossRef]
36. Lockwood, J.L.; Hoopes, M.A.; Marchetti, M.P. *Invasion Ecology*; Blackwell Publishing: Malden, MA, USA, 2007.
37. Liebhold, A.M.; Tobin, P.C. Population ecology of insect invasions and their management. *Annu. Rev. Entomol.* **2008**, *53*, 387–408. [CrossRef] [PubMed]

forests

MDPI

Article

Occurrence Prediction of the Citrus Flatid Planthopper (*Metcalfa pruinosa* (Say, 1830)) in South Korea Using a Random Forest Model

Dae-Seong Lee [1], Yang-Seop Bae [2], Bong-Kyu Byun [3], Seunghwan Lee [4], Jong Kyun Park [5] and Young-Seuk Park [1,6,*]

1 Department of Biology, Kyung Hee University, Seoul 02447, Korea
2 Division of Life Sciences, Incheon National University, Incheon 22012, Korea
3 Department of Biological Science and Biotechnology, Hannam University, Daejeon 34054, Korea
4 Division of Entomology, Seoul National University, Seoul 08826, Korea
5 Department of Applied Biology, Kyungpook National University, Sangju 37224, Korea
6 Department of Life and Nanopharmaceutical Sciences, Kyung Hee University, Seoul 02447, Korea
* Correspondence: parkys@khu.ac.kr; Tel.: +82-2-961-0946

Received: 6 June 2019; Accepted: 9 July 2019; Published: 12 July 2019

Abstract: Invasive species cause a severe impact on existing ecosystems. The citrus flatid planthopper (CFP; *Metcalfa pruinosa* (Say, 1830)) is an invasive species in many countries. Predicting potential occurrence areas of the species related to environmental conditions is important for effective forest ecosystem management. In this study, we evaluated the occurrence patterns of the CFP and predicted its potential occurrence areas in South Korea using a random forest model for a hazard rating of forests considering meteorological and landscape variables. We obtained the occurrence data of the CFP in South Korea from literature and government documents and extracted seven environmental variables (altitude, slope, distance to road (geographical), annual mean temperature, minimum temperature in January, maximum temperature in July, and annual precipitation (meteorological)) and the proportion of land cover types across seven categories (urban, agriculture, forest, grassland, wetland, barren, and water) at each occurrence site from digital maps using a Geographic Information System. The CFP occurrence areas were mostly located at low altitudes, near roads and urbanized areas. Our prediction model also supported these results. The CFP has a high potential to be distributed over the whole of South Korea, excluding high mountainous areas. Finally, factors related to human activities, such as roads and urbanization, strongly influence the occurrence and dispersal of the CFP. Therefore, we propose that these factors should be considered carefully in monitoring and surveillance programs for the CFP and other invasive species.

Keywords: hazard rating; invasive species; surveillance; forest ecosystem management; prediction model; species distribution model; random forest

1. Introduction

The occurrence of invasive species has increased worldwide due to global warming and international trade [1–3]. At the beginning of the invasion process, the community containing the invasive species does not have appropriate species interactions, such as prey–predator, host–parasite, and competition between species to stabilize the community dynamics [4–6]. Therefore, they can cause a severe impact on the existing ecosystem, resulting in notorious pests in many countries [7], for example, the gypsy moth (*Lymantria dispar* L.), the Asian long horn beetle (*Anoplophora glabripennis* Motschulsky), the emerald ash borer (*Agrilus planipennis* Fairmaire), found in the USA [8], and the pine wood nematode (*Bursaphelenchus xylophilus* Steiner and Buhrer), which is found in many countries, including Korea and Japan [9,10].

The citrus flatid planthopper (CFP; *Metcalfa pruinosa* Say) (Hemiptera, Auchenorrhyncha: Flatidae) is one invasive species which causes several problems in many countries. This species, originally from North America [11], has spread widely throughout America [12]. In Europe, it was first reported in Italy in 1979 [13] and is now found throughout Europe [14–20]. Meanwhile, in Asia, the presence of this species was first reported in South Korea in 2005 [21] and later in 2009 [22]. Recently, its distribution area has expanded to a nationwide scale [23].

The CFP damages its host plants by feeding on sap and ejecting honeydew, which induces a sooty mold disease [22,24]. The CFP is a polyphagous species feeding on a wide range of host plants in Korea, including woody trees, such as *Robinia pseudoacacia* L., *Castanea crenata* Siebold et Zucc., *Styrax japonicus* Siebold et Zucc., *Diospyros lotus* L., and *Acer palmatum* Thunb., and shrubs and herbaceous plants, such as *Hibiscus syriacus* L., *Rhamnus davurica* Pall., and *Lycium chinense* Mill. [22,23].

The occurrence and dispersal of the CFP are highly related to human activities such as roads, highways, and movement of vehicles on a local scale [25–27]. However, the occurrence and dispersal of invasive species are influenced by various other factors, including meteorological and geographical conditions, landscape, and host plants [28–30]. To effectively manage invasive pests, the evaluation of the occurrence patterns related to environmental conditions and the estimation of the potential occurrence area of the species are essential to provide information to efficiently identify current or future hazardous conditions [28,29,31].

Therefore, various hazard rating systems have been developed for different forest pests. In the early stage of hazard rating, qualitative and statistical methods such as discriminant analysis are used [32–34]. Further, recent machine-learning algorithms, such as self-organizing maps, multi-layer perceptron, classification and regression trees, and random forest models, can been used [28,29,35,36].

Despite the importance of hazard ratings for controlling the CFP, there are limited studies on hazard ratings using the CLIMEX model [37] which is used for the prediction of species distribution [38–41] and for risk assessment of invasive species [42,43]. CLIMEX is a climate-specific model to predict potential distribution and relative abundance of species by considering climatic conditions [44]. However, it does not consider other factors, such as road and landscape conditions. Meanwhile, MaxEnt [45] as a species distribution model is also frequently used to predict the potential distribution of invasive species based on presence-only species records [46]. MaxEnt cannot be applied to abundance data, only to presence data. Recently, studies to predict the distribution of (potential) invasive species, including CFP, using CLIMEX and MaxEnt were reviewed by evaluating their effectiveness [39].

The CFP is one of the quarantine pests listed by the Animal and Plant Quarantine Agency, Republic of Korea (https://www.qia.go.kr/) and its distribution area has been increasing since the first occurrence in Korea. Therefore, it has a high potential as a key insect pest on various plants in agriculture and forests [22]. In this study, we aimed to evaluate the occurrence patterns of the CFP and to predict its potential occurrence area for a hazard rating of forests, considering meteorological and landscape variables in Korea.

2. Materials and Methods

2.1. Data Collection

We obtained occurrence data of the CFP (*Metcalfa pruinosa*) from 105 presence (observed) sites in South Korea from the literature [23,27,47] and government documents from websites of the Gyeonggi-do Agricultural Research Services (https://nongup.gg.go.kr) and the Ministry of Agriculture, Food, and Rural Affairs (http://www.mafra.go.kr) (Figure 1). Our CFP occurrence data included only presence (i.e., observed) sites and we assumed that CFP was in fact absent at the unobserved sites at that moment because the field surveys were extensively conducted to monitor the occurrence of CFP. To evaluate the environmental conditions of the CFP occurrence sites, we selected absence sites with the same number of presence sites randomly with distances over 1.5 km from the presence sites and over 5 km from other selected absence sites to avoid similar environmental conditions. Finally, our dataset comprised 210 sites (105 sites each for presence and absence).

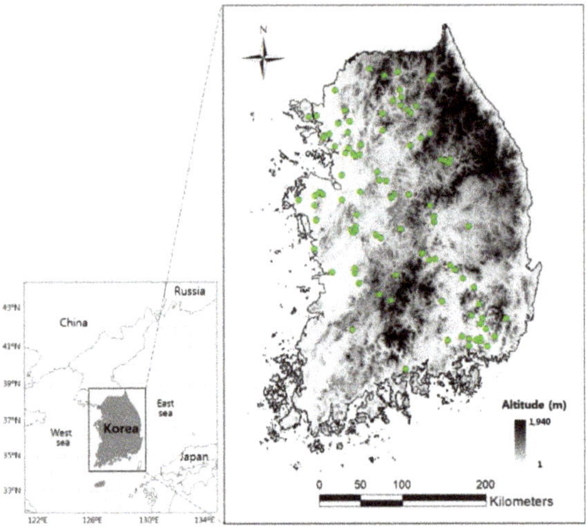

Figure 1. Occurrence (presence) sites of *Metcalfa pruinosa* based on the literature [23,27,47].

At each selected site (for both presence and absence), we extracted seven environmental variables (i.e., altitude, slope, and distance to road as geographical variables and annual mean temperature, minimum temperature in January, maximum temperature in July, and annual precipitation as meteorological variables) and the proportion of land cover types across seven categories (urban, agriculture, forest, grassland, wetland, barren, and water) from digital maps using a Geographic Information System (ArcGIS 10.1, ESRI). Altitude and slope were extracted from the digital elevation model data obtained from the National Geographic Information Institute (http://www.ngii.go.kr). Distance to road indicated a distance from the observation site to a nearest road and it was measured from a road map by combining road data from the National Spatial Data Infrastructure Portal (http://www.nsdi.go.kr) and the Intelligent Transport Systems of Standard Node Link (http://nodelink.its.go.kr). Meteorological variables were extracted from the digital map of Climatological Normals from the Climate Information Portal of the Korea Meteorological Administration (http://www.climate.go.kr) and Applied SERS A1B scenario data from the Digital Agro-climate Map Database for Impact Assessment of Climate Change on Agriculture (http://www.agdcm.kr) of the Rural Development Administration, Korea. The Environmental Geographic Information Service (https://egis.me.go.kr) operated by the Ministry of Environment, Korea, provided the digital land cover map, which was used to extract land cover area. We used raster digital maps with a resolution of 30 m × 30 m cell size.

2.2. *Analysis of Occurrence Patterns*

The number of occurrence regions, such as districts (i.e., "si", "gun", or "gu" level in Korea) and the area damaged by the CFP were evaluated for different years (Figure 2). The Mann–Whitney U test was used to evaluate the statistically significant differences between the presence and absence sites for environment and land cover conditions. The Mann–Whitney U test was performed with a "stats" package in R [48].

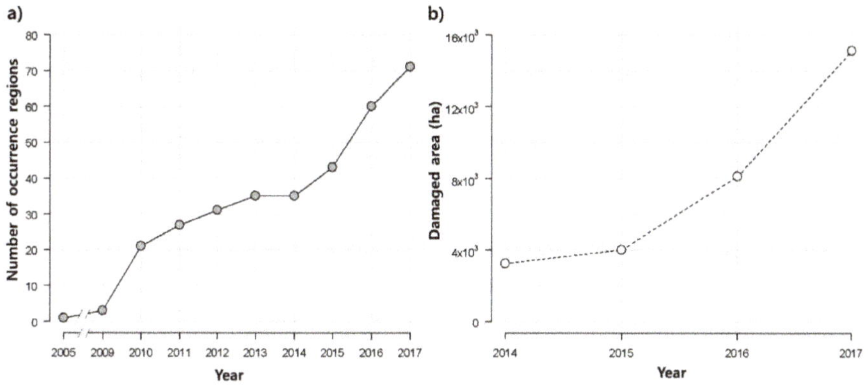

Figure 2. Change in (**a**) the number of occurrence regions and (**b**) area damaged by *Metcalfa pruinosa* in South Korea. Data were from databases (Gyeonggi-do Agricultural Research Services (https://nongup.gg.go.kr) and the Ministry of Agriculture, Food, and Rural Affairs (http://www.mafra.go.kr)). Damaged area data in Figure 2b were only available from 2014.

2.3. Prediction of Potential Occurrence

To estimate the occurrence risk of the CFP in the forests on a nationwide scale in Korea, we predicted the potential occurrence area (presence: 1; absence: 0) of the CFP based on the environmental conditions at the occurrence sites using a random forest (RF) model. The RF is an ensemble machine-learning model and it is a non-parametric method of predicting and assessing the relationships among a large number of potential predictor variables and a response variable [49]. In the RF model, we used five environmental variables (altitude, slope, and distance to road (geographical) and minimum temperature in January and maximum temperature in July). We excluded annual mean temperature to avoid collinearity among variables (Figure 3) and annual precipitation, which was not significantly different ($p < 0.05$) between the presence and absence sites. Although altitude and minimum temperature had high correlation coefficients, we included them as predictors because altitude strongly influences the distribution and abundance of insects [50] and the minimum temperature in January greatly affects the mortality of insects during the winter [6]. We calculated a mean decrease in node impurity by the residual sum of squares to evaluate the relative importance of each variable to determine the presence or absence status of CFP occurrence in the RF model.

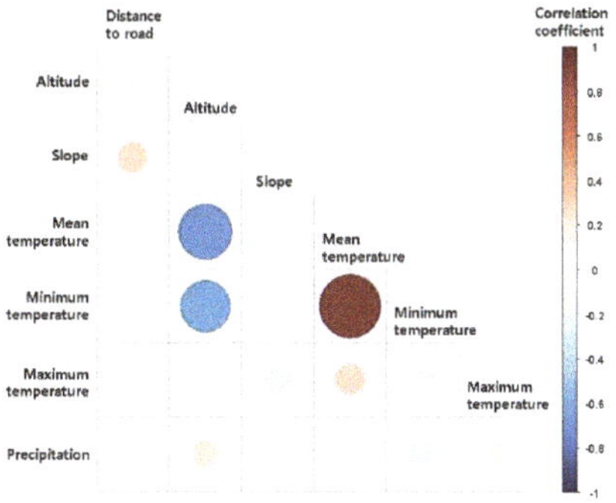

Figure 3. Correlation coefficient between environmental variables measured at 210 presence or absence sites.

We standardized all environmental variables a priori the modeling procedure. We divided our dataset in two parts with a ratio of 8:2 with respect to training and testing the models. We repeated the procedure for the RF model 10 times and averaged the calculated values. The RF models were performed using the package "randomForest" [51] in R. Partial dependence plot function (partialPlot) in "randomForest" and generalized additive model (GAM) in the "mgcv" package [52] were used to present the marginal effect of environmental variables in our model. The model performance was evaluated by accuracy and area under the receiver operating characteristic (AUROC) curve value using the "ROCR" package [53] in R.

3. Results

3.1. Occurrence Patterns of CFP

The number of occurrence regions (districts as "si", "gun", or "gu" in Korea) increased as a function of year after the first occurrence. It was less than five districts in 2009, but increased greatly to over 20 districts in 2010. From 2010 to 2015, the rate of increase was relatively low. However, in 2016, the number of occurrence regions abruptly increased from 43 to 60 districts, showing an exponential growth pattern. The damaged area displayed similar patterns regarding the number of occurrence regions from 2014 to 2017 ($r = 0.95$, $p < 0.05$) when the data were available.

Geographical and meteorological conditions were significantly different between the presence and absence sites (Mann–Whitney U test, $p < 0.05$) (Figure 4). The CFP was mostly observed in areas with low altitude (103.2 m ± 8.9 standard error (SE)), low slope (5.0° ± 0.6 SE), and short distance to the road (24.2 m ± 4.0 SE). Annual mean temperature and maximum temperature in July were higher at the presence sites than at the absence sites (Mann–Whitney U test, $p < 0.01$), while the minimum temperature in January was lower at the presence sites (Mann–Whitney U test, $p < 0.01$). Annual precipitation was not significantly different between the presence and absence sites.

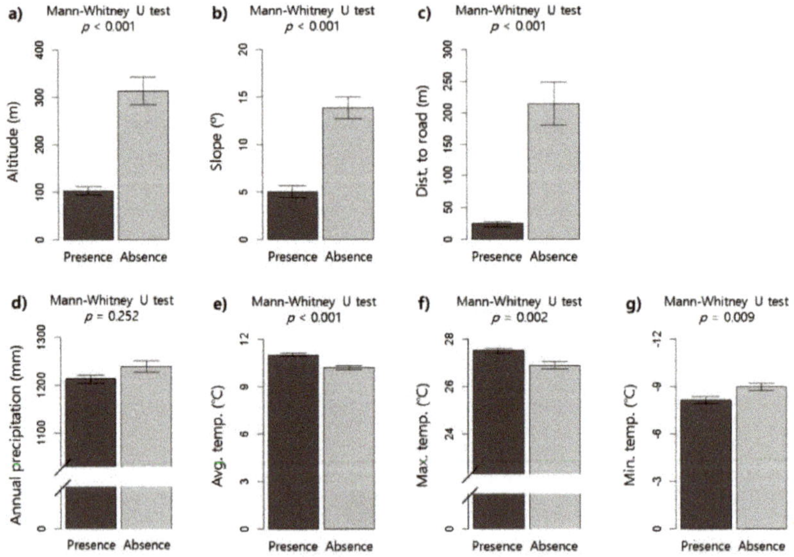

Figure 4. Differences in environmental conditions between the presence and absence sites of *Metcalfa pruinosa*. Bars denote averages and standard errors. The *p*-value (*p*) was calculated based on the Mann–Whitney U test. (**a**) Altitude, (**b**) slope, (**c**) distance to road, (**d**) annual precipitation, (**e**) annual average temperature, (**f**) maximum temperature in July, and (**g**) minimum temperature in January.

The ratios of all land cover types, except wetland, were significantly different between the presence and absence sites (Mann–Whitney U test, $p < 0.01$) (Figure 5). The presence sites were mostly observed in forests (33.8% ± 2.2 SE), agricultural land (31.6% ± 2.3 SE), and urban areas (19.5% ± 1.5 SE). The proportion of presence in forests was lower than that of absence, whereas the proportion of urban and agricultural areas of presence sites were higher than those of the absence sites.

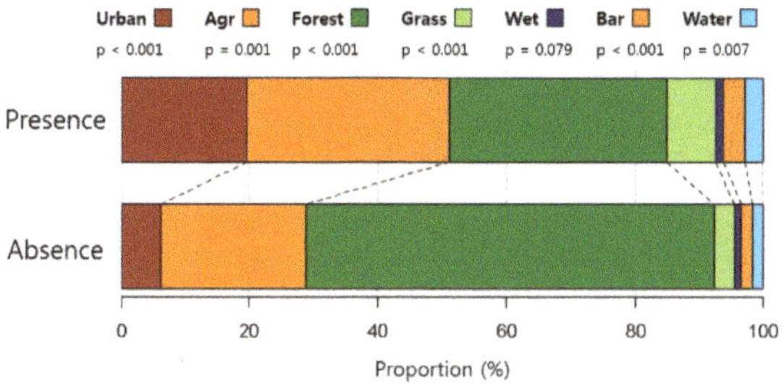

Figure 5. Differences in land cover types between presence and absence sites of *Metcalfa pruinosa*. The *p*-value (*p*) was calculated based on the Mann–Whitney U test. Urban: Urban area; Agr: Agricultural land; Wet: Wetland; and Bar: Barren.

3.2. Prediction of Potential Occurrence Area

The RF model predicted the potential occurrence areas of the CFP based on the geographical and meteorological conditions. The model showed high predictability: 0.78 (±0.01 SE) accuracy and 0.80 (±0.01 SE) AUROC curve value (Figure 6a). The model displayed that the occurrence probability was mostly high in lowland and urban areas (Figures 1 and 6b).

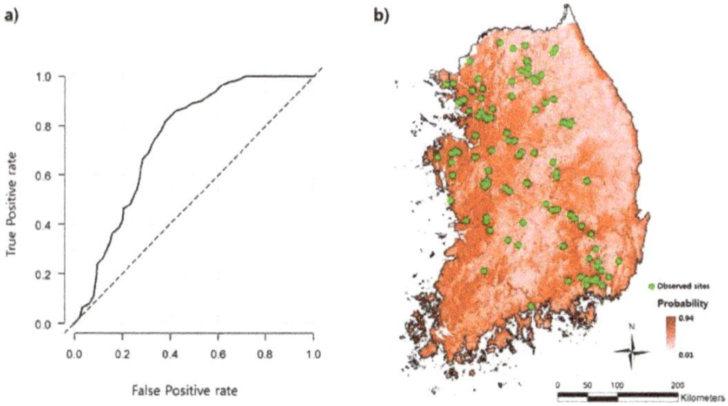

Figure 6. Prediction of potential distribution area of *Metcalfa pruinosa* with the random forest (RF) model. (**a**) Mean receiver operating characteristic (ROC) curve of the random forest (RF) model with 10 replications and (**b**) predicted potential distribution with different occurrence probability and actual observed sites with green circles.

The value of mean decrease in node impurity presented the distance to road as the most important variable for the prediction of the CFP occurrence sites in the RF model, followed by altitude and slope (Figure 7). This was reflected in the partial dependence plot of occurrence probability responding to the changes in the dependent variables (Figure 8). The occurrence probability of the CFP mostly changed as a function of distance to road and altitude, in particular, at low values of distance to road (mostly <50 m) and altitude (mostly <200 m), which displayed high occurrence probability. The probability was not highly influenced at high altitude and long distance to road.

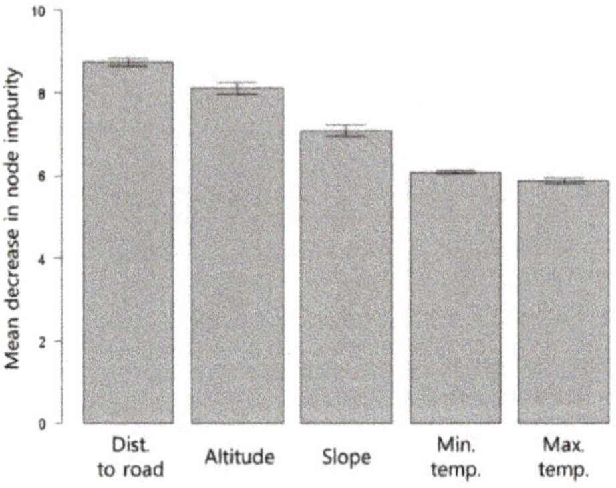

Figure 7. Relative importance of independent variables in the RF model for projecting presence or absence of *Metcalfa pruinosa*. Importance was calculated using node impurity in the RF model. Error bars indicate averages and standard errors. Dist. to road: distance to road, Min. temp.: minimum temperature in January, and Max. temp.: maximum temperature in July.

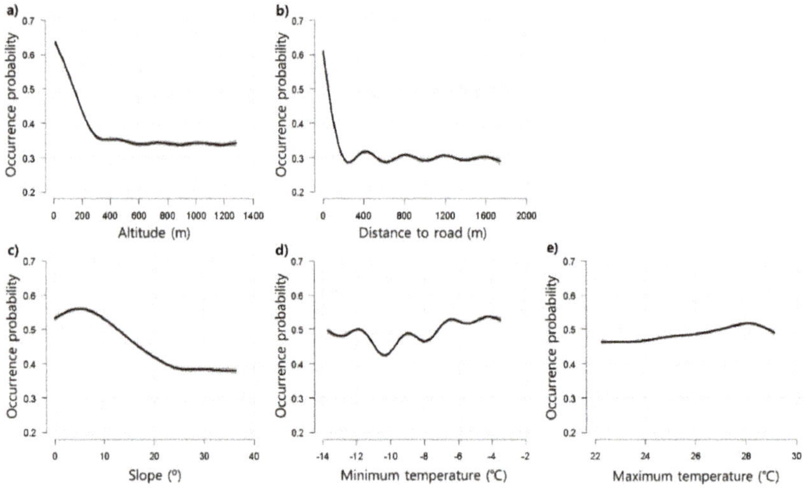

Figure 8. Partial dependence of occurrence probability of *Metcalfa pruinosa* against the changes of each environmental variable in the RF model. (**a**) Altitude, (**b**) distance to road, (**c**) slope, (**d**) minimum temperature in January, and (**e**) maximum temperature in July. Solid lines represent averages and gray areas display the 95% confidence interval.

4. Discussion

4.1. Occurrence Patterns of CFP

In South Korea, CFPs were first observed as adults only in Hanrim of Gimhae-si in the southern part of South Korea in 2005. The nymphs and adults were observed in Jinyoung of Gimhae-si, which is the town next to Hanrim, in 2008 [21,22]. Meanwhile, the CFP was discovered at several locations in the southern to central regions of the Korean Peninsula, including Seoul and Suwon in 2009 [22], and its occurrence and damage area increased greatly from 2010 (Figure 2). Currently, the CFP is widely distributed in South Korea. Thus, considering the distribution area of the CFP over time, the dispersal of the CFP was limited, with a small dispersal area, at the early stage of the invasion process from the first occurrence in 2005 until 2009. This phenomenon is commonly observed in the dispersal of invasive species [4,54]. During the early stage of invasion, the species might colonize and increase its population size, then disperse to the neighboring areas [8].

Regarding the increase in occurrence area, we have to note that there was a big difference between 2009 and 2010 (Figure 2). We infer that the actual distribution of the CFP in 2009 was wider than that displayed by the recorded data. The observed number of locations in 2009 might be smaller than that of the actual ones because the CFP was not well recognized as an invasive species and a pest insect by entomologists and policy decision-makers in South Korea until 2009. Meanwhile, the distribution area of the CFP increased exponentially after 2015, showing a stepwise dispersal pattern in the number of occurrence areas. This pattern was also reported in the dispersal of invasive species such as pine wilt disease in both Korea [9] and Japan [10]. This may be related to the increase in exponential population growth. However, the mechanism of this dispersal pattern was not clear. Therefore, further study is needed to understand the dispersal patterns of invasive species to effectively manage them.

4.2. Invasion Route and Factors Promoting Dispersals

Two possible routes were proposed for the CFP invasion into South Korea: From North America, which is the origin of the CFP, or from European countries [22], which would imply that the CFP arrived in South Korea through the imports of horticultural nursery stock. Similarly, the possible means of the CFP invasion were mostly by passive transport of the CFP adults or eggs with infested plants of many countries, including the Czech Republic and Romania [55,56]. Guillemaud, et al. [57] reported two main factors promoting biological invasion: Human activities and the biological adaptation of species. Human activities, including international trade and movement of vehicles among local regions, are highly responsible for recent biological invasions by moving invasive species and causing long distance dispersal [4,5,9,14,25,27,58–62]. Human activities also promote the dispersal of invasive species by modifying and disturbing natural habitats and providing the possibility of the introduction of new species into ecosystems [63].

Our results also revealed that human activities are the main influencing factors for the occurrence and dispersal of the CFP in South Korea. The CFP was observed mostly in areas close to the road with low altitude and slopes (Figure 4). This was also supported by the difference in occurrence patterns in different land use types. The occurrence of the CFP was observed mainly in urban and agricultural areas (Figure 5), which are highly influenced by human activities. Meanwhile, the proportion of presence sites in forest areas was smaller than that of absence sites. However, the occurrence proportion in forest areas was still the highest (over 33.8%) among land use types. This indicates that the occurrence areas in the forest are mostly in lowland areas which are close to the road and urban and agricultural areas.

Meanwhile, flight activities of the CFP adults contribute to local dispersal [14,56]. The annual dispersal distance of the CFP was estimated at 0.2–0.5 km in Austria [64]. The dispersal distance in natural conditions can be variable depending on various environmental conditions, such as temperature, precipitation, landscape, geology, and distribution of host plants [4,5,8,14,58–60], although human activities play a substantial role in their dispersal. Therefore, an integrated approach for the evaluation of the dispersal of invasive species is recommended with the consideration of environmental factors in several different categories, if data are available.

4.3. Host Plants

The high values of CFP occurrence in urban, agricultural, and forest areas reflect the diversity of both their habitats and host plants. The CFP is polyphagous; it has a variety of host plants and can inhabit very different environmental conditions. Therefore, the CFP can damage various types of plants, including ornamental or agricultural plants and herbaceous or woody plants [22,23,56]. It has a high probability of dispersing to new areas with commercial trade and transportation. Meanwhile, in this study, we did not include the distribution of host plants in the analysis and in the model for the potential occurrence prediction of the CFP because of its polyphagous nature and the difficulty in characterizing the distribution of host plants on the digital map. From another aspect, polyphagous species have a wide distribution of host plants. Therefore, our results would not be influenced by the exclusion of host plants distribution.

4.4. Hazard Rating with the CFP

The hazard rating of forests regarding susceptibility to insect pests and diseases provides information to efficiently identify current and future hazardous conditions in the concerned forests [28]. Therefore, it is very important in forest ecosystem management [29]. In this study, we evaluated the hazard rate of forests on a nationwide scale in Korea through the prediction of the potential occurrence of the CFP, showing that most areas in the South Korean territory, excluding high mountain areas, have a high occurrence probability of the CFP (Figure 5). Distance to road displayed the highest contribution regarding predicting the occurrence of CFP, followed by altitude and slope (Figure 7). The marginal effect of each variable to the model output displayed the influence of each variable to the hazard rating (Figure 8). The occurrence probability of the CFP showed a wide variation according to the distance to road, altitude, and slope. The occurrence probability was high at low values of these variables, while it remained at low values above certain values of each variable (300 m, 300 m, and 20° for the distance to road, altitude, and slope, respectively). The importance of roads in the invasion of species was supported by previous studies [25–27], showing that roads that allow human activities contribute highly to the early invasion and dispersal of species, because sites near roads are more likely to receive invasive species. These results indicate that these variables greatly affect the dispersal and occurrence of the CFP. Therefore, we propose that these factors should be considered carefully in the monitoring and surveillance programs for the CFP and other invasive species.

Meanwhile, minimum and maximum temperatures in January and July induced little variation in the occurrence probability. However, both temperature and occurrence probability displayed a positive relationship. This indicates that most areas in South Korea have a suitable temperature for the CFP. Although temperature is a fundamental factor for the development and distribution of species, it did not display an influential impact on the model output. This might be due to the relatively small differences among study sites where the South Korean territory was not big enough to display temperature differences. Therefore, if this was applied in a much larger area, temperature factors could have been influential on the distribution of the species. In this regard, the CLIMEX model has been frequently used for the prediction of the potential distribution area based on temperature conditions. Strauss [26] predicted the CFP's potential geographical distribution and identified areas at risk in Austria using CLIMEX®. The potential distribution of CFP and its dispersal were evaluated in South Korea using CLIMEX according to the climate change scenario [38]. Similar to our study, their results showed that CFP has spread in all regions of South Korea under the current climate. Both CLIMEX and MaxEnt are frequently used for the prediction of potential distribution of various species with different objectives, such as conservation of endangered species, control of invasive species, etc. However, there are some limitations in their applications. CLIMEX is based on climate-oriented data, whereas MaxEnt is not applicable to abundance data, only presence/absence data. Meanwhile, the RF model used in this study can be applied to presence/absence data as well as abundance data with various types of predictors, although presence/absence data were used in this study.

Although our models predicted the occurrence probability and potential distribution areas based on various environmental factors, the actual occurrence probability and hazard rating could be influenced by the adaptation ability of invasive species, which was not considered in our study. Species have a great ability to adapt to new environments, including changes in temperature and habitats, and this promotes the colonization of invasive species and dispersal to new regions. Therefore, further study is needed on the influence of biological adaptation of invasive species on dispersal and settlement.

5. Conclusions

The CFP occurrence areas were mostly located at low altitudes and near roads and urbanized areas. The RF model predicted the potential distribution areas of CFP with five environmental variables, showing that the CFP has a great potential to be distributed over the whole of South Korea, excluding high mountainous areas. Finally, factors related to human activities, such as roads and urbanization, strongly influenced the occurrence and dispersal of the CFP. Therefore, these variables should be considered carefully in monitoring and surveillance programs for the CFP and other invasive species to effectively control the CFP.

Author Contributions: Conceptualization, D.-S.L. and Y.-S.P.; data curation, D.-S.L., Y.-S.B., B.-K.B., S.L., and J.K.P.; methodology, D.-S.L. and Y.-S.P.; writing—original draft preparation, D.-S.L. and Y.-S.P.; writing—review and editing, D.-S.L., Y.-S.B., B.-K.B., S.L., J.K.P., and Y.-S.P.

Funding: This study was supported by the 'R&D Program for Forest Science Technology (FTIS 2017042A00-1823-CA01) provided by Korea Forest Service (Korea Forestry Promotion Institute) and by the National Research Foundation of Korea (NRF) funded by the Korean government (MSIP) (grant number NRF-2016R1A2B4011801).

Acknowledgments: The authors are grateful to survey members involved in the project supported by the 'R&D Program for Forest Science Technology provided by Korea Forest Service for their participation in the field survey.

Conflicts of Interest: The authors declare no conflict of interest. The funders had no role in the design of the study; in the collection, analyses, or interpretation of data; in the writing of the manuscript, or in the decision to publish the results.

References

1. Hughes, L. Biological consequences of global warming: Is the signal already apparent? *TREE* **2000**, *15*, 56–61. [CrossRef]
2. Walther, G.R.; Post, E.; Convey, P.; Menzel, A.; Parmesan, C.; Beebee, T.J.C.; Fromentin, J.M.; Hoegh-Guldberg, O.; Bairlein, F. Ecological responses to recent climate change. *Nature* **2002**, *416*, 389–395. [CrossRef] [PubMed]
3. Hulme, P.E. Trade, transport and trouble: Managing invasive species pathways in an era of globalization. *J. Appl. Ecol.* **2009**, *46*, 10–18. [CrossRef]
4. Choi, W.I.; Song, H.J.; Kim, D.S.; Lee, D.-S.; Lee, C.-Y.; Nam, Y.; Kim, J.-B.; Park, Y.-S. Dispersal patterns of pine wilt disease in the early stage of its invasion in South Korea. *Forests* **2017**, *8*, 411. [CrossRef]
5. Lee, D.-S.; Nam, Y.; Choi, W.I.; Park, Y.-S. Environmental factors influencing on the occurrence of pine wilt disease in Korea. *Korean J. Ecol. Environ.* **2017**, *50*, 374–380. [CrossRef]
6. Choi, W.I.; Jeon, M.J.; Park, Y.S. Structural dynamics in the host-parasitoid system of the pine needle gall midge (*Thecodiplosis japonensis*) during invasion. *PeerJ* **2017**, *5*, e3610. [CrossRef]
7. USDA. *Major Forest Insect and Disease Conditions in the United States: 2015*; U.S. Department of Agriculture: Washington, DC, USA, 2017; p. 45.
8. Liebhold, A.M.; Tobin, P.C. Population ecology of insect invasions and their management. *Annu. Rev. Entomol.* **2008**, *53*, 387–408. [CrossRef]
9. Choi, W.I.; Park, Y.-S. Dispersal patterns of exotic forest pests in South Korea. *Insect Sci.* **2012**, *19*, 535–548. [CrossRef]
10. Togashi, K.; Shigesada, N. Spread of the pinewood nematode vectored by the Japanese pine sawyer: Modeling and analytical approaches. *Popul. Ecol.* **2006**, *48*, 271–283. [CrossRef]
11. Wilson, S.W.; McPherson, J.E. Life histories of *Anormenis septentrionalis*, *Metcalfa pruinosa* and *Ormenoides venusta* with description of immature stages. *Ann. Entomol. Soc. Am.* **1981**, *74*, 299–311. [CrossRef]
12. Metcalf, Z.P.; Bruner, S.C. Cuban Flatidae with new species from adjacent regions. *Ann. Entomol. Soc. Am.* **1948**, *41*, 63–118. [CrossRef]
13. Zangheri, S.; Donadini, P. Appearance in the Venice district of a Nearctic bug: *Metcalfa pruinosa* Say (Homoptera, Flatidae). *Redia* **1980**, *63*, 301–305.
14. Preda, C.; Skolka, M. Range expansion of *Metcalfa pruinosa* (Homoptera: Fulgoroidea) in Southeastern Europe. *Ecol. Balk.* **2011**, *3*, 79–87.
15. Balakhnina, I.V.; Pastarnak, I.N.; Gnezdilov, V.M. Monitoring and control of *Metcalfa pruinosa* (Say) (Hemiptera, Auchenorrhyncha: Flatidae) in Krasnodar Territory. *Entomol. Rev.* **2014**, *94*, 1067–1072. [CrossRef]
16. Malumphy, C.; Baker, R.; Cheek, S. *Citrus Planthopper, Metcalfa pruinosa*; Plant Pest Notice No. 19; Central Science Laboratory: Aberdeen, UK, 1994.

17. Drosopoulos, A.; Broumas, T.; Kapothanassi, V. *Metcalfa pruinosa* (Hemiptera, Auchenorrhyncha: Flatidae) an undesirable new species in the insect fauna of Greece. *Ann. Benaki Phytopathol. Inst.* **2004**, *20*, 49–51.
18. Nickel, H. Arrival of the citrus flatid planthopper *Metcalfa pruinosa* (Say, 1830) in Germany and northern Switzerland. *Entomo Helv.* **2016**, *9*, 129–136.
19. Della Giustina, W. *Metcalfa pruinosa* (Say 1830), new for French fauna (Hom.: Flatidae). *Bulletin de la Société Entomologique de France* **1986**, *91*, 89–92.
20. European and Mediterranean Plant Protection Organization (EPPO). *EPPO Global Database*; European and Mediterranean Plant Protection Organization: Paris, France, 2018.
21. Lee, H.S.; Wilson, S.W. First report of the Nearctic flatid planthopper *Metcalfa pruinosa* (Say) in the Republic of Korea (Hemiptera: Fulgoroidea). *Entomol. News* **2010**, *121*, 506–513. [CrossRef]
22. Kim, Y.; Kim, M.; Hong, K.-J.; Lee, S. Outbreak of an exotic flatid, Metcalfa pruinosa (Say) (Hemiptera: Flatidae), in the capital region of Korea. *J. Asia-Pac. Entomol.* **2011**, *14*, 473–478. [CrossRef]
23. Kim, D.-E.; Kil, J. Occurrence and host plant of *Metcalfa pruinosa* (Say) (Hemiptera: Flatidae) in Korea. *J. Environ. Sci. Int.* **2014**, *23*, 1385–1394. [CrossRef]
24. Wilson, S.W.; Lucchi, A. Feeding activity of the flatid planthopper *Metcalfa pruinosa* (Hemiptera: Fulgoroidea). *J. Kans. Entomol. Soc.* **2007**, *80*, 175–178. [CrossRef]
25. Pantaleoni, R.A. The ways in which Metcalfa pruinosa (Say, 1830) (Auchenorrhyncha Flatidae) invades a new area. *Bollettino dell'Istituto di Entomologia Guido Grandi della Università degli Studi di Bologna* **1989**, *43*, 1–7.
26. Strauss, G. Pest risk analysis of Metcalfa pruinosa in Austria. *J. Pest Sci.* **2010**, *83*, 381–390. [CrossRef]
27. Kil, J.H.; Lee, D.H.; Hwang, S.M.; Lee, C.W.; Kim, Y.H.; Kim, D.E.; Kim, H.M.; Yang, H.S.; Lee, J.C. *Detailed Studies on Invasive Alien Species and Their Management (VI)*; National Institute of Environmental Research: Incheon, Korea, 2011.
28. Park, Y.-S.; Chung, Y.-J. Hazard rating of pine trees from a forest insect pest using artificial neural networks. *For. Ecol. Manag.* **2006**, *222*, 222–233. [CrossRef]
29. Park, Y.-S.; Chung, Y.-J.; Moon, Y.-S. Hazard ratings of pine forests to a pine wilt disease at two spatial scales (individual trees and stands) using self-organizing map and random forest. *Ecol. Inform.* **2013**, *13*, 40–46. [CrossRef]
30. Futai, K. Pine wilt in Japan: From first incidence to the present. In *Pine Wilt Disease*; Zhao, B.G., Futai, K., Sutherland, J.R., Takeuchi, Y., Eds.; Springer: Tokyo, Japan, 2008; pp. 5–12.
31. Mason, G.N.; Lorio, P.L., Jr.; Belanger, R.P.; Nettleton, W.A. *Rating the Susceptibility of Stands to Southern Pine Beetle Attack. Integrated Pest Management Handbook*; Agriculture Handbook No. 645; USDA, Forest Service: Washington, DC, USA, 1985.
32. Valentine, H.T.; Houston, D.R. A discriminant function for identifying mixed-oak stand susceptibility to gypsy moth Lymantria dispar defoliation. *For. Sci.* **1979**, *25*, 468–474.
33. Hicks, R.R., Jr.; Howard, J.E.; Coster, J.E.; Watterston, K.G. Rating forest stand susceptibility to southern pine beetle in east Texas using probability functions. *For. Ecol. Manag.* **1980**, *2*, 269–283. [CrossRef]
34. Lorio, P.L., Jr.; Sommers, R.A. Gulf coastal plain (Lousiana). In *Site, Stand, and Host Characteristics of Southern Pine Beetle Infestations*; USDA Forest Service, Tech. Bull., 1612. Comb. For. Pest Res. Devlop. Prog.; Coster, J.E., Searcy, J.L., Eds.; USDA Forest Service: Pineville, LA, USA, 1980.
35. Nam, Y.; Koh, S.-H.; Jeon, S.-J.; Youn, H.-J.; Park, Y.-S.; Choi, W.I. Hazard rating of coastal pine forests for a black pine bast scale using self-organizing map (SOM) and random forest approaches. *Ecol. Inform.* **2015**, *29*, 206–213. [CrossRef]
36. Baker, F.A.; Verbyla, D.L.; Hodges, C.S., Jr.; Ross, E.W. Classification and regression tree analysis for assessing hazard of pine mortality caused by *Heterobasidion annosum*. *Plant Dis.* **1993**, *77*, 136–139. [CrossRef]
37. Sutherst, R.W.; Maywald, G.F.; Bottomley, W.; Bourne, A. *CLIMEX for Windows, Version 2 User's Guide*; CRC for Tropical Pest Management: Brisbane, Australia, 2004.
38. Byeon, D.-H.; Jung, J.-M.; Lohumi, S.; Cho, B.-K.; Jung, S.; Lee, W.-H. Predictive analysis of Metcalfa pruinosa (Hemiptera: Flatidae) distribution in South Korea using CLIMEX software. *J. Asia-Pac. Biodivers.* **2017**, *10*, 379–384. [CrossRef]
39. Byeon, D.-H.; Jung, S.; Lee, W.-H. Review of CLIMEX and MaxEnt for studying species distribution in South Korea. *J. Asia-Pac. Biodivers.* **2018**, *11*, 325–333. [CrossRef]
40. Vanhanen, H.; Veteli, T.O.; Paivinen, S.; Kellomaki, S.; Niemela, P. Climate change and range shifts in two insect defoliators: Gypsy moth and nun moth—A model study. *Silva Fenn.* **2007**, *41*, 621. [CrossRef]

41. Sridhar, V.; Verghese, A.; Vinesh, L.S.; Jayashankar, M.; Jayanthi, P.K. CLIMEX simulated predictions of Oriental fruit fly, Bactrocera dorsalis (Hendel) (Diptera: Tephritidae) geographical distribution under climate change situations in India. *Curr. Sci.* **2014**, *106*, 1702–1710.
42. Logan, J.A.; Régnière, J.; Powell, J.A. Assessing the impacts of global warming on forest pest dynamics. *Front. Ecol. Environ.* **2003**, *1*, 130–137. [CrossRef]
43. Kumar, S.; Neven, L.G.; Zhu, H.; Zhang, R. Assessing the global risk of establishment of *Cydia pomonella* (Lepidoptera: Tortricidae) using CLIMEX and MaxEnt niche models. *J. Econ. Entomol.* **2015**, *108*, 1708–1719. [CrossRef]
44. Canberra Australia: Commonwealth Scientific and Industrial Research Organisation (CSIRO). *CLIMEX Version 4: Exploring the Effects of Climate on Plants, Animals and Diseases*; CSIRO: Canberra, Australia, 2015.
45. Phillips, S.J.; Anderson, R.P.; Schapire, R.E. Maximum entropy modeling of species geographic distributions. *Ecol. Model.* **2006**, *190*, 231–259. [CrossRef]
46. Elith, J.; Phillips, S.J.; Hastie, T.; Dudík, M.; Chee, Y.E.; Yates, C.J. A statistical explanation of MaxEnt for ecologists. *Divers. Distrib.* **2011**, *17*, 43–57. [CrossRef]
47. Korea Forestry Promotion Institute (KOFPI). *Development of Prediction Models for Dispersal and Population Change of Forest Exotic Insects (2nd Year Report)*; Korea Forestry Promotion Institute: Seoul, Korea, 2018.
48. R Core Team. *R: A Language and Environment for statistical Computing*; R Foundation for Statistical Computing: Vienna, Austria, 2017. Available online: https://www.R-project.org/ (accessed on 23 June 2018).
49. Breiman, L. Random forests. *Mach. Learn.* **2001**, *45*, 5–32. [CrossRef]
50. Hodkinson, I.D. Terrestrial insects along elevation gradients: Species and community responses to altitude. *Biol. Rev.* **2005**, *80*, 489–513. [CrossRef]
51. Liaw, A.; Wiener, M. Classification and Regression by randomForest. *R News* **2002**, *2/3*, 18–22.
52. Wood, S.N. *Generalized Additive Models: An Introduction with R*; Chapman and Hall/CRC: Boca Raton, FL, USA, 2017.
53. Sing, T.; Sander, O.; Beerenwinkel, N.; Lengauer, T. ROCR: Visualizing classifier performance in R. *Bioinformatics* **2005**, *21*, 3940–3941. [CrossRef] [PubMed]
54. Lee, S.D.; Park, S.; Park, Y.-S.; Chung, Y.-J.; Lee, B.-Y.; Chon, T.-S. Range expansion of forest pest populations by using the lattice model. *Ecol. Model.* **2007**, *203*, 157–166. [CrossRef]
55. Lauterer, P. Citrus flatid planthopper—*Metcalfa pruinosa* (Hemiptera: Flatidae), a new pest of ornamental horticulture in the Czech Republic. *Plant Prot. Sci.* **2002**, *38*, 145–148. [CrossRef]
56. Grozea, I.; Gogan, A.; Virteiu, A.M.; Grozea, A.; Stef, R.; Molnar, L.; Carabet, A.; Dinnesen, S. *Metcalfa pruinosa* Say (insecta: Homoptera: Flatidae): A new pest in Romania. *Afr. J. Agric. Res.* **2011**, *6*, 5870–5877. [CrossRef]
57. Guillemaud, T.; Ciosi, M.; Lombaert, É.; Estoup, A. Biological invasions in agricultural settings: Insights from evolutionary biology and population genetics. *Comptes Rendus Biol.* **2011**, *334*, 237–246. [CrossRef] [PubMed]
58. Auffret, A.G.; Cousins, S.A.O. Humans as long-distance dispersers of rural plant communities. *PLoS ONE* **2013**, *8*, e62763. [CrossRef] [PubMed]
59. Wilson, J.R.; Dormontt, E.E.; Prentis, P.J.; Lowe, A.J.; Richardson, D.M. Something in the way you move: Dispersal pathways affect invasion success. *Trends Ecol. Evol.* **2009**, *24*, 136–144. [CrossRef] [PubMed]
60. Lozon, J.D.; MacIsaac, H.J. Biological invasions: Are they dependent on disturbance? *Environ. Rev.* **1997**, *5*, 131–144. [CrossRef]
61. Cox, G.W. *Alien Species and Evolution: The Evolutionary Ecology of Exotic Plants, Animals, Microbes, and Interacting Native Species*; Island Press: Washington, DC, USA, 2004.
62. Strauss, G. Environmental risk assessment for *Neodryinus typhlocybae*, biological control agent against *Metcalfa pruinosa*, for Austria. *Eur. J. Environ. Sci.* **2012**, *2*, 102–109. [CrossRef]
63. Lake, J.C.; Leishman, M.R. Invasion success of exotic in natural ecosystems: The role of disturbance, plant attributes and freedom from herbivores. *Biol. Conserv.* **2004**, *117*, 215–226. [CrossRef]
64. Kahrer, A.; Strauss, G.; Stolz, M.; Moosbeckhofer, R. Beobachtungen zu Faunistik und Biologie der vor kurzem nach Österreich eingeschleppten Bläulingszikade (*Metcalfa pruinosa*). *Beiträge zur Entomofaunistik* **2009**, *10*, 17–30.

MDPI
St. Alban-Anlage 66
4052 Basel
Switzerland
Tel. +41 61 683 77 34
Fax +41 61 302 89 18
www.mdpi.com

Forests Editorial Office
E-mail: forests@mdpi.com
www.mdpi.com/journal/forests

www.ingramcontent.com/pod-product-compliance
Lightning Source LLC
LaVergne TN
LVHW071947080526
838202LV00064B/6699